Safety Science Research

Evolution, Challenges
and New Directions

Safety Science Research

Evolution, Challenges and New Directions

Edited by
Jean-Christophe Le Coze

CRC Press
Taylor & Francis Group
Boca Raton London New York

CRC Press is an imprint of the
Taylor & Francis Group, an **informa** business

CRC Press
Taylor & Francis Group
6000 Broken Sound Parkway NW, Suite 300
Boca Raton, FL 33487-2742

International Standard Book Number-13: 978-0-8153-9268-2 (Hardback)

Library of Congress Cataloging-in-Publication Data

Names: Le Coze, Jean-Christophe, editor.
Title: Safety science research : evolution, challenges and new directions / edited by Jean-Christophe Le Coze.
Description: Boca Raton : Taylor & Francis, 2019. | Includes bibliographical references.
Identifiers: LCCN 2019013614| ISBN 9780815392699 (pbk. : alk. paper) | ISBN 9780815392682 (hardback : alk. paper) | ISBN 9781351190237 (e-book)
Subjects: | MESH: Accidents, Occupational--prevention & control | Safety | Occupational Health
Classification: LCC HD7260 | NLM WA 485 | DDC 363.11--dc23
LC record available at https://lccn.loc.gov/2019013614

Visit the Taylor & Francis Web site at
http://www.taylorandfrancis.com

and the CRC Press Web site at
http://www.crcpress.com

Contents

Preface..vii

Editor..ix

Contributors..xi

Introduction..xv

Section I New Generation

1 Standardisation and Digitalisation: Changes in Work as
  Imagined and What This Means for Safety Science.....................3
  Petter G. Almklov and Stian Antonsen

2 The Interaction between Safety Culture and National Culture.........21
  Tom W. Reader

3 Governance for Safety in Inter-Organisational
  Project Networks...39
  Nadezhda Gotcheva, Kirsi Aaltonen and Jaakko Kujala

4 Coping with Globalisation: Robust Regulation and Safety in
  High-Risk Industries...55
  Ole Andreas Engen and Preben Hempel Lindøe

5 On Ignorance and Apocalypse: A Brief Introduction to
  'Epistemic Accidents'...75
  John Downer

6 Revisiting the Issue of Power in Safety Research.......................87
  Stian Antonsen and Petter Almklov

7 Sensework...103
  Torgeir Haavik

8 Drift and the Social Attenuation of Risk....................................119
  Kenneth Pettersen Gould and Lisbet Fjæran

9 Safety and the Professions: Natural or Strange Bedfellows?..........133
  Justin Waring and Simon Bishop

10 **Visualising Safety** 151
    Jean-Christophe Le Coze

11 **The Discursive Effects of Safety Science** 173
    Johan Bergström

12 **Investigating Accidents: The Case for Disaster Case Studies in
    Safety Science** 187
    Jan Hayes

13 **Towards Actionable Safety Science** 203
    T. Reiman and K. Viitanen

14 **Safety Research and Safety Practice: Islands in a Common Sea** 223
    Steven T. Shorrock

Section II   Pioneers

15 **Safety Research: 2020 Visions** 249
    Rhona Flin

16 **The Gilded Age?** 263
    Erik Hollnagel

17 **Observing the English Weather: A Personal Journey from
    Safety I to IV** 269
    Nick Pidgeon

18 **A Conundrum for Safety Science** 281
    Karlene H. Roberts

19 **Some Thoughts on Future Directions in Safety Research** 289
    Paul R. Schulman

20 **Redescriptions of High-Risk Organisational Life** 299
    Karl E. Weick

21 **Skin in the Game: When Safety becomes Personal** 311
    Ron Westrum

Index 321

Preface

Safety-critical systems such as offshore platforms, hospitals, aircraft, nuclear power plants, refineries, bridges, dams, mines, and so on, all rely on a myriad of artefacts and actors to operate safely. It is an admirable political, technological, social and economic endeavour re-enacted everyday all over the world. But sometimes, when a bridge collapses, a building burns, an offshore platform explodes, a nuclear reactor melts down, a train derails, a ship sinks or a plane crashes, we are reminded how precarious such successes are, and we are also reminded of the diversity of practices across countries, sectors and companies. To study these operations, to understand their complexities, to develop practical solutions, safety science has matured over 40 years.

And, because of their immense complexity, high-risk systems have been investigated through the lenses of different research traditions. The 1980s and 1990s were foundational decades in this respect; many of the core programs within safety science have been structured ever since by influential networks of researchers. They have developed ideas, methods and concepts including among others, safety culture, cognitive engineering, safety regulations, high-reliability organisations or normal accidents. These networks of authors created the lenses through which we grasp safety-critical systems; their mode of operating, but also their ways of failing.

If traditions have created, framed and shaped the contours of safety science, they are also historically situated in a different era, the end of the 20th century. What about their relevance for a world that has tremendously changed at the turn of the 21st century? The intention of this book is to explore this question; to step back, and to connect these established traditions with the insights of a new generation of researchers within, across and beyond these traditions. These writers indeed apply, adapt, extend, challenge, combine or reformulate these traditions when confronted with our increasingly digitalised, globalised and networked world.

In turn, their ideas are reflected upon by some of the pioneers of safety science, representing a diversity of research traditions, and bringing exclusive perspectives based on their long-time involvement in the maturing of the field. As a result, this book offers a unique opportunity to consider the growing and evolving field of safety science and its relevance for contemporary challenges, its adaptation to ongoing evolutions and its new directions.

Editor

Jean Christophe Le Coze is a safety researcher (PhD, Mines ParisTech) at INERIS, the French national institute for environmental safety. His activities combine ethnographic studies and action research in various safety-critical systems with an empirical, theoretical, historical and epistemological orientation.

Contributors

Kirsi Aaltonen, PhD, is associate professor of project management and complex systems at the University of Oulu, Finland. Her current research focusses on the governance and management of complex projects with a particular focus on integrated project deliveries.

Petter Almklov, PhD, holds a PhD in organisational anthropology and also has an MSc in engineering geology. He is currently employed as research professor at Norwegian University of Science and Technology (NTNU) Social Research, Norway.

Stian Antonsen, PhD, holds a PhD in organisational sociology from NTNU, Norway. He is currently market director of safety research at SINTEF and adjunct professor in safety management at NTNU.

Johan Bergström is a reader and senior lecturer at Lund University, Sweden. He is involved in both teaching and research introducing a critical approach to the discourse of safety and risk. He is also frequently invited to give external lectures on topics of safety, resilience, complexity and risk.

Simon Bishop is associate professor of organisational behaviour at the University of Nottingham, England. His research considers how social institutions are inhabited, contested and refashioned through the micro-sociological practices of social actors.

John Downer currently works at the School of Sociology, Politics and International Studies (SPAIS), University of Bristol, England. There he teaches on various subjects pertaining to Science and Technology Studies (STS) and risk and spends the rest of his time perennially failing to finish a book that explores the epistemology of safety-critical technologies.

Ole Andreas Engen is professor at the University of Stavanger, Norway. His research interests are focussed on the relationships between policy, risk, technology and safety. Empirically, his research has mainly addressed risk governance, safety regulation and HSE in the petroleum sector.

Lisbet Fjæran is a research fellow in risk analysis at the University of Stavanger, Norway. She has previously worked as an environmental consultant and her main research topics include risk assessment, risk regulation and food and environmental risks, as well as foundational issues in risk analysis.

Rhona Flin is professor of industrial psychology, Aberdeen Business School, Robert Gordon University, Scotland and emeritus professor of applied psychology, University of Aberdeen, Scotland.

Nadezhda Gotcheva, PhD, is senior scientist at VTT Technical Research Centre of Finland Ltd. She holds a PhD from Tampere University of Technology, Finland. Her current research work focusses on safety culture evaluation and leadership for safety.

Kenneth Pettersen Gould is an associate professor in risk management and societal safety at the University of Stavanger, Norway. His main research interests include organisational risk, safety management and foundational theories and methods in safety science.

Torgeir Kolstø Haavik, PhD, has a background in geological engineering, social geography and sociology. He holds positions as a senior researcher at NTNU Social Research, Norway and adjunct associate professor at NTNU, Norway.

Jan Hayes, PhD, has 30 years of experience in safety and risk management. She holds an associate professor appointment at RMIT University, where she is program leader for the social science research activities of the Energy Pipelines Cooperative Research Centre. Her current activities cover academia, consulting and regulation.

Erik Hollnagel is senior professor of patient safety at Jönköping University, Sweden and visiting professorial fellow, Macquarie University, Australia. Throughout his career, he has worked at universities, research centres and in industries in many countries and with problems from a variety of domains.

Jaakko Kujala, PhD, is professor in the Industrial Engineering and Management Research Unit at the University of Oulu, Finland. His current research interests include coordination of complex project networks, stakeholder management and simulation as a research method.

Preben Hempel Lindøe is professor emeritus at the University of Stavanger, Norway. For the last 30 years, he has worked within applied research, action research methodology, occupational health and safety, safety management and risk regulation.

Nick Pidgeon is director of the Understanding Risk Research Group at Cardiff University, Wales, and professor of environmental risk. His research looks at risk perception and communication for environmental and technology risks, as well as the causes of organisational accidents. Following the untimely death of his colleague Barry Turner in 1995, he completed the preparation of the second edition of the classic book, *Man-Made Disasters* (1997).

Tom W. Reader, PhD, is an associate professor at the London School of Economics, England. His research explores the influence of organisational culture upon safety outcomes in high-risk industries (e.g. aviation, energy and healthcare).

Teemu Reiman, PhD, works as a safety culture manager at the Finnish nuclear power company Fennovoima, and also does part-time research and consultancy. He has a doctoral degree in psychology from the University of Helsinki, Finland, and is an adjunct professor at Aalto University, Finland.

Karlene H. Roberts, PhD, is professor emerita at the Haas School of Business, University of California, Berkeley. She received her bachelor's degree from Stanford University and her PhD in psychology from the University of California, Berkeley. Her research is on the design and management of organisations in which errors can have catastrophic consequences.

Paul Schulman is professor emeritus of government at Mills College, Oakland, California, and senior research fellow at the Center for Catastrophic Risk Management at the University of California, Berkeley.

Steven Shorrock, PhD, is a chartered ergonomist and human factors specialist and a chartered psychologist with over 20 years of experience. He is a human factors and safety specialist in air traffic control in Europe with EUROCONTROL, Brussels. He is also adjunct associate professor at the University of the Sunshine Coast, Australia.

Kaupo Viitanen, MA (psych), is currently working as a research scientist at VTT Technical Research Centre of Finland Ltd. His research focusses on topics such as safety culture evaluation and change, organisational resilience, organisational learning and safety management and leadership.

Justin Waring is professor of organisational sociology at the University of Nottingham, England. His research examines the social organisation of public services with a particular focus on the evolving governance of quality and safety in health and care services.

Karl Weick is emeritus professor of organisational behaviour and psychology. His research interests include collective sensemaking under pressure, medical errors, handoffs in extreme events, high-reliability performance, improvisation and continuous change.

Ron Westrum is emeritus professor of Sociology at Eastern Michigan University, Michigan. He has frequently written on systems safety, communication and corporate cultures. He is preparing a book on *The Cultures of Information Flow*.

Introduction

In this introduction, I present the context of this book and come back to safety science developments and research traditions over the past decades, focussing on the 1980s and 1990s characterised as a 'golden age'. I then discuss the contours of safety research considering its multidimensions and, finally, argue about the interest of building a safety research narrative of new evolutions, directions and challenges considering the significant changes witnessed in the past two decades, which followed the 'golden age'.

This introduction proceeds with a certain degree of simplicity. It cannot delve into the nuances that would be required to give credit to so many authors, ideas, domains, research outcomes, debates or related safety practices and current micro and macro transformations. This is an editorial choice, designed to capture the salient features of the field, which are believed to be useful to situate, orient and stimulate research.

I.1 Presentation of This Book's Initiative

What is now the book that you are reading, based on 22 contributions including this introduction by a 'newer' and 'older' generation of safety researchers, started as a single review article project. I intended, around 2013, to provide a presentation of what I thought were novel insights or concepts introduced by a generation of authors who, during the mid-2000s were PhD students who wrestled with their own topics, but shared a similar situation: they started within established research traditions in safety and had to develop their ideas from there.

They were in this respect in quite a different position from a first generation of authors who institutionalised these traditions in the 1980s, some 30 to 40 years ago. At the time of the project, writing this review article, concepts such as those of power (Antonsen, 2009), sensework (Haavik, 2011) or epistemic accidents (Downer, 2011) seemed to me to be examples, among others, of contributions identifying what these authors considered to be gaps in existing traditions and explored them to formulate original contributions.

Some of these researchers happened to meet from time to time in a rather dispersed fashion in the late 2000s and early 2010s, at conferences or workshops organised by different networks or communities within the safety field. I thought, therefore, that it would probably be more exciting to have these researchers sharing and debating their ideas together rather than associating them, more artificially, in a review article written from an office desk.

This is how the idea of organising a workshop came, to discuss the situation with others, for instance in 2014 during the Working on Safety Conference in Glasgow. I made a partial and subjective selection of authors that I knew of, and for some of them, met before, and we gathered in June 2017 in Paris. Many of the authors who contributed to the first part of this book participated in the event and engaged in lively debates in the course of these two days.

From there, an additional move consisted of connecting these new ideas or orientations of recent researchers to the work of the pioneers of safety research by collecting their reactions to these 'self-proclaimed' new orientations or ideas. Selecting pioneers was a contentious task and the decision was made to restrict the invitation to researchers who had already started writing articles or books in the 1980s.

Proceeding this way excludes important researchers and writers of the 1990s onwards. There are so many names that it would be unfair to even start a list. It is not the intention of this book to draw frontiers between generations of writers, to decide who is in and who is out or to establish a network of a new generation of writers. The initiative is far more modest, far more reflexive and far more inclusive and open-minded than this project might be perceived.

The request formulated to those pioneers who accepted to join the project was to reflect back over 40 years of research based on the reading of the chapters of the first part of the book. Guidance was minimal, and their contribution was left to their appreciation to reflect as they wished, by either selecting one or several, or even perhaps all of the chapters. This is part two.

I would like in this respect to thank Rhona Flin, Erik Hollnagel, Nick Pidgeon, Karlene Roberts, Paul Schulman, Karl Weick and Ron Westrum for accepting the invitation and finding the time to read and then write a chapter. Their contribution is a great added value, which highlights one central argument of this book.

The argument is that safety, as a research topic, is not limited to or the preserve of one research tradition but is investigated and explored from a diversity of perspectives, which will never exhaust the complexity of the topic (Dupré, Le Coze, 2014), although it is important to try to keep a big picture in sight (Le Coze, 2016).

In what follows, research traditions in safety are identified and succinctly presented, then new ideas are briefly introduced and further developed in the chapters of the first part of the book. The second part is based on reflections by some of the pioneers of this field of safety who established strong directions for safety research from the 1980s onwards.

I.2 The 1980s–1990s: The Advent of (High-Risk Systems) Safety Research

Safety research, and more specifically in relation to the project of this book, industrial safety research,* as a scientific field, is a product of the late 20th century. It is a moment of history during which sociotechnical systems such as civil aviation, nuclear power plants and weapons following World War II, raised increasing concern about their complexity and the potential consequences associated with losing control (La Porte, 1975; Winner, 1978).

The historian of technology, Hughes, makes the following comment about this period during which the feeling of an increasingly technological-driven society became a fundamental issue: 'In "Men, machines, and modern times" (1966), Elting Morison, an MIT history professor, eloquently expresses with a biblical cadence his views about large systems. He holds that they have acquired an intricacy, mass, scale, and rate of change making it extremely difficult for individuals to cope' (Hughes, 2005, 88).

Clearly, safety practices and concerns for safety existed for a much longer period of time before the 1970–1980s, as for instance in the railway, mining or chemical industries. These industries were already fairly complex sociotechnical systems with a potential for large-scale disasters. Catastrophes such as the *Titanic* (maritime, UK, 1908, 1125 casualties) or La Courrière (mining, France, 1907, 1099 casualties) are but a few of these events that indicate their reality a century ago.

Historians have documented the interactions between states, technology, science, companies and citizens, creating debates and expectations for preventing accidents and promoting safety during the decades of the Industrial Revolution in the 19th century (Fressoz, 2012; Jarrige, 2013). The production of laws, the development of technical standards, the inspections of factories and the work and bureaucratic safety arrangements were already part of the growing set of engineering, regulatory and management methods, principles, tools and practices associated with the control of hazardous operations.

But it was in the late 20th century that a multidisciplinary scholarship with specialised books, scientific journals and conferences developed specifically on this topic. Engineers, psychologists, ergonomists, sociologists and political scientists dedicated a growing interest to the issue of hazardous operation safety in various domains, such as aviation, the (petro)chemical industry or with nuclear weapons, although one can obviously find traces of these academic developments earlier (Swuste et al., 2010; Swuste et al., 2012).

---

* An arbitrary distinction is made here between risk studies, occupational safety, crisis management, disaster research and industrial safety to make it manageable. Readers understand that there are strong links between these fields, and that many books are available in these various domains.

In the 1960s and 1970s, methods for technical risk assessment such as HAZOP (hazard operability study) or FMAE (failure mode and effect analysis) were created by chemical or reliability engineers for analysing weapons, aircraft, chemical or pyrotechnic processes during this period. The study of the relationship between human behaviours, work situation design and safe performance was consolidated by engineering psychologists, human factors engineers and ergonomists. Safety management systems were also a topic of interest in this period with, for instance, the developments of accident investigation tools such as the management oversight and risk tree (MORT) produced by scientists, engineers and experienced practitioners and managers.

Technical, human and organisational perspectives on safety were therefore produced in parallel over the years, with an increase following World War II, to tackle families of problems associated with the operation of a diverse range of large-scale and hazardous technical systems. But it is a little later in this second half of the 20th century that one book acquired a special academic and popular status in this area.

Triggered by Three Mile Island in 1979, Perrow's *Normal Accidents,* released in 1984, convincingly argued about and contributed to the advent of the category of high-risk systems. It spurred much interest far beyond the confines of individuals who started specialising in the topic of industrial safety because of its resonance with the debates of the time over the development of technology.

Its timeliness in relation to disasters (e.g. Bhopal in 1984, *Challenger* in 1986, Chernobyl in 1986) and argument created a momentum and a platform for sustained debates on the newly proposed category of high-risk systems. An international and multidisciplinary event held by the World Bank (Rasmussen and Batstone, 1989) illustrates the growing interest among a community of researchers. In retrospect, the publication of the book *Normal Accidents* is certainly one of the key moments of the institutionalisation of safety research.

It most likely helped reinforce the need for sustained investment from industries, states and universities in this area. Of course, there were many other books that reflected the fact that a momentum was taking place when, in the United States and Europe, research traditions developed around concepts, authors and safety-critical domains (some of the important ones are listed in Table I.1 below). The 1980s and the 1990s were indeed very active decades for the development of an academic domain specialising in the topic of high-risk systems safety, based on a range of research traditions.

I.3 What Are These Traditions?

These traditions are anchored both in disciplines (e.g. engineering, psychology, sociology) that were established before safety became an academic

TABLE I.1

Research Traditions in Safety Science

Research tradition	Authors (not exhaustive) 1980s/1990s
Incubation, Safety Culture and Learning	Barry Turner, Nik Pidgeon, David Blockley
Normal Accidents and the Critical Perspective	Charles Perrow, Paul Shrivastava, Lee Clarke, Carl Sagan, Scott Snook
High-Reliability Organisations	Karlene Roberts, Todd La Porte, Paul Schulman, Gene Rochlin, Karl Weick
Safety Regulations and Socio-Legal Perspectives	Aaron Wildavsky, Bridget Hutter, John Braithwaite, Joseph Rees, Andrew Hopkins
Socio-Constructivist Perspectives	Ron Westrum, Robert Gephart, Trevor Pinch, Brian Wynne, Diane Vaughan
Safety Climate, Leadership and Management	Dov Zohar, Ian Glendon, Andrew Hale, Rhona Flin, Amy Edmondson, Eduardo Salas, Patrick Hudson
Human Error, Interface Design and System Safety	Jens Rasmussen, James Reason, Donald Norman, David Woods, Erik Hollnagel, Nancy Leveson, Gary Klein

domain as mentioned above and in younger ones, sometimes still in a process of institutionalisation in the 1970s at various stages (e.g. ergonomics, cognitive engineering, management, political sciences, science and technology studies).

Research traditions in high-risk or safety-critical systems can be identified around concepts often created by one (or a network of) author(s) from different scientific backgrounds who then developed an empirical, methodological and theoretical agenda, sometimes in interaction with other traditions through debates or circulation of concepts (Le Coze, 2016).

Table I.1 introduces seven of these traditions along with their main authors: incubation, safety culture and learning; normal accident and the critical perspective; high-reliability organisations; safety regulations and the socio-legal perspective; socio-constructivist perspectives; safety climate, leadership and management; and human error, interface design and system safety. I now introduce them briefly.

I.4 Research Traditions, Concepts and Authors of the 1980s and 1990s

I.4.1 Incubation, Safety Culture and Learning

Without claiming to be exhaustive or to be able to systematically discriminate sharply between research traditions, concepts and authors because they overlap in many ways (Table I.1), one can mention the development of the Incubation Model of Barry Turner. In close association with researchers David Blockley and Nick Pidgeon in the UK and in a multidisciplinary context (i.e. engineering, psychology, sociology), this model has developed strong links to early conceptualisations of the notion of safety culture and learning.

The innovation of this model back in the late 1970s and during the 1980s was to extract a common pattern across a diversity of disaster reports, and to show how individuals could be likely to miss signals because of the complex social processes exhibited by bureaucracies, including prominently their cultural dimensions. Disasters appeared to be not the products of a mistake of a single individual, but the products of individuals interacting and interpreting their world in organisationally and socially shaped contexts instead.

This tradition was one of the first to connect with the safety culture debate in the late 1980s when the concept was introduced. By providing early inputs to the topic and advocating a careful take on this ambiguous notion, Turner and Pidgeon explored the implications of the empirical and theoretical dimension of the incubation model. Another important concept that seemed to derive quite naturally from the incubation model was learning, which also became in the following years an additional core idea in the safety field.

I.4.2 High-Reliability Organisations

Another tradition is the body of work around High-Reliability Organisations in the United States during the 1980s with authors such as Karlene Roberts, Todd La Porte, Paul Schulman, Gene Rochlin and also Karl Weick. The difference with the previous tradition is the focus on daily operations instead of a focus on past events, but the interdisciplinary nature of the research is also a common point (e.g. organisational psychology, political science, social psychology, management).

What high-reliability researchers delivered were the first ethnographic descriptions and conceptualisations of the daily operations of these specific organisations for which the social and technological environment was unforgiving. They looked for some of these social, organisational and managerial properties that could help explain their ability to succeed in these particularly unforgiving contexts, within this requirement of operating in nearly error-free systems.

Several features appeared to be connected to the ability of maintaining reliable operations including redundancy within, among and between teams, namely the possibility for tasks to be performed while being checked by several individuals. The description of a property such as *having the bubble* was another of these features, which meant that some individuals in managerial positions had a broad view of operations allowing them to keep the big picture in mind. The concept of collective mindfulness further theorised during the 1990s some of these properties into what became a successful practical and theoretical proposition.

I.4.3 Normal Accidents and the Critical Perspective

In the U.S. context, this tradition was in dialogue with the approach, style and intellectual posture followed and conveyed by *Normal Accidents* in the United States by Charles Perrow. As a very influential book it was debated and advocated in different ways by authors such as Paul Shrivastava, Lee Clarke or Scott Sagan during the 1990s. The value and innovation of Perrow's book was first its synthetic angle, second, its provocative thesis and third, its collection of cases.

The synthetic angle was the ability to identify, associate, classify and then visualise a new category of organisations: high-risk systems. The provocative thesis declared some accidents to be inevitable in certain kinds of systems, and implied that nuclear power plants had to be abandoned for this very reason. The collection of cases or stories of accidents provided an overview of the diversity of ways accidents could happen in various hazardous processes, based on a critical stance.

The initial tensions, which existed between the message of Perrow, based on retrospective cases and what the high-reliability organisation researchers looked into through fieldwork of daily operations during the course of the 1980s, was pictured as two opposite perspectives to choose from by Sagan who, following Perrow, finally believed that accidents were inevitable. The debate never really settled, and extended versions of the normal accident thesis were developed, as for instance by Snook at the end of the 1990s.

I.4.4 Safety Regulations and the Socio-Legal Perspective

In different countries in the 1980s and 1990s, safety regulations and socio-legal views also delineated the contours of a research tradition among sociologists, political scientists and legal scholars such as Aaron Wildavsky, Bridget Hutter, John Braithwaite, Andrew Hopkins or Joseph Rees, who studied how laws, public policies, regulations and inspections frame the conditions of safe performances. Andrew Hopkins, over the years, developed his own specific analytical lenses, combining descriptive and normative sociological interpretations beyond this socio-legal angle, into widely read accounts of disasters, from the end of the 1990s onwards.

This tradition explored and showed how the state, through laws and inspections, played a key role in designing the conditions surrounding the practices in safety-critical organisations. The multiplicity of options and their evolution over time when it comes to shaping companies' expectations through legal requirements, but also through principles and implementation of control of these requirements by inspectorates of agencies, translate societies' degree of concern for safety through their ideological and political orientations (e.g. command and control, self-regulation).

The notion of regulation regimes developed at the end of the 1990s aggregated various of these dimensions into a broad concept. Interpreting accident and safety as a result of how well designed, implemented and controlled legal expectations are, such regulation regimes must be developed to achieve the right combination of resources and skills but also the right balance of persuasion and sanction. In this ambition, regulation regimes face the risk of regulatory capture of private interests, which might attempt to limit the degree of control and oversight to which they are exposed.

I.4.5 Socio-Constructivist Perspectives

It makes sense to distinguish what could be called a socio-constructivist orientation of accidents and safety with authors in the United States and the UK such as Ron Westrum, Robert Gephart, Trevor Pinch, Brian Wynne and Diane Vaughan because of their connections with an important new wave of social studies approaching science and technology from very innovative angles. These angles consisted in challenging the much taken for granted idea that science and technology come out of rational minds disconnected from the historical and social contexts within which they exert their socio-cognitive skills.

In this view of science and technology, uncertainties, controversies and values have a stronger role than in the previous interpretation of science and technology. In this view, social and technical realities are tightly intertwined. Applied to disasters and safety, such a view implies a careful analysis of the complexities associated with the design, interpretation and handling of hazardous artefacts.

Within this tradition, their uncertainties, ambiguities and messiness are part of the picture, and generate this potential for, sometimes, deceiving expectations and predictions. Understanding safety or accidents based on these ideas leads to a nuanced perspective on the practices of engineers and managers, on decision-making processes and on the possibility of avoiding surprises or not. Within this tradition, knowledge is not about the discovery of an outside world, but the construction of a temporary, limited and unstable understanding of the world.

I.4.6 Safety Climate, Leadership and Management

From a more psychological, and very often combined system perspective, areas such as safety climate, safety leadership, teamwork and safety management systems can constitute other examples of important themes in the safety field to be associated with a research tradition found in many countries with the work, for instance, of Rhona Flin, Eduardo Salas, Patrick Hudson, Amy Edmondson, Dov Zohar, Andrew Hale and Ian Glendon.

Understanding the psychology of individuals in relation to different kinds of contexts, whether in teams or emergency situations, has made a difference in the way safety could be thought and promoted in organisations. Conceptualising safety climate through psychological and psychosociological insights allowed researchers and practitioners to identify key dimensions that favour interactions conducive to safe performances in the way people interact, speak up and feel empowered to do so.

Describing principles of team leadership that promote safety and translating this knowledge into crew resource management programmes are some examples of important contributions, which were developed in the 1980s and 1990s with the help of this tradition (and the next one). Some of these authors extended their perspective from individuals to systemic, safety management systems and directions, to include a broader view, with links to the safety culture concepts.

I.4.7 Human Error, Interface Design and System Safety

Closely connected to this psychological and psychosociological tradition but with a stronger link to an engineering mindset and ecological perspectives of cognition are traditions exploring human error, interface design and the system view of safety/accidents. Cognitive (system) engineering, naturalistic decision-making but also system safety as promoted by authors such as Jens Rasmussen, James Reason, Donald Norman, David Woods, Erik Hollnagel, Gary Klein and Nancy Leveson innovated in the field of safety in many ways.

This tradition made a difference with its conceptualisation of human error. Clarifying this important issue through cognitive models of how the brain (and body) operate in complex environments opened new preventive strategies and options. Instead of immediately blaming individuals, the tradition argued that it was best to look into contexts in relation to strengths and limitations of cognition. Expanded through systemic models of safety and accidents, and often served by appealing and heuristic visuals, this new understanding of how individuals perform in real-life situations proved invaluable to practitioners.

The engineering or design rationale of this tradition indeed tremendously contributed to its practical relevance as in the field of interface design for which guidance was needed in order to incorporate models of cognition that would increase the expected reliability of the coupling between operators, displays and the world. But prevention practices in companies, such as learning from events, also greatly beneficiated from these developments by targeting measures beyond individuals, including design of technology or organisation.

I.4.8 Beyond the Anglo-Saxon World, a Few Words …

Beyond the Anglo-Saxon context (i.e. United States, UK, Australia), in other European countries like France, Germany, Belgium, the Netherlands, Spain, Italy, Sweden, Norway, Denmark* or Finland, some research traditions in safety were also developed, but it is beyond the scope and the current knowledge of the author to embrace this wealth of works in a systematic way. As an indication, for a country (that I know sufficiently) such as France, some distinctive research traditions in the 1980s and 1990s can be identified.

They are based on psychology, ergonomics, sociology and political sciences with authors such as Claude Gilbert (political science), Jacques Leplat, François Daniellou, René Amalberti (psychological and cognitive ergonomics) or Gilbert De Terssac and Mathilde Bourrier (organisation and work sociology). Explicitly linked to existing French research traditions and translated in the safety field, they were also, but in different ways, in dialogue with the Anglo-Saxon traditions as with Jacques Leplat, then René Amalberti or Mathilde Bourrier. These authors established bridges or connections between French and Anglo-Saxon writers, while remaining anchored in a French background of human and social sciences.

I.5 Research Traditions as Frames

Clearly, this attempt at delineating research traditions can be perceived as simplistic and quite imperfect because authors could be classified in or connected with several of these traditions and not exclusively in a single one. For instance, an author such as Weick could fit in other traditions aside from high-reliability organisations, including the socio-constructivist perspective but also safety climate, leadership and management traditions. Another writer such as Hopkins, initially connected to the safety regulation and socio-legal tradition could be associated with the incubation, safety culture and learning tradition, but also with the high-reliability organisation tradition.

Another simplification is that some categories are distinguished whereas it would make sense to group them as for instance the traditions studying disasters (or errors) rather than studying daily operations (reliability, culture or resilience). This ordering, which classifies authors, models and topics, nevertheless, allows us to make four important points, which structure the background of this book.

First, it appears that the study of safety as an object has always strongly depended on a point of departure, which consists of an established discipline

---

* Of course, cognitive engineering and system safety developed with the input of Risøe research (Denmark), with writers like Rasmussen and Hollnagel.

or an existing body of knowledge bringing its own specific conceptual, methodological, analytical and empirical background to the field. The boundaries created by this process operate through a cognitive closure, they cut a slice of the world through their methodological and conceptual lenses, which is reinforced by networks, funding, publications or academic careers. It is important to keep this idea in mind for later discussions.

Second, it is important not to underestimate the diversity of conceptual, methodological and ideological assumptions behind these traditions, and the impossibility of reducing one to the other ... they cannot be articulated easily, they are based on diverse options and are not always compatible. Some see it as a problem, others as a virtue. I see it as an expression of the irreducible complexity of the world, and of the need to design research strategies to cope with it.* In this respect, this book celebrates diversity while recognising the importance of combining perspectives, although a very challenging endeavour.

Third, it is important to notice that the traditions identified above are often a mix of different disciplines as for instance with the HRO research in which political scientists, organisational psychologists and social psychologists worked together. A similar description could be made for other schools as with human factors, cognitive-system engineering and system safety. Contrasting the disciplinary backgrounds of high-reliability organisation and resilience engineering shows differences as well as commonalities, with the possibility of connections and hybridisations (Le Coze, 2019).

Fourth, these traditions have as a consequence an interest in different kinds of actors involved in high-risk systems, from front-line operators (e.g. a pilot or a process operator) to regulatory ones (e.g. an inspector). An option, which was considered for this introduction at one point, was to classify traditions between micro, meso or more macro levels of analysis targeting various actors, but this type of distinction also has its limitations.

The fifth point is further developed next and concerns these 10 to 20 years, roughly between the 1980s and the 1990s, during which the development of these research traditions shaped the core assumptions, concepts and directions in safety science for the following decades.

I.6 The 1980–1990s: A Golden Age of Safety Research?

Indeed, considered from this perspective, the 1980s and 1990s were a kind of 'golden age' for safety research, which exerted a strong influence throughout the first decades of the 21st century. A wave of high-profile accidents in the

* One explicit formulation of this research problem was expressed by Rasmussen (1997).

1980s already mentioned (e.g. Chernobyl, 1986; Bhopal, 1984; *Challenger*, 1984, etc.) ensured a sustained attention and propelled a generation of researchers who shaped the field, providing its thrust and academic legitimacy.

Key books were published by many whom could now be called pioneers, establishing through their research, publications and social activities, the above-mentioned research traditions (Turner, 1978; Braithwaite, 1984; Perrow, 1984; Rasmussen, 1986; Hale and Glendon, 1987; Wildavsky, 1988; Reason, 1990; Hollnagel, 1993; Roberts, 1993; Rees, 1994; Woods et al, 1994; Vaughan, 1996; Flin, 1996; Reason, 1997; Hopkins, 1999; Hoods et al, 2000; Rasmussen and Svedung, 2000; Weick and Sutcliffe, 2001). So, safety research is a recent product (from an academic point of view) shaped by 20 years, the 1980s and 1990s, of intense publications establishing the traditions mentioned above.

Journals were created in the late 1970s and early 1980s, such as *Safety Science* (formerly the *Journal of Occupational Accidents*), *Reliability Engineering and System Safety, Journal of Hazardous Material, Journal of Loss Prevention in the Process Industry, Accident Analysis and Prevention, Journal of Contingencies and Crisis Management* or *Journal of Risk Research*, which hosted some of these debates, and created a background of dynamic and active research.

To this, we can add many professional journals in domains such as mining, pipelines, aviation, nuclear, and so on, in which safety is regularly discussed very often and understandably so from a practical point of view. The influence of these different traditions can be found in this literature. But safety can also be discussed in journals that are not dedicated safety journals.

One can find important contributions in the *Journal of Applied Psychology; Ergonomics; Human Factors; Cognition, Technology and Work; Administrative Science Quarterly; Law and Society; California Review of Management; Organization Science, Regulation and Governance; Journal of Organizational Behaviour* (this is far from being an exhaustive list).

Debates ensured the visibility and tensions needed for a scientific field to be (feel) alive. The opposition between two schools, on the one side Normal Accident and, on the other, High-Reliability Organisation framed by Sagan was probably one of the most referred to debates in the safety literature. Debates about the relationship between safety climate and safety culture are another example. But successful safety models, in particular the acclaimed Swiss Cheese Model of James Reason, which was one of the most visible, visual and practical of these models propelled the field as a new one during this 'golden age'.

This period constitutes therefore a beginning for most of the notions currently discussed, high-reliability organisations, safety culture or resilience engineering (and related ideas such as the new view versus old view of human error, safety differently, Safety I and II), which have been quite successful and popular over the years. And they all have their roots in the 1980s and 1990s. Even Resilience Engineering (Hollnagel, Woods and Leveson, 2006), which could be initially perceived as a recent development of the early 21st century pursues the thread of ergonomics, cognitive engineering and the system view of safety, which forms one of the research traditions identified above (Le Coze, 2019).

I.7 Where Are We Now?

One question is 'Where are we now?' The combination of globalisation, climate change, technological innovations and the recurrence of high-profile accidents in safety-critical organisations (e.g. Fukushima Daïchi, 2011; Deep Water Horizon, 2010) during the past two to three decades shaped a different context in the first decades of the 21st century. The usual suspects associated with these trends include standardisation, financialisation, ecological degradation, climate change, commodification, neo-liberalism, digitalisation, disruption, self-regulation, and so on (Le Coze, 2017, 2018).

One question is how has safety research adapted to this new context? An option to find out is to follow the research traditions, which have exerted an enduring influence through their networks of writers, practitioners and researchers over the years. Smith and Hoffmann (2017) for human error, interface design and system safety; Flin, Youngson and Yule (2015) or Edmondson (2019) for safety climate, leadership and management; Ramanujam and Roberts (2018) for high-reliability organisations; or Drahos (2017) for the safety regulations and socio-legal perspective, are typical examples of books in which one can see how research traditions consolidate their basis, and evolve to incorporate new insights.

Another option followed in this book is to search the literature and look for new empirical insights, concepts, ideas and research within and across existing research traditions. Instead of remaining within one established tradition, the strategy is to think simultaneously inside, across and beyond these traditions while attempting to keep a big picture of safety research.

Such a strategy relies on the contention that the topic of safety is multidimensional, requires to be thought from various perspectives, and can be associated with multiple approaches to performing research. There is not only one way of progressing in safety research. When following this idea, it appears that a new generation of researchers, sometimes themselves students of some of the pioneers mentioned earlier, have specialised in the context of these established safety research traditions.

They pursued the pioneers' lead through PhDs and then publications (articles, books) at the end of the first decade and during the second decade of the 21st century (2000s–2010s). They occupy research positions within a diversity of institutions and are involved in the study of a range of safety-critical activities (e.g. petroleum industry, chemical industry, nuclear power plants, hospitals, civil aviation, gas transport, air traffic control, finance, railways, etc.).

A selection of some of these writers that I could think of were invited to participate in the Paris workshop or invited later to write a chapter. As explained earlier, at the beginning of this introduction, this is not a systematic and exhaustive coverage but one that offers an interesting perspective of the field when approached from this second option, which consists in departing from the first option of considering one research tradition.

I.8 Within, Across and Beyond Research Traditions

What characterises the work of these authors in the context of the changes over the past 20 to 30 years mentioned above when looking inside, across, beyond and throughout existing research traditions? What do these writers do? They do a number of different things. First, they apply, develop, expand, adapt or challenge existing concepts developed during the 'golden age' within the research traditions (e.g. high-reliability organisations, cognitive engineering, human factors and system safety, normal accident, safety climate/culture, etc.).

Second, they do so by introducing new disciplinary insights from different fields translated in the context of safety research. These include sociology of science and technology; philosophy of science; visual studies; computer-supported coordinated work; sociology of profession; project management; actor-network theory; distributed, embodied or extended cognition; and globalisation studies.

But they also, with the help of these disciplines, address contemporary issues, represented by some of the trends enumerated above (e.g. standardisation, digitalisation, outsourcing, globalisation), and/or they develop new concepts/models (e.g. sensework, epistemic accidents, etc.).

And, finally, they apply models, concepts and ideas from safety research in concrete practical situations but are also reflexive about their use and are sometimes critical about the success and limits of such attempts. They wonder about the increase of safety as both a practice and a profession, and the interaction between research and practitioners in the context of an explosion of safety as a consulting market.

Although each chapter in Section I of this book is a mix of these four ingredients to various degrees, one proposal is to organise them to emphasise what I believe to be interesting evolutions, challenges and new directions in safety research. Empirical, theoretical, critical or practical dimensions of the chapters are distinguished in this respect.

The empirical aspect targets contributions clearly addressing practices, transformations or trends currently observed while studying safety in high-risk systems. The theoretical aspect concerns chapters discussing, formulating, replacing or adding new concepts in the field, from one or sometimes several research traditions through a kind of mixture typical of the interdisciplinary nature of safety research.

The critical and practical aspects distinguish chapters wondering about the dynamics of safety research in relation to practices, actors and institutions, the former exploring its performative dimensions (wondering about a potential negative side of safety research development), the latter investigating the possibilities, limits and conditions of its relevance and usefulness for practitioners.

This second option of understanding safety research not only operates within but also across and beyond established research traditions because each can be questioned from either an empirical, theoretical, critical or practical point of view, and possibly from all of these points of view. If one takes high-reliability organisation as an example of tradition, empirical and theoretical studies can pursue the programme of the tradition, but this tradition can also be investigated using the critical or practical approach.

Proceeding this way consists of considering safety research from two broad angles cutting across traditions: *safety as a research topic* (from an empirical or theoretical angle) or *safety research as a topic* (from a critical or practical viewpoint). In the former, safety is the object which researchers describe, explore, conceptualise or explain, in the latter, the dynamics of safety research are the object which researchers study.

Taken together, the following chapters offer a portion of some of the challenges of safety research in our current times, expressed from empirical, theoretical, critical or practical perspectives within, across and beyond research traditions. Let us briefly introduce the chapters of Section I.

I.8.1 Safety as a Research Topic

I.8.1.1 Studying (Fairly) New Empirical Realities

On the empirical side of safety research, the increasing combination of standardisation and digitalisation in relation to work practices is explored and questioned by Almklov and Antonsen in Chapter 1: *Standardisation and Digitalisation: Changes in Work as Imagined and What This Means for Safety Science*. Whereas standardisation is not a new topic in bureaucracies, its amplification through globalised trends and its new shape as algorithms penetrate deeply organisational processes raises new positive and negative potentialities when it comes to safety.

Reader in Chapter 2: *The Interaction between Safety Culture and National Culture* explains how safety culture as a very popular construct used for assessing and managing safety across countries must incorporate a national component into its rationale. But separating what is the influence of a specific geographical context from what is the influence of a particular organisation might not prove to be as easily done as what was initially thought.

Gotcheva, Aaltonen and Kujala in Chapter 3: *Governance for Safety in Inter-Organisational Project Networks* introduce the project dimension into the field of safety, stressing the importance of investigating the specific nature, dynamics and properties of these networks of actors and organisations. Because of their temporary but also highly extended properties through a myriad of contracted organisations, these configurations represent a real challenge for managers and regulators.

Engen and Lindøe in Chapter 4: *Coping with Globalisation. Robust Regulation and Safety in High-Risk Industries*, address the new globalised landscape and technological innovations that safety regulatory regimes face and how such institutions adapt to cope with these changes. Discussing the property of robust regulations, they introduce and explore issues of regulatory design principles, but also of trust and power that structure the conditions under which high-risk systems are regulated.

I.8.1.2 Developing (Fairly New) Conceptual Lenses

On the theoretical side of safety research, Downer introduces, in Chapter 5: *On Ignorance and Apocalypse: A Brief Introduction to 'Epistemic Accidents'*, the idea of an epistemic accident, derived from a philosophical argument that there are intrinsic limits to the human pretention of reaching a true picture of reality, and concludes with a provocative statement about the impossibility of preventing accidents, which differs from the classic view of *Normal Accidents*, particularly in the context of (socio)technical innovations.

Antonsen and Almklov in Chapter 6: *Revisiting the Issue of Power in Safety Research*, emphasise the need to consider power as one important perspective to conceptualise the problem of safety, after decades of (safety) culture as a dominant topic across industries. They see power as a ubiquitous phenomenon, albeit through various forms, pervading the conditions of safe operations.

Through Chapter 7: *Sensework*, Haavik considers that the way safety is portrayed in ethnographic descriptions of safety critical organisations should take into account more carefully than done so far, the embedded aspect of daily practices of the diversity of actors populating these systems. Communication, coordination and cooperation are cognitive, social and material sensemaking processes, which should be better understood as interwoven levels of explanation.

In Chapter 8: *Drift and the Social Attenuation of Risk*, Pettersen Gould and Fjæran propose to understand drift from a wider perspective than usually understood and they analytically distinguish for this purpose between practical, organisational and societal layers of drift. Moreover, they articulate these three levels of descriptions with a substitute to the amplification of risk: the attenuation of risk. Risk attenuation allows for relaxed regulations, increase in production and misplaced trust of public opinion in capitalist forces.

I.8.2 Safety Research as a Topic

I.8.2.1 Critical Reflection on the Performative Side of Safety Research

From what is considered in this book as a critical angle of analysis, Waring and Bishop explore in Chapter 9: *Safety and the Professions: Natural or Strange Bedfellows?* the interaction between safety in professions and safety as a profession. For professionals, safety is a core aspect of their expertise, particularly in safety-critical organisations (e.g. surgeons, pilots). Their legitimacy is found in their ability to perform safely. But safety has also slowly become in itself an independent expertise (i.e. experts in human error, cognition and system safety), independent of existing established professions (e.g. surgeons, pilots), and is likely to provide insights outside the precincts of older professions, creating new social interactions, cultural alterations and power issues.

In Chapter 10: *Visualising Safety*, Le Coze introduces the importance of thinking of safety research as a product of visualisations. Going through a selection of some of the most iconic ones in the field, he moves on to criticise both the lack of a better grasp of their centrality in safety research and the lack of an update or alternatives to some of these most revered ones in relation to the conceptual and empirical changes of the past years. Some new visualisations are modestly suggested.

Finally, in Chapter 11: *The Discursive Effects of Safety Science*, Bergström continues this line of critical studies by questioning the performative role of safety research when concepts travel from articles or books to regulations, using resilience and just culture as examples. The distortion, departure or unintended effects of what researchers initially wished to contribute to is analysed through a Foucauldian, discursive analysis. When migrating to regulatory contexts, concepts can become an absence of something (e.g. resilience or just culture), which can then be turned into reasons for blaming, rather than improving practices, as originally expected.

I.8.2.2 Practical Concerns on the Relevance of Safety Research

Slightly differing from the critical angle, the practical one explicitly considers the possibilities, limits and conditions for an effective link between research and practices in safety-critical systems. In this respect, Hayes discusses, in Chapter 12: *Investigating Accidents: The Case for Disaster Case Studies in Safety Science*, the value of studying major events instead of smaller-scale events (e.g. near misses, incidents) often approached in the context of daily operations.

Rather than opposing two modes of studying safety (after an event, or in daily operations), Hayes insists on the heuristic value of telling stories, of writing narratives, in which professionals can reflect about their own practices. Relying on the works of several academics who made retrospective analyses a core ingredient of safety research, Hayes argues for the necessity of maintaining a strong commitment to these types of case studies.

Reiman and Viitanem also wonder in Chapter 13: *Towards Actionable Safety Science*, about the conditions of a good use of safety research for practitioners, and the other way around. They bring insights into the various relationships one can experience between the two positions, insisting on the need to consider what separates the two, mainly discussing their respective assumptions about what constitutes a valuable outcome for research or practice (i.e. knowledge advancement versus practical relevance). Referring to the processes of adaptation or translation from research to practice and practice to research, they illustrate their argument with two different methods and concepts: AcciMaps and safety culture.

Shorrock also elaborates in Chapter 14: *Safety Research and Safety Practice: Islands in a Common Sea*, on the relationship between safety research and safety practice through a systematic exploration of which tasks dominate the daily activities of researchers (e.g. reading and publishing articles, writing research proposals) and practitioners (e.g. conducting audits, writing safety policies) while conveying a structural interpretation of why the two are so frequently separated, with only limited cooperation. Shorrock concludes with some avenues for remedying the situation, and advocates for the need to develop a genuine interest in this problem.

I.9 A Safety Research Narrative of Evolutions, Challenges and New Directions

These chapters, classified according to a framework distinguishing theoretical, empirical, critical and practical aspects could have been presented differently, using different classification principles, for instance, referring to their connection, use or combination of existing research traditions. But the interest of the chosen option is to provide a kind of order to the (messy) dynamics of a field, its complex ramifications, hybrid character and mixture of directions while addressing contemporary issues.

As introduced earlier, the idea behind the book is to situate, orient and stimulate. And this proposed order in the chapters can help tell a story. They can allow a narrative of the evolution of an academic field, its new directions but also its challenges, to be built. One possible narrative is as follows.

In the past two decades, it has become more important in safety research within, across and beyond established traditions, to consider the issue of power (Chapter 1, Almklov and Antonsen) now transformed by the new configuration of global actors and institutions which currently shape the landscape of high-risk systems, also creating new challenges for safety regulations (Chapter 4, Engen and Lindøe). But these increasingly global and networked activities of high-risk systems across continents also require for safety management concepts and tools, such as safety culture, to be adapted to the different national contexts where they are used (Chapter 2, Reader).

Moreover, propelled by digitalisation and standardisation among other drivers (e.g. financialisation), these globalised processes create new resources, constraints and problems for safety practices, including a new phase of standardising work (Chapter 1, Almklov and Antonsen) and new environments shaping the sensemaking of sharp-end actors (Chapter 7, Haavik). Considering the extent, speed and disruptive realities behind these reconfigurations of high-risk systems, epistemic accidents (Chapter 5, Downer) might be lurking around the corner more than ever but might also be creating multiple opportunities for global drift, creating a path for disasters (Chapter 8, Pettersen Gould and Fjaeran).

Building the capabilities for learning from major events (Chapter 12, Hayes) is and will remain therefore of great importance, an approach also valuable to improve our understanding of the network properties of current safety-critical organisations but also complex projects for which adapted safety practices and models need further developments than are available today (Chapter 3, Gotcheva, Aaltonen and Kujala). This, more generally, is about the ability of safety research to connect its outcomes and results (e.g. knowledge, tools) to practitioners' needs (Chapter 14, Shorrock).

However, the ideal of a research that is adequately linked to practices (and the other way around) is surely not a straightforward process, requiring translation (Chapter 13, Reiman and Viitanem) with cases of unintended effects of the introduction of safety models which have to be acknowledged and perhaps better anticipated when it comes to regulations (Chapter 11, Bergström) or professions (Chapter 9, Waring and Bishop). Keeping the big picture in mind, from power, through standardisation to unintended effects of safety research (among other challenges), within, across and beyond existing traditions, for the safety management of high-risk systems, might be supported in the future by visualisations powerful enough to replace the most iconic ones in the field (Chapter 10, Le Coze).

This short and simple narrative is not meant to be a closure. It is one among many other possibilities. With the help of pioneers' reflections in Section II (Chapter 15, Flin; Chapter 16, Hollnagel; Chapter 17, Pidgeon; Chapter 18, Roberts; Chapter 19, Schulman; Chapter 20, Weick; and Chapter 21, Westrum), the risk of reducing this wealth of research to a unique ordering scheme and narrative is now limited in a way that would not have been possible without their long-time experience, expertise, wisdom and knowledge of the field translated in each of their chapters. Again, a great thanks to them.

References

Antonsen, S. (2009). Safety culture and the issue of power. *Safety Science*, Volume 47, 183–191.

Braithwaite, J. (1984). *To Punish or Persuade: Enforcement of Coal Mine Safety*. Albany: State University of New York Press.

Braithwaite, J. (1984). *Corporate Crime in the Pharmaceutical Industry*. London and Boston: Routledge & Kegan Paul.

Downer, J. (2011). 737-Cabriolet: The limits of knowledge and the sociology of inevitable failure. *American Journal of Sociology*, Volume 117, 725–762.

Drahos, P. (ed.). (2017). *Regulatory Theory: Foundations and Applications*. Acton: ANU Press.

Dupré, M., Le Coze, J.C. (2014). *Réaction à risque. Regards croisés sur la sécurité dans la chimie*. Paris: Lavoisier. [Risky reactions. Multiple perspectives on safety in the chemical industry].

Edmondson, A.C. (2019). *The Fearless Organization. Creating Psychological Safety in the Workplace for Learning, Innovation, and Growth*. Hoboken, NJ: John Wiley & Sons.

Flin, R. (1996). *Sitting in the Hot Seat. Leaders and Teams for Critical Incident Management*. Chichester, UK: John Wiley & Son.

Flin, R., Youngson, G.G., Yule, S. (2015). *Enhancing Surgical Performance: A Primer in Non-Technical Skills*. Boca Raton, FL: CRC Press.

Fressoz, J.B. (2012). *L'apocalypse joyeuse. Une histoire du risque technologique*. Paris: Le Seuil.

Haavik, T.K. (2011). On components and relations in sociotechnical systems. *Journal of Contingencies and Crisis Management*, Volume 19, 99–109.

Hale, A.R., Glendon, A.I. (1987). *Individual Behavior in the Control of Danger*. Amsterdam: Elsevier.

Hoffman, R.R., Smith, P. (2017). *Cognitive Systems Engineering: The Future for a Changing World*. Boca Raton, FL: Taylor & Francis/CRC Press.

Hollnagel, E. (1993). *Human Reliability Analysis: Context and Control*. London: Academic Press.

Hoods, C., Rothstein, H., Baldwin, R. (2001). *The Government of Risk. Understanding Risk Regulation Regimes*. Oxford University Press.

Hollnagel, E., Woods, D., Leveson, N. (2006). Resilience engineering. Concepts and precepts. Aldershot UK: Ashgate.

Hopkins, A. (1999). *Managing Major Hazards: The Lessons of the Moura Mine Disaster*. Sydney: Allen & Unwin.

Hughes, T. (2005). *Human Built World. How to Think about Technology and Culture*. Chicago: University of Chicago Press.

Jarrige, F. (2013). *Technocritique*. Paris: La découverte.

La Porte, T.R. (ed.). (1975). *Organized Social Complexity: Challenge to Politics and Policy*. Princeton, NJ: Princeton University Press.

Le Coze, J.C. (2018). An essay: societal safety and the global 1,2,3. *Safety Science*, Volume 110(C), 23–30.

Le Coze, J.C. (2017). Globalization and high-risk systems. *Policy and practice in health and safety*, Volume 15(1), 57–81.

Le Coze, J.C. (2016). *Trente ans d'accidents. Le nouveau visage des risques sociotechnologiques*. Toulouse: Octarès. [Thirty years of accidents. The new face of socio-technological risks].

Le Coze, J.C. (2019). Vive la diversité. High reliability organisation and resilience engineering. *Safety Science*, Volume 117, 469–478.

Perrow, C. (1984). *Normal Accidents, Living with High-Risk Technologies* (1st ed.). Princeton: Princeton University Press.

Ramanujam, R., Roberts, K.H. (eds.). (2018). *Organizing for Reliability: A Guide for Research and Practice.* Stanford, CA: Stanford University Press.

Rasmussen, J. (1997). Risk management in a dynamic society: A modeling problem. *Safety Science*, Volume 27, 183–213.

Rasmussen, J. (1986). *Information Processing and Human–Machine Interaction.* North-Holland, Amsterdam.

Rasmussen, J., Svedung, I. (2000). *Proactive Risk Management in a Dynamic Society.* Karlstad: Swedish Rescue Service Agency.

Reason, J. (1990). *Human Error.* Cambridge University Press.

Reason, J. (1997). *Managing the Risk of Organizational Accidents.* Aldershot, Hampshire, England: Ashgate.

Rees, J. (1994). *Hostages of Each Other: The Transformation of Nuclear Safety Since Three Mile Island.* Chicago: University of Chicago Press.

Roberts, K. (ed.). (1993). *New Challenges in Understanding Organisations.* New York: Macmillan.

Swuste, P., van Gulijk, C., Zwaard, W. (2010). Safety metaphors and theories, a review of the occupational safety literature of the US, UK and The Netherlands, till the first part of the 20th century. *Safety Science*, Volume 48(8), 1000–1018.

Swuste, P., van Gulijk, C., Zwaard, W., Oostendorp, Y. (2012). Occupational safety theories, models and metaphors in the three decades since World War II, in the United States, Britain and the Netherlands: A literature review. *Safety Science*, Volume 62, 16–27.

Turner, B, A. (1978). Man-made disaster. *The Failure of Foresight.* Wykeham Science Press, London.

Vaughan, D. (1996). *The Challenger Launch Decision: Risky Technology, Culture and Deviance at NASA.* Chicago: University of Chicago Press.

Weick, K., Sutcliffe, K. (2001). *Managing the Unexpected.* San Francisco: Jossey-Bass.

Wildavsky, A. (1988). *Searching for Safety.* New Brunswick, NJ: Transaction Publishers.

Winner, L. (1978). *Autonomous Technology: Technics-out-of-Control as a Theme in Political Thought.* Cambridge, MA: MIT Press.

Woods, D., Cook, R., Johanessen, L., Sarter, N. (1994). *Behind Human Error.* Farnham, Surrey: Ashgate.

Section I

New Generation

1

Standardisation and Digitalisation: Changes in Work as Imagined and What This Means for Safety Science

Petter G. Almklov and Stian Antonsen

CONTENTS

1.1 Standardisation: From Weber and Taylor to Standards and Digitalisation 5
1.2 An Epistemology of Representation and Standards................................6
1.3 Standards, Accountability and Control...7
1.4 Standardisation as a Part of Neo-Liberal Organising 9
1.5 Accountability 10
1.6 Situated Work: Standardisation and the Particularity of Situations 11
1.7 Digitalisation of Accountability...13
1.8 Conclusion 15
References..16

Introduction

In the last decade, the difference between work as it is imagined and work as it is actually done has been important in the literature on safety and resilience.* The salience and analytical power of this seemingly trivial distinction lies in the fact that it challenges organisational epistemes (dominant discourses of work within organisational and management literature) by underscoring that there is more to work than our descriptions of it. Work, as it is actually done in practice, can never be fully described or prescribed. There is always more to work in context than can be captured by formal descriptions. So while the perfect description of work as it is actually done is more of an aspiration than an achievable goal, we have found it useful in our own research to address situated work, and to challenge the distanced,

---

* See Dekker (2006) and Nathanael and Marmaras (2006). See also Hollnagel (2015) and Haavik (2014) for further discussions of WAI and WAD.

stereotyping descriptions of work as generated from a distance by researchers, management systems and managers (Suchman, 1995). Understanding work as it is performed in real-life contexts is particularly important for understanding the variability of normal work that we associate with resilience.

We argue that it is necessary to follow the call of Barley and Kunda (2001) to 'bring work back in' to organisational theory; to base our theories on empirical studies that seek to understand work as it is performed in real-life situations, with all of their material, social and temporal particularities. In doing this, we are inspired by strands of research from outside the realm of safety science, found across the social sciences more generally.* It is important to critically reflect on the discourses of work within organisations, not only to find out whether or not they are good and true representations, but to discover how their pragmatics influence their organisational decision processes. For example, representations can make some types of work (as well as aspects of work organisation) visible, while obscuring or even suppressing others. Also, as argued by Suchman (1987), it is necessary to understand the role that representations of actions (i.e. procedures, plans) have as *resources for* situated action.†

In this chapter, we will address changes in the dominating discourses of work and their consequences for safety. First and foremost, we argue that there is an increasing tendency towards an ever more detailed standardisation. Timmermans and Epstein, drawing on Bowker and Star (2000), define standardisation as a 'process of constructing uniformities across time and space, through the generation of agreed-upon rules' (Timmermans and Epstein, 2010, p. 71). While we concur with this general definition, we also see a need to add one further aspect: when standardisation meets specialisation across organisational boundaries, it introduces a logic where work processes are increasingly seen as discrete operations, consisting of atomistic products to be delivered, rather than a dynamic flow of actions. This is accompanied with regimes of accountability, of ever more detailed reporting and control, which are also based on standards.

Digitalisation is a critical catalyst for the increased ubiquity and increased level of detail of standardisation. Digitalisation here means making use of digital technology to support the execution and control of work processes. Information infrastructures facilitate more detailed control through descriptions of work as consisting of atomistic standardised entities (see Hanseth and Monteiro, 1997; Bowker and Star, 2000; Almklov et al., 2014b). Moreover,

---

* Much inspiration can be found in the work of anthropologists at Xerox's Palo Alto Research Institute in the 1980–1990s (Suchman, 1987; Orr, 1996). Other inspirational works outside the safety research community are Barley's (1996) detailed study of what technicians actually do, and Strauss et al.'s (1987) studies of medical work.
† In Antonsen et al. (2010), we sought to investigate some of the different roles procedures have in work situations. This was also investigated in a space research context in Johansen et al. (2016).

the digital systems have a performativity of their own. They put constraints on action, and exercise power in ways that a procedure on paper simply cannot do, thereby changing the dynamics between work as it is imagined and work as it actually done.

In the following, we will discuss the trends towards increasing standardisation in organisations, as an element of the dominating discourses of management, as well as a general development in modern life. We will then discuss how this meets the particularities of situated action. Thereafter we will discuss the role of digital technologies, both as enablers of detailed control and carriers of a standardising discourse, as well as the possibilities they present for new forms of situated action. On the surface, standards seem neutral and technical, but they nevertheless have politics.[*] They constrain the leverage for work as it is actually done. The developments we discuss affect these politics in the sense that they increase the level of detail in which this can be done. Also, we will argue that the way in which standardised descriptions of work are themselves part of the transactional coordination of work (e.g. in organisations relying on outsourcing of operational work) and the way they are inscribed in the digital systems through which work is performed, skews these power balances in ways that need attention from researchers interested in reliability, safety and resilience.

1.1 Standardisation: From Weber and Taylor to Standards and Digitalisation

Standards and trans-contextual agreements arise when different localities are connected by infrastructures. Consider, for example, the indication of time at a given place. Even though ways of measuring time had already been around for centuries, there was no definite standard between the many different towns of pre-industrial Europe. The church tower gave the local time, and that was sufficient. However, when trains and railways connected the towns up to one another, the need for timetables necessitated the coordination of measures of time. Common rules or conventions of how to represent the world, through standards or measurement, are ubiquitous features of modern societies. Processes of decontextualisation, commensuration[†] (Espeland and Stevens, 1998; Larsen, 2009) and measurement (Porter, 1995; Crosby, 1997) are intrinsic features of modern life, and their proliferation is a key aspect of globalisation. While anthropologists describe a myriad of ways in which different societies categorise the world, there is, as part

---

[*] See for example Winner (1980), Suchman (1994) and Busch (2011).
[†] Commensuration means to describe different phenomena as belonging to the same class, typically by reference to an external standard or measure.

of modernity and globalisation, a tendency towards description based on atomistic standardised entities (see e.g. Larsen, 2009).

This is the dominant ideology within which organisations are shaped. When we discuss standardisation here, we describe developments that are continuations of developments from the onset of bureaucracy. Developments that, arguably, manifest themselves in new ways within a neo-liberal society. All but the most rudimentary of organisations are based on standardisation of work processes to some extent (Mintzberg, 1993). This is also the core of bureaucracy as described by Weber, and of Taylorism in organisations (Antonsen et al., 2012). When we discuss trends towards standardisation, we discuss the tendency towards a more detailed control of operational work, through procedures and reports based on measurements and standardised entries, moving into work situations where professional discretion has traditionally had leverage. The term 'standards' is often associated with third-party standards: international or industry-wide standards, such as ISO standards, developed and regulated by dedicated public or private agencies. Though these are certainly relevant for our discussion of the standardisation of work processes in organisations, we will refer more generally to a certain epistemology within which work is described and governed.*

1.2 An Epistemology of Representation and Standards

As already indicated, standardisation consists of descriptions that depend on context-overriding 'rules of abstraction' (Almklov, 2008). Work can be described in several ways, but standardised descriptions shape it into certain classes of objects according to rules that are not specific to the singular context. Representations of work adhering to such 'rules' or formats produces decontextualised descriptions that are mobile and comparable across contexts. If you describe your workday there is an infinite possibility of its aspects that you could choose to highlight; you could use a myriad of categories to describe each individual task. Using the predefined codes generated by the administration may not say that much to you, but it transforms them into entities that can be compared and combined with the descriptions of others. Such representations are the key levers of control for modern day management. In the same way that the railroads connecting towns required a harmonisation of measures of time, the management's information system requires that I report my work according to categories and measures that follow certain standards.

Representation always depends on a subject choosing to highlight certain aspects and ignoring others. A standardised description is one where some

---

* See Almklov et al. (2014a) for a discussion of international standards in maritime transport and their relationship to situated practice.

predefined, context-independent aspects are highlighted. The relevant categories are constructed independent of the specifics of the context of work. Thus, by adhering to some sort of rules, your work is described in a way that makes it commensurable (Espeland and Stevens, 1998) with other workdays and the work of other employees. Moreover, it can be enrolled in systems of standardisation-based audit. Standardisation is, in this sense, a way or logic in which the local is described according to global* categories. The reach of the standard, and how far away the rules are accepted, produces the mobility of the representations.[†]

If standardisation is, as argued above, description and prescription according to trans-contextual rules, then it follows that measurement and quantification are forms of standardised representation by which these rules of representation are mathematical. Both standardisation and quantification depend on commensuration (Espeland and Stevens, 1998): to count one must construct entities as belonging to the same class. If a hospital counts the number of medical interventions, or a police district counts their number of solved cases, it also places these in a class of comparable entities, even though each arrest or medical intervention is unique. In the case of management by objectives, such counting has effects on what is done and what is not done. Bevan and Hood (2006) argue that what gets measured is what matters, observing that the obsession with counting has produced a drift towards satisfying measurable goals in the English National Health Service (NHS).[‡] Historian Theodore Porter (1996, p. 4) argues that numbers convey a 'mechanical objectivity'. They are harder to dispute or criticise, as the rules by which they are constructed and manipulated are generally accepted as 'neutral'. In many cases, though, the commensurability of the entities they represent may be problematic. Numbers, due to their associations with neutrality, often obscure the contextual variation from which they are drawn.

1.3 Standards, Accountability and Control

Standards are in one sense information infrastructures (II's) along which certain entities, those who adhere to the rules of abstraction, are highly mobile.[§] II's are networked systems through which structured types of information

---

* These 'global' categories are not neutral either, rather developed in other settings with their own situational constraints, e.g. for managerial control.
† See Bowker and Star (2000). The discussion of the mobility of standardised representations in Almklov (2008) is also inspired by Latour's discussion of 'immutable mobiles' (Latour, 1995).
‡ See also Dekker (2017) for a discussion of how measurement turned into targets cease to be measurements and result in gaming practices.
§ II's are, in practice, typically networked digital systems, but formally a structured list of standardised items can also function as an information infrastructure (Bowker and Star, 1999).

are mobile (Star and Ruhleder, 1996; Bowker and Star, 2000). Importantly, the literature on II's has underscored their sociotechnical nature. They are infrastructures for practice because certain practices are intertwined with them. Hanseth and Monteiro (1997), in a study of ICT systems at hospitals, argued that standards are the ways behaviour is inscribed into II's. II's are typically (but not necessarily) digital systems providing mobility to standardised representations, in our case of work.

The mobility of standardised information in II's also means that control can be exercised along the same infrastructures. In the audit society (Power, 1997; Shore and Wright, 2015) control is exercised by auditing standardised representations of work. Standards have politics; the aspects that are highlighted and mobilised are not neutral. The standards governing work in an organisation makes some aspects of work, and certain types of work, more visible than others (see for example Lampland and Star, 2009; Bowker and Star, 2000). Moreover, these politics are often invisible as well-functioning infrastructures tend to sink into the background and remain unchallenged (Bowker and Star, 2000).

Standardisation in organisations and bureaucracies is not something new. However, when we argue that there is a trend towards more standardisation, we also refer to *changes* in forms of standardisation, its granularity and how it moves into new domains. Professions that have had more leverage for contextual adaptation (to an increasing extent) see their work and its outcomes rendered in terms of standardised descriptions.* These developments can, among other things, be attributed to new organisational forms, external and internal governance by accountability-based structures and audit regimes, as well as the increasing digitalisation of work.

Standards and measurements are cornerstones of managerialism (Pollitt, 1993; Røyrvik, 2011), the generic approach to management decoupled from the specifics of the actual operation to be managed. Within this line of thinking the manager does not need to understand the work itself. Thus, he becomes reliant on decontextualised, standardised and measurable descriptions of work, and its outcomes. He bases his control, also of risk and safety, on systems of audit (Power, 1997, 2007). When we describe an increasing standardisation within organisations and work, it is intimately wed to an ideology of control through such structures. Increasing demands for more standardised descriptions come from actors that are removed from the context in which the actual work is performed. When descriptions are decontextualised, they can be audited by third parties, and controlled from afar. What is lost in terms of context is countered by ever more detailed descriptions (Røyrvik and Almklov, 2013).

Standards and measurement, as II's, tend to sink into the background when work is performed and managed. They become trivialised and seem

---

* See for example Timmermans and Berg (2010) for a discussion of changes in the medical profession.

neutral as descriptions, thus making it harder to address what they lift to the foreground and what they suppress. While this is a general concern for organisational theory, our main concern here is how it might affect aspects of work with regard to safety.

1.4 Standardisation as a Part of Neo-Liberal Organising

In our study of how New Public Management changed the infrastructure sectors in Norway (Almklov and Antonsen, 2010, 2014), we noted that one of the key features of the deregulation and outsourcing-based organisational changes was that coordination increasingly relied on market-based or transactional standardisation. There were different strategies in different critical infrastructure industries we studied, but in all sectors operational work tended to be described in procedures and reporting systems as standardised product-like entities; delimited standardised tasks with a price. This was both the case in companies that relied on external suppliers for their operations, and for those that had internal pseudo-markets. When one organisation bought operative work from another in a task-by-task fashion, the procedure became not only a description of how the work should be performed, but a definition of a product. Work was subject to transactions in which it was conceived as standardised, atomistic tasks. Both in interviews and strategic documentation, even brochures (see Almklov and Antonsen, 2010), the network companies described how standardising technological components facilitated the standardisation of work operations, and how the comparability of tasks made it easier for them to control costs and quality. Operational work in infrastructure sectors, which earlier could be described as some sort of continuous caretaking, was transformed to the execution tasks according to predefined prices and quality demands.

Standards serve to reduce transaction costs, in the words of Busch (2000, p. 275) by 'making certain terms of (always incomplete) contracts transparent to all parties'. Standards gave transparency and control to the buyer and reduced the leverage for the supplier to operate with their own measures of quality, and of how the work should be executed. This also concurred with our observations in the interviews in all infrastructure sectors: the outsourcing-based strategy depended on describing technical components and the tasks to be performed on the system in a standardised manner, broken down into detailed task descriptions. The buyer of services sought to break down the work of maintaining and operating the grid into atomistic tasks. To compare prices and to generally have more control, it was preferable to describe these atomistic tasks in standardised ways. We called this commoditisation. It made work auditable and controllable from afar. And it actively suppressed variability. Doing more is wasteful.

Doing less, or something else rather than what is prescribed is a breach of contract.* Commoditisation of work can thus be seen as a reification of the procedure.

1.5 Accountability

Organisations operating in the audit society (Power, 1997) need to account for their activities in a manner that is transparent for external and internal audits. This also goes for the management of risk (Power, 2007). The auditor usually does not inspect the work or physical systems themselves, but rather whether the systems of internal control are sound. For this to work, such systems must be comparable while relying on standardised ways of representing work. In the Norwegian petroleum industry, for example, the authorities only seldom inspect the facilities themselves. Their role is to monitor that the internal control of the companies is sufficient, that they have updated procedures, statistics over incidents and remedies to address vulnerabilities, or even that they address the informal aspects of safety, such as safety culture (Antonsen et al., 2017). This is also the case in the shipping industry, where the 'International Safety Management Code' (ISM Code) requires all ships to have a Safety Management System (SMS) that follows certain criteria. The auditors and inspectors rarely address practice, but typically concern themselves with inspecting the systems that govern practice (Størkersen et al., 2018; Almklov et al., 2014a). A typical trait in the neo-liberal governance of risk is thus that it relies on a paper trail of risk management. For our power network operators, this was also the case. Their company, the supplier of operative services, would not get paid unless the work was executed according to procedures and documented accordingly. The network operator would then exercise their control based on standardised reports on the performed tasks. Situational variability is of little interest. What counts, and what gets counted, is the execution of the items on the check list. Governance of work and governance of risk through systems of audit-based accountability fits neatly with a modern doctrine of management, where management is an independent discipline. Even a non-expert manager can monitor performance based on measurements and standardised checklists. By means of scorecards, indicators and benchmarks, and irrespective of his knowledge of the actual system, he can monitor work execution and risk (Power, 2007).

This type of management and governance favours standardised representations of work. It also directs resources and attention towards certain types

---

* In principle, that is. In practice, the competent buyer still understood and accepted some deviance, but there was a clear tendency towards less leverage.

of work. In a recent study of changes in the Norwegian justice sector after the 22 of July terrorist attack in 2011 (Nilsen et al., 2015; Almklov et al., 2017), we observed that there was a clear tendency to prioritise measures that gave visible results that could be checked off and counted. Several of our informants stated that the urgent need to 'show results' harmed their ability to undertake strategic efforts directed at more fundamental problems.

The tendency to regard work as the production of auditable results also makes it difficult, but necessary, to make certain types of work and aspects of work organisationally visible. When the operations of a power system, a processing plant or policing are organised as the production of a set of countable predefined tasks, the situated nature of work is not accounted for.

1.6 Situated Work: Standardisation and the Particularity of Situations

In a project on a large petroleum processing facility we asked an expert on these plants what the main role of the operators was. 'It is hard to say exactly' she said, 'they seem to be just driving around in their cars, but they are crucial. The plant wouldn't run without them'. We realised during the study that their role, much as the operators of critical infrastructures such as the power grid, was about keeping the system running by navigating situational variability. Key parts of their work were to facilitate, coordinate and monitor the work of others (contractors and specialists) on the running facilities up against situational constraints (technical, temporal, weather and other contingencies). In both these cases, understanding the situated nature of work was a key to understand reliability and resilience of the system.

Although studying personnel whose key tasks are situational coordination is particularly useful to illustrate how organisationally invisible situational aspects of work are important for reliability and safety, all forms of work count as situated practice. It is action undertaken in the context of 'particular, concrete circumstances' (Suchman, 2007, p. 26). There are always differences between work as it is imagined and work as it is actually done.

Based on our research within various industries* we have observed some characteristics of situatedness that can be hard to capture in standardised representations of work. In the following paragraphs, we will delineate three such characteristics of situated work.

---

* Critical infrastructures (Almklov and Antonsen, 2010, 2014), societal safety (Almklov et al., forthcoming), space research operations (Johansen et al., 2016), petroleum processing plants (Kongsvik et al., 2015), offshore petroleum production (Antonsen et al., 2012), digitalisation of the petroleum industry (Almklov et al., 2014b) and shipping (Almklov et al., 2014a).

1. *The importance of social and personal knowledge*: In communities of practice there are often networks of practical knowledge and repositories
   of social knowledge (Orr, 1996). These grow out of individuals' and
   collectives' experience with the system and with each other. Though
   procedures may contain formal requirements for competence, the
   tacit knowledge of individuals and the social resources they might
   draw on is hard to standardise. For example, fitters and plant operators might have a broad system knowledge that is based on experience from working on other plants, or working within other roles.
   They might also have personal networks of people who are able to
   help them with particular situations.

2. *Temporality and historicity*: Procedures and plans are seldom able to
   capture all temporal dimensions of work. They typically do not
   reflect – at least not to a full extent – the implications of the fact that
   work is situated in time. To name but a few, what does it mean that
   the same task has been conducted several times before; that it comes
   at the beginning or end of a shift, or the general workload at that particular moment; what does it mean that there are parallel activities?
   Work, when it is executed, must be aligned with ongoing activities:
   even the most detailed procedures, such as the pre-planned experiments on the International Space Station discussed in Johansen et al.
   (2016), will need to be implemented in temporal conditions where the
   trajectories of other activities at the station could interfere with their
   operations. Another temporal dimension is the aggregating history
   of the technical system and the work done on it. Aging systems, be
   it a power grid or a processing plant, gradually deviate their design;
   both by the work performed on them and by external perturbations
   that change them in ways that are not possible to fully document.

3. *Materiality of the system and its surroundings*: A transformer station
   might be a square on the technician's chart and the control room
   operator's screen. But each transformer station is different, in terms
   of where they are located and what goes on around them. One fitter
   exemplified this by discussing how stations might be in the base-
   ments of buildings and/or situated next to schools or other institutions, and what that mean for their work with them. Also, as
   mentioned, the material condition of the system changes over time,
   increasing the variability of the material conditions over time.

A related point, combining the latter two items, is the aggregating sociomaterial entanglement* between workers and the materiality of their context. By sociomaterial entanglement we refer to the way that knowledge,

---

* See Barad (2007); Suchman (2007) Orlikowski and Scott (2008), Almklov et al. (2014b),
Niemmima (2017).

practices and technology (and other material aspects of the context) grow together over time, becoming inseparable to the extent that they should be understood by relational perspectives only. The history of previous work is sedimented both in the materiality of the technology and the knowledge and practices of the workers. The current state of an aging infrastructure is the result of the interventions of workers, and the current state of the worker's knowledge of it is the result of a continued interaction with the system. They have grown together.

These are aspects of situatedness making work as imagined different from work as it is performed in concrete contexts. Standardised descriptions of work tend to fail to represent them in a useful way. It is not impossible, but it is hard. Thus, the discourses of work based on standardisation highlight some aspects of work and obscure others. In the following we will argue that the inscription of standards in digital systems further strengthens the contrast between situated work and standardised work as imagined.

1.7 Digitalisation of Accountability

The regimes of ever more detailed accountability in the governance of work would be forbiddingly bureaucratic if it were not for developments in digital systems. Though control by measurement and standards has been the kernel of bureaucracies for centuries, the increasing detail in which tasks can be described and followed up from afar depends on digital II's. Thus, digitalisation is a catalyst for detailed standardisation. In our discussion of the commoditisation of operational work into atomistic standardised tasks, it was striking to observe that the developments were always accompanied with new ICT's. The buyer's coordination of these tasks was made possible by the increasing level of detail in the available data.

However, in addition to just increasing the ubiquity and detail of standardisation, digitalisation carries with it a specific performativity of its own.

Digital compliance and accountability, or 'the tyranny of the drop-down menu': While a procedure, written on paper, accompanied with some admonitions about adhering to them may influence practice, a digital procedure inscribed in a digital system can have a stronger performativity. Work as it is imagined is now often inscribed in the systems *by which we conduct work*. We have, for example, seen how caseworkers in the social services (Almklov et al., 2017) are guided through procedural steps. You just can't go to the next step before the proceeding one is filled out. While a paper procedure is a separate representation of work as imagined, procedures inscribed in the technical systems are inescapable parts of the execution of work. This, of course, can have many advantages in terms of safety during normal operations. For instance, it would be easy to install some software in your car so that it would not start before you put on your seatbelt. It influences practice in a much stronger way than a rule simply telling you to do it would. But such inscriptions also have the potential for reducing the leverage for situational adaptation. Digitalisation of work-control mechanisms will involve transferring some functions and decision-making authority to predefined algorithms, or machine learning, where human decision-making is reduced to a minimum. This may be a good idea for operations and decisions taken under conditions of very low situational variability, and a highly limited number of available action strategies. It can, however, be a very bad idea under conditions where this is not the case. As safety researchers, we know that situational adaptation is key to performing successful operations. We also know that there is sometimes a fine line between normal situations and highly complex crisis situations, and that the tipping point between these two states is highly dependent on the active intervention of human beings (e.g. Roe and Schulman, 2008). If we take as a premise that digitalisation can be a game changer for compliance and surveillance, the well-known trade-off between standardisation and improvisation (e.g. Antonsen et al., 2012; Grøtan, 2015) becomes more pertinent than ever. If sharp-end decision-making in normal operations is replaced with 'computer says no' scenarios, the room for gaining experience and training on situational adaptation may be very limited. In a similar manner, work can be automatically monitored in new ways. Digital traces of your work can and do provide the basis of a new form of surveillance and accountability. See also Antonsen and Almklov (this volume, Chapter 6), for a discussion of the power dimensions of digital accountability.

Digital situatedness: The ever more atomistic standardised descriptions of work may be seen to reduce the leverage for situational adaptation. One may imagine that, particularly in practical work, there will be less leverage for humans to handle the dynamic variability of situations. However, there is an upside of digitalisation, also in terms of adaptive capabilities. The increased amount of all sorts of data can form the basis for new forms of situationally adaptive work, where information from diverse sources is dynamically managed. Typically, we have seen this in control rooms earlier (e.g. Roe and Schulman, 2008), but the control room types of work are proliferating into new settings. In Almklov et al. (2014b), we discussed how new forms of situated practice emerge in settings that rely heavily on sensors and II's. Situatedness becomes, in many ways, trans-local, and situated in settings with a digital materiality.* For example, we observed how the remote monitoring of offshore production data produced new forms of knowledge in an emergent manner. The interpretation processes, interventions, tinkering with displays and ways of representing and aggregating data all led to new understandings: new knowledge that was situated in these non-local webs of interaction. Importantly, even though these kinds of work are heavily based on standardised data flows, the work processes typically consist of extensive interpretative work in which the technologies and the data are interrogated, shuffled, extrapolated, tweaked and refined. New forms of situated knowledge emerge in these interactions and are inseparable from it. Thus, understanding situated practice in such settings depends on addressing the relations between humans and their technological tools. The 'human factor' is inextricably entwined with the technological.[†]

1.8 Conclusion

Standardisation is not new. It is a cornerstone of organisation, and a central element in any analysis of modernity and globalisation (Le Coze, 2017). Standardisation has consequences for how work is managed and the leverage for the situational adaptation that is necessary for resilience. In addition to the increasing level of detail and ubiquity of standardisation, there are at least two ways that standardisation has become more powerful, in which the performative power of standards has become stronger. These are partly interrelated.

First: Standards form the basis for a mode of organising based on a transactional product logic, what we have referred to here as a commoditisation of work. This means that the standardised description of work, its procedures or specifications, are not only 'recipes for reality' (Busch, 2011), they are a 'tender for reality as a product'. Fulfilling procedural demands *is* the product. This reduces the leverage for doing something else and makes it wasteful to do more than what is prescribed.

Second: When work as it is imagined is inscribed in the technical systems and monitored by the systems through which work is governed, it also constrains the leverage for situational adaptability in new ways. The balance between 'Working to rule or working safely?' (Hale and Borys, 2013), is

---

[*] Haavik (this volume, Chapter 7) also discusses some of the situated sociotechnical nature of resilience in highly mediated settings. See also Niemimaa (2017) for a discussion of the sociomaterial nature of operational work in digitalised electricity infrastructures.

[†] A similar relational argument is put forward in the literature on distributed situational awareness (e.g. Stanton et al., 2006).

not a choice in the same way as it was before. The rationalistic discourse where work is imagined as consisting of standardised tasks is inscribed in technical systems and sunk into the infrastructure that enables work to be done.

Digitalisation leads to new challenges in safety research and management: the vulnerability to cyber-attacks, remote control and monitoring of processes, intractable couplings and dependencies between systems, to name but a few. These are all important topics. Our main concern here is how digitalisation changes the ways that work is governed by rationalistic description and prescription, increasing the level of detail possible and the power exercised by governing systems. Our warning about digitalisation has to do with the potential of allowing for unprecedented levels of standardisation and a powerful way of 'hard coding' compliance into the technologies that support the execution and audit-based control of work processes. Weber's dystopian 'iron cage of rationality' would today certainly be a digital cage of rationality.

We are not blind to the upside of digitalisation, but as safety researchers it is our job to take the role as 'party poopers' in innovation processes and major societal changes. As part of the next generation of safety researchers, it is also our job to view major societal and technological changes in light of our existing theoretical frameworks, considering the match between established theories and new developments. In this case, we find that the classic tension between compliance and situational adaptation is more relevant than ever. The role of operational history that we find relevant for maintaining safety in practice is also relevant for theory. While it might be a sign that the new generation of safety researchers is already growing old, we underline that our quest for new theories and perspectives on safety should be wary of tapping old wines into new bottles. Understanding the present is always contingent on our understanding of the past, and this should be borne in mind by all generations of safety researchers.

References

Almklov, Antonsen, Bye, Øren (forthcoming). Organisational culture and societal safety. Collaborating across boundaries. In review for *Safety Science*. (Special issue on societal safety).

Almklov, P. G. (2008). Standardised data and singular situations. *Social Studies of Science,* Volume 38(6), 873–897.

Almklov, P. G., and Antonsen, S. (2010). The commoditisation of societal safety. *Journal of Contingencies and Crisis Management,* Volume 18(3), 132–144.

Almklov, P. G., and Antonsen, S. (2014). Making work invisible: New public management and operational work in critical infrastructure sectors. *Public Administration,* Volume 92(2), 477–492.

Almklov, P. G., Rosness, R., and Størkersen, K. (2014a). When safety science meets the practitioners: Does safety science contribute to marginalisation of practical knowledge? *Safety Science,* Volume 67, 25–36.

Almklov, P. G., Østerlie, T., and Haavik, T. K. (2014b). Situated with Infrastructures: Interactivity and Entanglement in Sensor Data Interpretation. *Journal of the Association for Information Systems,* Volume 15(5), 263–286.

Almklov, P. G. Ulset, G., and Røyrvik, J. (2017). Standardisering og måling i barnevernet [Standardisation and measurement in Child Welfare Services]. In T. Larsen and E. A. Røyrvik (Eds.), *Trangen til å telle. Objektivering, måling og standardisering som samfunnspraksis.* Oslo: Spartacus.

Antonsen, S., Skarholt, K., and Ringstad, A. J. (2012). The role of standardisation in safety management–A case study of a major oil and gas company. *Safety Science,* Volume 50(10), 2001–2009.

Antonsen, S., Nilsen, M., and Almklov, P. G. (2017). Regulating the intangible. Searching for safety culture in the Norwegian petroleum industry. *Safety Science,* Volume 92, 232–240.

Barad, K. (2007). *Meeting the universe halfway: Quantum physics and the entanglement of matter and meaning.* Duke University Press.

Barley, S. R. (1996). Technicians in the workplace: Ethnographic evidence for bringing work into organisational studies. *Administrative Science Quarterly,* 404–441.

Barley, S. R., and Kunda, G. (2001). Bringing work back in. *Organisation Science,* Volume 12(1), 76–95.

Bevan, G., and Hood, C. (2006). What's measured is what matters: targets and gaming in the English public health care system. *Public administration,* Volume 84(3), 517–538.

Bowker, G. C., and Star, S. L. (2000). *Sorting things out: Classification and its consequences.* MIT Press.

Busch, L. (2000). The moral economy of grades and standards. *Journal of Rural Studies,* Volume 16(3), 273–283.

Busch, L. (2011). *Standards: Recipes for reality.* MIT Press.

Crosby, A. W. (1997). *The measure of reality: Quantification in Western Europe, 1250–1600.* Cambridge University Press.

Dekker, S. (2017). *The safety anarchist: Relying on human expertise and innovation, aareducing bureaucracy and compliance.* Routledge.

Espeland, W. N., and Stevens, M. L. (1998). Commensuration as a social process. *Annual Review of Sociology,* Volume 24(1), 313–343.

Grøtan, T. O. (2015). Organizing, thinking and acting resiliently under the imperative of compliance On the potential impact of resilience thinking on safety management and risk consideration. PhD Thesis. Norwegian University of Science and Technology.

Haavik, T. K. (2014). Sensework. *Computer Supported Cooperative Work (CSCW),* Volume 23(3), 269–298.

Hale, A., and Borys, D. (2013). Working to rule, or working safely? Part 1: A state of the art review. *Safety Science,* Volume 55, 207–221.

Hanseth, O., and Monteiro, E. (1997). Inscribing behaviour in information infrastructure standards. *Accounting, Management and Information Technologies,* Volume 7(4), 183–211.

Hollnagel, E. (2015). Why is work-as-imagined different from work-as-done. *Resilience in everyday clinical work.* Farnham, UK: Ashgate, 249–264.

Johansen, J. P., Almklov, P. G., and Mohammad, A. B. (2016). What can possibly go wrong? Anticipatory work in space operations. *Cognition, Technology and Work*, Volume 18(2), 333–350.

Kongsvik, T., Almklov, P., Haavik, T., Haugen, S., Vinnem, J. E., and Schiefloe, P. M. (2015). Decisions and decision support for major accident prevention in the process industries. *Journal of Loss Prevention in the Process Industries*, Volume 35, 85–94.

Lampland, M., and Star, S. L. (2009). *Standards and their stories: How quantifying, classifying, and formalising practices shape everyday life*. Cornell University Press.

Larsen, T. (2010). Acts of entification. In Rapport, N. (ed) *Human Nature as Capacity: Transcending Discourse and Classification*, New York: Berghahn books, 154–178.

Larsen, T. (2013). Introduction: "Objectification, measurement and standardisation". Culture unbound. *Journal of Current Cultural Research*, Volume 4(4), 579–583.

Latour, B. (1995). The "Pédofil" of Boa Vista. *Common Knowledge*, Volume 4(1), 144–187.

Le Coze, J. C. (2017). Globalisation and high-risk systems. *Policy and Practice in Health and Safety*, Volume 15(1), 57–81.

Mintzberg, H. (1993). *Structure in fives: Designing effective organisations*. Prentice Hall.

Nathanael, D., and Marmaras, N. (2006). The interplay between work practices and prescription: A key issue for organisational resilience. In *Proceedings of the 2nd Resilience Engineering Symposium* (pp. 229–237).

Niemmimaa, M. (2017). *Performing continuity of/in smart infrastructure: Exploring entanglements of infrastructure and actions* (PhD thesis). Turku School of Economics, Finland.

Nilsen, M. Almklov, P. Albrechtsen, E., and Antonsen, S. (2015). Risk governance deficits revealed by the Oslo terror attacks. *Proceedings of ESREL 2015*.

Orlikowski, W. J., and Scott, S. V. (2008). 10 sociomateriality: Challenging the separation of technology, work and organisation. *The Academy of Management Annals*, Volume 2(1), 433–474.

Orr, J. E. (1996). *Talking about machines: An ethnography of a modern job*. Cornell University Press.

Pollitt, C. (1993). *Managerialism and the public services: Cuts or cultural change in the 1990s?* Blackwell Business.

Porter, T. M. (1996). *Trust in numbers: The pursuit of objectivity in science and public life*. Princeton University Press.

Power, M. (1997). *The audit society: Rituals of verification*. Oxford: OUP.

Power, M. (2007). *Organised uncertainty: Designing a world of risk management*. Oxford University Press on Demand.

Roe, E., and Schulman, P. R. (2008). *High reliability management: Operating on the edge*, (Volume 19). Stanford University Press.

Røyrvik, E. A. (2011). *The allure of capitalism: An ethnography of management and the global economy in crisis*. Berghahn Books.

Røyrvik, J., and Almklov, P. G. (2013). Towards the gigantic: Entification and standardisation as technologies of control. *Culture Unbound: Journal of Current Cultural Research*, Volume 4(4), 617–635.

Shore, C., and Wright, S. (2015). Audit culture revisited. *Current Anthropology*, Volume 56(3), 421–444.

Stanton, N. A., Stewart, R., Harris, D., Houghton, R. J., Baber, C., McMaster, R., ... Linsell, M. (2006). Distributed situation awareness in dynamic systems: Theoretical development and application of an ergonomics methodology. *Ergonomics*, Volume 49(12–13), 1288–1311.

Star, S. L., and Ruhleder, K. (1996). Steps toward an ecology of infrastructure: Design and access for large information spaces. *Information Systems Research*, Volume 7(1), 111–134.

Størkersen, K. V. (2018). Bureaucracy overload calling for audit implosion: A sociological study of how the International Safety Management Code affects Norwegian coastal transport. PhD Thesis the Norwegian Univsersity of Science and Technology.

Suchman, L. (1995). Making work visible. *Communications of the ACM*, Volume 38(9), 56–64.

Suchman, L. (2007). *Human-machine reconfigurations: Plans and situated actions*. Cambridge University Press.

Suchman, L. A. (1987). *Plans and situated actions: The problem of human-machine communication*. Cambridge University Press.

Suchman, L. (1994). Do categories have politics? *Computer Supported Cooperative Work (CSCW)*, Volume 2(3), 177–190.

Timmermans, S., and Berg, M. (2010). *The gold standard: The challenge of evidence-based medicine and standardisation in health care*. Temple University Press.

Timmermans, S., and Epstein, S. (2010). A world of standards but not a standard world: Toward a sociology of standards and standardisation. *Annual Review of Sociology*, Volume 36, 69–89.

2

The Interaction between Safety Culture and National Culture

Tom W. Reader

CONTENTS

2.1 What Is Safety Culture? 22
2.2 Safety Culture and National Culture...24
    2.2.1 Power Distance..26
    2.2.2 Collectivism 27
    2.2.3 Uncertainty Avoidance 27
    2.2.4 Masculinity...27
    2.2.5 Short-Term Orientation 27
2.3 Implications for Practice 28
2.4 Safety Culture in Air Traffic Management 28
    2.4.1 Developing a Methodology for Investigating Safety
        Culture Internationally 29
    2.4.2 Establishing a Reliable Safety Culture Tool.........................30
    2.4.3 Safety Culture Is Associated with National Culture...........31
    2.4.4 Safety Culture Benchmarking Should Take into Account
        National Culture 31
    2.4.5 The Influence of National Culture on Safety Culture
        Varies According to Employee Role 32
2.5 Conclusion 33
References...34

Introduction

In this chapter we explore the interaction between safety culture and national culture. Safety management in high-risk industries often crosses national boundaries: for example, within multi-national companies (e.g. oil and gas), industries that operate globally (e.g. aviation, shipping) and organisations that must interact to maintain safety (e.g. suppliers, operating companies). Furthermore, workforces are increasingly diverse in nature, with people from different backgrounds and cultures interacting together to ensure safety (e.g. healthcare, where teams are often multi-national in nature). Safety culture is often identified as an organisational characteristic critical for avoiding accidents, as it relates to norms and values for managing risk. However, institutional norms and values are a product of wider societal factors (Erez and Gati, 2004), and thus investigating the interaction between safety culture and national culture is important. For example, in terms of understanding how safety culture develops, how it can be reliably measured and how policies can be developed for improving safety in diverse operating environments.

In the following sections we define safety culture, and then consider in further detail the requirement to consider national culture into its conceptualisation. We then explore potential interactions between theories of national culture and safety culture – particularly through reference to work by Hofstede (1983). Finally, we examine this through drawing on empirical observations from one of the largest ongoing international safety culture investigations in the world: the measurement and monitoring of safety culture in European Air Traffic Management.

2.1 What Is Safety Culture?

Safety culture is a facet of organisational culture, and refers to the norms, values and practices shared by groups in relation to safety and risk (Cooper, 2000; Guldenmund, 2000). It came to prominence after the Chernobyl, *Challenger* and Piper Alpha disasters, in which institutional pressures to prioritise 'production' over 'safety' negatively influenced values, practices and systems for managing safety (Paté-Cornell, 1993; Pidgeon, 1998; Reason, 2000). Conceptualisations of safety culture can be somewhat ambiguous, and this reflects its diverse empirical and theoretical origins.

Consistent with the broader organisational culture literature, two general theorisations emerge (Schneider et al., 2013). First, quantitative approaches that focus more on the norms and values held by organisational members (e.g. values and attitudes towards safety), and second, more qualitative approaches that examine the assumptions and everyday practices that embody safety culture.

Early research on organisational culture primarily drew on the latter form of conceptualisation, with case study investigations of large-scale accidents, ethnographies, observations and interviews being used to understand the psychological and behavioural manifestations of the concept, and its relationship with safety outcomes (Guldenmund, 2000). From this perspective, safety culture is an emergent and shared property that is symbolised by the stories and narratives that permeate an organisation (Berger and Luckmann,

1991), and developed through safety norms and taboos that are constructed around issues of professionalism, identity, group boundaries and conflicts of interest (Antonsen, 2009; Collinson, 2003; Douglas, 1992).

However, research on safety culture has increasingly moved towards the realm of quantitative psychometric measurement (i.e. the use of surveys). This is, in part, due to surveys better operationalising safety culture in terms of providing quantitative data (e.g. average survey response scores) on the state of safety, clearly distinguishing between the different forms of thinking and practice on safety; being psychometrically robust, easy to use and associable with other data points (e.g. outcome variables). Surveys also have benefits in terms of being less time-consuming to run, and easy to repeat and compare. However, the use of surveys to measure safety culture has been critiqued, as they do not necessarily provide insight into the subtle or particularly problematic aspects of safety culture (Guldenmund, 2007). Qualitative methods are better suited for achieving this (i.e. arriving at nuanced understandings of safety-related activities), and ideally approaches from the psychometric and anthropological traditions should be used together when assessing safety culture (Mearns et al., 2013). The role of safety culture assessments remains debated overall, with safety culture being argued as poorly understood in terms of its relationship with safety systems and processes (Reiman and Rollenhagen, 2014), and the process of developing culture towards a desired end state being argued as more important than assessments (Guldenmund, 2017).

Within the psychometric tradition, various models have been used to describe safety culture, and the dimensions that are used to characterise safety culture vary. Common dimensions include: management commitment to safety, collaborating for safety, incident reporting, communication, colleague commitment to safety, safety support, compliance, training and safety rules (Guldenmund, 2007; Mearns et al., 2013). Perhaps most crucially, safety culture remains a highly relevant topic to organisational scientists, as it is consistently identified as a causal problem in recent mishaps within the aviation, energy, chemical, rail, financial and energy industries (Leaver and Reader, 2017; Nordlöf, Wiitavaara, Winblad, Wijk and Westerling, 2015; Reader and O'Connor, 2014; Strauch, 2015).

The continued relevance of safety culture means that it is a construct that it is continually measured and monitored in high-risk industries. This work, often using safety culture surveys, examines and evaluates the 'strength' of a given culture. A 'strong' rating is where beliefs and activities in relation to safety are both positive and shared, thereby leading to a reduced likelihood of organisational mishaps. Conversely, a 'weak' safety culture is where perceptions, beliefs and activities for safety are poor and fragmented, increasing accident probability. Empirical research testing the association between safety culture and safety performance supports this. For example, in terms of safety behaviours and incidents (Choudhry, Fang and Mohamed, 2007; Christian, Bradley, Wallace and Burke, 2009; Clarke, 2006). Thus, and despite its contested and somewhat vague nature, safety culture is identified as an organisational characteristic that is measurable, attainable and desirable (Hollnagel, 2014).

Discussions on safety culture inevitably focus on its distinction from safety climate. In this chapter, we conceptualise the two constructs as highly overlapping, with safety climate embedded within safety culture. Safety climate scholars have described it as reflecting 'surface features of the safety culture... at a given point in time' (Flin, Mearns, O'Connor and Bryden, 2000, p. 178), with survey measures investigating and aggregating individual perceptions in terms of organisational prioritisation of safety (Zohar, 2010). Safety culture draws on this conceptualisation too, while also drawing on a wider set of constructs and methodologies. For example, the values and practices that characterise a safety culture are broad, ranging from safety communications, collaboration on safety, capturing and learning from incident reporting, resourcing and managerial prioritisation of safety. Methodologies to study safety culture often rely on surveys; yet can also include case studies, interviews, focus groups and incident analysis. Thus, safety climate is understood as more changeable and rooted in employee perceptions of management commitment to safety, whilst organisational culture refers to a more dynamic construct than safety culture (DeJoy, 2005; Guldenmund, 2007).

A critical difference between safety culture and safety climate is the idea that it is shaped and formulated by group, organisational and societal cultural practices (Guldenmund, 2000; Schein, 1992). This is consistent with conceptualisations of organisational culture. For example, Erez and Gati (2004) describe culture within a multi-level theorisation, whereby the different levels of culture are each nested within a hierarchical structure (e.g. individual → group → organisational → national → global). According to this theorisation, organisational culture is changed through both the leadership and the membership of a given organisation, as well as in the wider environment in which that organisation is situated (e.g. economic, cultural). Safety culture research has often examined the role of leaders and groups in its formation, but the role of national culture is less explored.

2.2 Safety Culture and National Culture

The requirement to adopt a more international approach to studying safety culture is emphasised by the increasingly globalised nature of many safety-critical industries, and the national variations in safety performance and culture that have been observed.

In particular, research has shown that differences in safety culture between international organisations (e.g. prioritisation of safety, communication) that coordinate on safety (e.g. aviation, energy) have contributed to organisational mishaps (Mearns and Yule, 2009; Reader and O'Connor, 2014). Also, as work-forces become more mobile, safety-critical work is increasingly conducted by multicultural, co-located teams (Manzey and Marold, 2009), which can present challenges for safety management where national cultural tendencies (e.g. for challenging authority) create divergences in behavioural expectations and activities (e.g. correcting a supervisor's mistake) for safety practices (Manzey and Marold, 2009; Sivak, Soler, Tränkle and Spagnhol, 1989; Weber and Hsee, 1998). Furthermore, where multi-national organisations function in different regulatory and political environments, research has shown there to be diverse and conflicting (e.g. on safety protocols, inspections) safety objectives that shape safety culture (Aycan et al., 2001; Colakoglu, Lepak and Hong, 2006; Steinzor, 2011). Finally, research investigating safety management in international settings has shown the potential influence of the above issues. For instance, national variations in accident rates on construction projects crossing two countries (Spangenberg et al., 2003), variations in accident rates amongst seafarers who originate from different countries (Lu, Lai, Lun and Cheng, 2012), national cultural factors underlying communication problems (speaking up behaviours, adhering to protocols) in the cockpit (Merritt and Helmreich, 1996; Soeters and Boer, 2000) and national differences in safety procedures and management underlying national variations in medical error (Vincent, 2011).

Yet, until recently, empirically investigating potential links between national cultural norms and safety culture has been challenging. For example, due to safety culture models not functioning successfully in international environments (Bahari and Clarke, 2013), variations in response patterns according to the cultural background of participants (Cigularov, Lancaster, Chen, Gittleman and Haile, 2013), the lack of transferability of productive models used to associate safety culture with performance across cross-cultural environments (Hsu, Lee, Wu and Takano, 2008) and the sheer scale of safety culture data required to rigorously establish associations (Reader et al., 2015).

Nonetheless, a growing body of research has investigated the association between safety culture and national culture, often utilising Hofstede's (2001) national tendencies paradigm. This asserts that individuals from a nation will share a set of core values with fellow citizens, with these values emerging due to shared experiences in educational and cultural institutions (Hofstede and Minkov, 2010). Five key dimensions are reported: power distance, collectivism, uncertainty avoidance, masculinity and long-term orientation (Hofstede and Minkov, 2010; Schwartz, 1992).

Hofstede's conceptualisation of national culture is widely utilised. However, the model is much critiqued (McSweeney, 2002). National culture scholars have debated whether the above conceptualisation is either accurate or appropriate for describing national cultures (McSweeney, 2002). For example, in terms of whether national culture is homogenous, measurable (and distinguishable between societies) and associated with everyday behaviour. Furthermore, critiques have been made of the self-report approach used to distinguish national culture, and whether these can measure the values that constitute national culture. Specifically, if national culture represents implicit norms and values within a society, then the ability of self-report measures to access these is limited (i.e. people might be unaware of their own practices, or they might not know how to compare them to the practices of other, foreign societies). Other models, for example of societal tightness–looseness (which relates to the importance of social norms and extent of sanctioning within societies) (Gelfand, Nishii and Raver, 2006) or alternative conceptualisations of human values (Schwartz, 1992), have been suggested.

Despite these criticisms, there are a number of reasons to utilise Hofstede's national culture model in analyses of safety culture (Hofstede, 2002; Reader, Noort, Kirwan and Shorrock, 2015; Williamson, 2002). First, measurements have good psychometric properties, and are found to be reliable and robust (Hofstede, 2001) providing a strong platform for examining associations between national culture and safety practices. Second, a large body of evidence has emerged linking Hofstede's national culture model with other organisational variables, for example citizenship behaviour, organisational commitment, identification, team-related attitudes and job performance (Taras, Kirkman, and Steel, 2010). Thus, there is existing evidence of organisational practices in different national settings being influenced by the dimensions outlined in Hofstede's national culture model. Finally, many of the dimensions described by Hofstede are close correlates to the practices of interest in safety culture research. For example, safety culture research often examines speaking-up, for which power distance is clearly relevant. Uncertainty in managing risk and difficult social situations (e.g. challenging management) are also of interest to safety researchers, and the dimension of uncertainty avoidance appears to tap into these phenomena. Thus, Hofstede's model of national culture appears particularly relevant to safety culture because the links with safety practices are quite apparent.

To illustrate, we consider the potential relationships between safety culture and the five national culture dimensions outlined by Hofstede below.

2.2.1 Power Distance

Power distance relates to how people within a society think of, and relate to, hierarchies and authority in interpersonal contexts (Hofstede, 2003). A country that is shown to be high on power distance tends to have steep authority gradients between superiors and subordinates, as well as a greater acceptance of hierarchy. This shapes safety culture through influencing whether a junior person can correct an error by their superior, placing primacy on communication about safety to superiors, as well as influencing the likelihood of authority being challenged, and whether open discussion about safety issues (e.g. asking for more resources) between junior and senior organisational members is likely to occur (Mearns and Yule, 2008; Reason, 1997; Soeters and Boer, 2000).

2.2.2 Collectivism

This refers to the extent to which people in a society place primacy on group needs over individual needs, and whether they perceive themselves to be part of the collective. Collectivism has been argued to interact with safety culture in the following way: people in less collective (i.e. individualistic societies) are said to have reduced fear of endangering relationships through highlighting safety problems, reduced embarrassment for making and reporting an error themselves, as well as a greater willingness to critique or break unsafe group norms (Bakacsi, Sandor, András and Viktor, 2002; Soeters and Boer, 2000).

2.2.3 Uncertainty Avoidance

This refers to the national cultural tendencies for avoiding the anxiety caused by risky and ambiguous situations. Uncertainty avoidance has been argued to influence safety culture through shaping whether organisational employees place a greater reliance on technical solutions for managing safety, whether they are prepared to innovate on safety solutions, retaining flexibility within rules and procedures for unexpected scenarios (with employees being empowered to disregard a rule when required), as well as being willing to engage in situations that are socially ambiguous in terms of consequences (e.g. challenging a senior manager) (Helmreich, 1999; Waarts and Van Everdingen, 2005).

2.2.4 Masculinity

This refers to the extent to which a society prioritises caring, well-being and collaborative approaches, or focusses upon materialism, rewards and heroism. High masculinity is generally argued to be a negative influence upon safety culture, because it prioritises competitive and target-focussed behaviours (i.e. prioritising performance) at the expense of consensus, collaboration and institutional learning (e.g. on human error) (Mearns and Yule, 2009).

2.2.5 Short-Term Orientation

Finally, short-term orientation relates to whether a society generally focusses on immediate rewards and dividends, or whether it is constantly trying to adapt and pragmatically solve problems in order to retain long-term growth and sustainability. In general, tendencies for short-term orientation would appear to influence safety culture less positively, as they are likely to create pressures focussing on short-term goals (e.g. performance), plan less for the long-term and have less consideration for ongoing safety problems in general (a focus on 'firefighting').

2.3 Implications for Practice

The association between national culture and safety culture appears important for understanding how safety culture develops and is enacted. For example, if safety culture is in part reflective of how people within a society communicate and act (Aarts and Dijksterhuis, 2003; Berger and Luckmann, 1966), it is still not clear the extent to which safety-related practices and beliefs of employees can be influenced (Berry, 1997). Furthermore, considering the common practice of benchmarking in safety culture, interactions with national culture have implications for how data is interpreted and understood. Safety culture benchmarking is used to identify weak and strong performing units and organisations and recognise opportunities for inter-organisational learning where practices can be shared to improve safety (Mearns, Whitaker and Flin, 2003). Yet, when conducted in an international context, it is not established whether data from different societal environments are comparable: for example, to what extent are 'strong' cultures a product of national contexts rather than robust safety management systems? Finally, in terms of practical interventions for improving safety culture, methods identified as 'best' in one cultural setting might not necessarily transfer well to another. For example, research in multi-national industries shows that safety policy effectiveness is, in part, shaped by their appropriateness to the wider national cultural environment in which they are enacted (Janssens, Brett and Smith, 1995).

2.4 Safety Culture in Air Traffic Management

To explore the above issues around the association between safety culture and national culture, research from the international air traffic management (ATM) industry has been utilised. This has drawn on a ten-year project developing a model for examining safety culture in Air Traffic Services within European countries. ATM is an industry where National Air Navigation Service Providers (ANSPs) work to direct commercial aircraft during take-offs and landings, managing the flow of air traffic across and between countries. It is highly standardised in terms of technical functions, organisational structures and occupational roles, thereby lending itself to the study of interactions between safety culture and national culture.

The project in Europe has been funded and organised by EUROCONTROL, which coordinates the various ANSPs working within Europe. The purpose is to monitor safety culture in the ATM industry, find areas for local improvement and learning, and place the topic of safety culture as central to the work conducted in ANSP. The project was, in part, a response to the Überlingen mid-air collision in 2002 (caused due to ANSP-level problems and resulted in 71 fatalities). To date, more than 30 countries have been involved in the project, with over 20,000 participants in total. Through this research, the following insights on the relationship between safety culture and national culture have been observed.

2.4.1 Developing a Methodology for Investigating Safety Culture Internationally

Through a systematic literature review, interviews and focus groups with 52 ANSP staff in different countries, and the collection of questionnaire data from 1420 staff in four ANSPs, a methodology for investigating safety culture was developed (Mearns et al., 2013). This involved a safety culture survey, followed-up by a set of focus groups in order to understand the survey results. Initial testing of the survey instrument identified weak, but consistent, construct validity across different countries. The survey was further refined, through the results of the confirmatory factor analysis, using focus groups in different countries to examine the specific questionnaire items in greater detail. The items and themes were substantially culled in order to ensure that the questions could be understood and interpreted as equivalently as possible in all of the different cultures and languages. A final safety culture instrument was developed for examining safety culture in European ATM, with this being used as a mechanism to qualitatively explore and highlight safety issues at a local ANSP level (and to compare across ANSPs).

Based on this work, the method used to assess safety culture in different national contexts has remained stable for nearly 10 years. ANSPs are surveyed using the safety culture assessment tool, with item-level analyses (e.g. least-to-most positive responses to items), group-level analyses (e.g. comparing controllers to management, different centres), and dimension analyses (e.g. management commitment to safety, incident reporting) being used to profile safety culture. Surveys are translated into local languages, managed and analysed by a university, and the results are fed back to EUROCONTROL and the ANSP. The results are then presented to management (and groups of employees) in two-hour workshops, with qualitative insight on the practices and contexts underlying good or poor response patterns being generated. For example, the problems in ATM staff reporting and learning from incidents, or in communication between units. Common issues are synthesised into recommendations for the ANSP (e.g. on how to improve incident reporting systems, communication training), with EUROCONTROL supporting their implementation. Many sites have been surveyed multiple times, meaning that the effectiveness of safety recommendations can be assessed through changes in survey response patterns. For some ANSPs, the safety culture survey has become embedded into their safety management systems (e.g. they undertake a survey every three years), for others, safety culture assessments are conducted on a needs-based approach.

Crucially, the project has led to the first published methodology for investigating safety culture in different national contexts; inturn, this has opened up opportunities for investigations around the association between safety culture and national culture.

2.4.2 Establishing a Reliable Safety Culture Tool

Using the aforementioned survey instrument, an investigation was made into whether safety culture can be measured reliably in different societal contexts (Reader et al., 2015). Specifically, and employing multi-group confirmatory factor analysis, the psychometric properties of the safety culture tool and its underlying structure, were examined in 17 European countries from four culturally distinct regions of Europe (North, East, South and West). The sample included ATM operational staff (n = 5176) and management staff (n = 1230).

The analysis focussed on developing a measurement model for four different European regions: Northern Europe (Scandinavian countries, UK), Eastern Europe (former Eastern Bloc nations), Southern Europe (Mediterranean nations) and Western Europe (France, Germany and Benelux). These were clustered together due to their widely acknowledged similar cultural tendencies (Leung, Bhagat, Buchan, Erez and Gibson, 2010), geographic proximity, religious and linguistic backgrounds, socio-political and economic environments, as well as value systems, attitudes, values and work goals (Bakacsi et al., 2002; Brodbeck et al., 2000; Gupta, Hanges and Dorfman, 2002; Jesuino, 2002; Vinken, Soeters and Ester, 2004).

For the first time, this analysis established that safety culture can be measured with psychometric equivalence in different cultural environments (as well as across occupational groups) (Reader et al., 2015). The established model consisted of the following dimensions. First, management commitment to safety, which refers to the extent to which management prioritise safety. Second, collaborating for safety, which refers to group attitudes and activities for safety management. Third, incident reporting, which measures the extent to which respondents believe that it is safe to report safety incidents. Fourth, communication, examining the extent to which staff are informed about safety-related issues in the ATM system. Fifth, colleague commitment to safety, measuring beliefs about the reliability of a given colleague's safety-related behaviour. Sixth, measuring the availability of resources and information for safety management.

The work is significant, because it indicates that safety culture is understood similarly in a range of different cultural environments (i.e. with similar latent dimensions), and thus data can be compared between countries. This is a prerequisite to establishing associations between safety culture and national culture, as well as an attempt towards an international benchmark of safety culture.

2.4.3 Safety Culture Is Associated with National Culture

In the same study (Reader et al., 2015), associations were made between national culture and safety culture. Across the sample of 17 countries, a number of notable findings emerged. Utilising publicly available data on Hofstede's five national culture dimensions (collectivism, power distance, uncertainty avoidance, masculinity and uncertainty avoidance) for countries in Europe, the relationship with the dimensions of safety culture was examined.

A significant negative relationship was found between the safety culture dimensions (mean responses by staff to the safety culture survey at each ANSP) and independent data on collectivism, power distance, uncertainty avoidance, masculinity and short-term orientation for the country in which each ANSP was based. This indicated an association between safety culture and national culture. Further analysis found safety culture to be mostly positive in Northern Europe, somewhat less positive in Western and Eastern Europe and quite negative in Southern Europe. This indicates that national cultural traits may influence the development of organisational safety culture, with significant implications for safety culture theory and practice.

The associations between national culture and safety culture were hypothesised to occur due to high power distance reducing communication openness on safety, high collectivism creating tendencies to not endanger group harmony through challenging unsafe group activity, high uncertainty avoidance creating an over-reliance on rules and reluctance to encounter potentially threatening social situations and high masculinity and short-term orientation leading to the prioritisation of immediate (over long-term) gains.

2.4.4 Safety Culture Benchmarking Should Take into Account National Culture

To investigate the above observation in greater depth, a more detailed examination was made of the association between uncertainty avoidance and safety culture (Noort, Reader, Shorrock and Kirwan, 2016). Through an analysis of 13,616 employees in 21 ANSPs, the implications of an interaction between safety culture and national norms for uncertainty avoidance was considered in relation to organisational benchmarking.

Specifically, this interaction was explored through two mechanisms. First, that norms for uncertainty avoidance are likely to shape how people behave in relation to safety (e.g. willingness to try new safety solutions), influencing how safety culture is perceived (e.g. commitment to safety). Second, that norms for uncertainty avoidance will shape how staff perceive safety-related practices (e.g. social consequences of highlighting the error of a superior), which will also shape how safety culture is perceived (e.g. willingness to report an incident).

In both cases, these have implications for safety culture benchmarking. Specifically, attempts towards an international benchmark will likely always lead to a skewed set of results, whereby units based in low uncertainty avoidant countries will perform better than those in high uncertainty avoidant countries. However, the extent to which safety culture perceptions reflect the actual state of safety management (and not national cultural norms) is not clear: potentially providing a misleading picture on the state of safety across an international system. Also, if the purpose of safety culture is to encourage organisational learning, benchmarking internationally will lead to unidirectional learning, with it not being clear that the safety practices developed to solve a problem in one context will help solve those in another.

To overcome this issue, it is suggested that *safety culture against international group norms* (SIGN) scores are developed to statistically control for the influence of uncertainty avoidance upon safety culture data, and to benchmark within clusters (e.g. European regions) of similar countries rather than across those with highly distinct national cultures. Ignoring the relationship between uncertainty avoidance and safety culture means that, potentially, benchmarking exercises will ignore the structural influences in safety culture perceptions and practices.

2.4.5 The Influence of National Culture on Safety Culture Varies According to Employee Role

Finally, the influence of national culture upon perceptions of safety culture is not constant. Controllers and managers tend not to have differing perceptions of safety culture, whereby managers perceive it more positively (Reader et al., 2015).

To explore this further, interactions between national norms for power distance and the status associated with occupational roles have both been examined (Tear, Reader, Shorrock and Kirwan, in press). Drawing on the same sample used to explore uncertainty avoidance (Noort et al., 2016), managers in ANSP were found to have more positive perceptions of safety culture than those in less-powerful roles (e.g. controllers, engineers), and safety culture is shown to be less favourable in high-power distance countries. However, for the first, an interaction was shown between the two, whereby differences in the perception of safety culture between those in senior, and those in non-senior roles, was found to be deepened by national norms for power distance. This shows the way in which power distance can shape the dynamics between those in different organisational roles. Also, it shows that the effect of national culture upon safety culture perceptions and practices is not necessarily equivalent among all staff working within an organisation.

2.5 Conclusion

Models of organisational culture have long supposed a relationship between national culture and institutional values and practices (Erez and Gati, 2004). This has also been theorised for safety culture, but it has not been demonstrated until now.

Future investigations are required to qualitatively establish how national culture shapes safety-related practices and perceptions within organisations. Furthermore, issues such as acquiescence effects (where participants from a given culture have different response patterns to those from another culture), also appear important. Perhaps the most critical next step is the undertaking and design of longitudinal research: whereby the effect of interventions to improve safety are tested across multiple countries, with their net effect upon safety culture being assessed (and divergences between countries established). Finally, the data reported from the EUROCONTROL study featured different organisations (ANSPs): the influence of national culture upon safety culture has not been examined within a single organisation, and this would appear to be an essential next step in establishing how the interaction both manifests and impacts upon safety.

Nonetheless, the investigation of safety culture in the international ATM industry has revealed a number of important theoretical and practical issues. The theoretical issues focus on how safety culture develops; that it is not independent of societal environments, with national norms (particularly around power distance and uncertainty avoidance) shaping how safety is viewed and enacted. Safety culture does not exist in a vacuum: it is influenced by organisational safety management and the underlying value systems of a given society. This has practical implications for how safety culture is measured, monitored and changed. Although a one-size-fits all model can be taken to measuring safety culture, this might not be useful in identifying improvements, understanding the experiences of all staff groups or determining what is a weak or strong safety culture.

References

Aarts, H. and Dijksterhuis, A. (2003). The silence of the library: Environment, situational norm, and social behavior. *Journal of Personality and Social Psychology,* Volume 84(1), 18–28.

Antonsen, S. (2009). Safety culture and the issue of power. *Safety Science,* Volume 47, 183–191.

Aycan, Z., Kanungo, R., Mendonca, M., Yu, K., Deller, J., Stahl, G. and Kurshid, A. (2001). Impact of culture on human resource management practices: A 10 country comparison. *Applied Psychology,* Volume 49, 192–221.

Bahari, S. and Clarke, S. (2013). Cross-validation of an employee safety climate model in Malaysia. *Journal of Safety Research,* Volume 45, 1–6.

Bakacsi, G., Sandor, T., András, K. and Viktor, I. (2002). Eastern European cluster: Tradition and transition. *Journal of World Business,* Volume 37, 69–80.

Berger, P. L. and Luckmann, T. (1966). *The Social Construction of Reality: A Treatise in the Sociology of Knowledge.* London, England: Penguin.

Berger, P. L. and Luckmann, T. (1991). *The Social Construction of Reality: A Treatise in the Sociology of Knowledge.* London, England: Penguin.

Berry, J. W. (1997). Immigration, acculturation, and adaptation. *Applied Psychology,* Volume 46(1), 5–34.

Brodbeck, F. C., Frese, M., Akerblom, S., Audia, G., Bakacsi, G., Bendova, H. and Brenk, K. (2000). Cultural variation of leadership prototypes across 22 European countries. *Journal of Occupational and Organizational Psychology,* Volume 73, 1–29.

Choudhry, R. M., Fang, D. and Mohamed, S. (2007). The nature of safety culture: A survey of the state-of-the-art. *Safety Science,* Volume 45, 993–1012.

Christian, M., Bradley, J., Wallace, J. and Burke, M. J. (2009). Workplace safety: A meta-analysis of the roles of person and situation factors. *Journal of Applied Psychology,* Volume 94, 1103–1127.

Cigularov, K., Lancaster, P., Chen, P., Gittleman, J. and Haile, E. (2013). Measurement equivalence of a safety climate measure among Hispanic and White Non-Hispanic construction workers. *Safety Science,* Volume 54, 58–68.

Clarke, S. (2006). The relationship between safety climate and safety performance: A meta-analytic review. *Journal of Occupational Health Psychology,* Volume 11, 315–327.

Colakoglu, S., Lepak, D. P. and Hong, Y. (2006). Measuring HRM effectiveness: Considering multiple stakeholders in a global context. *Human Resource Management Review,* Volume 16, 209–218.

Collinson, D. L. (2003). Identities and insecurities: Selves at work. *Organization,* Volume 10, 527–547.

Cooper, M. (2000). Towards a model of safety culture. *Safety Science,* Volume 36, 111–136.

DeJoy, D. M. (2005). Behavior change versus culture change: Divergent approaches to managing workplace safety. *Safety Science,* Volume 43, 105–129.

Douglas, M. (1992). *Risk and blame: Essays in Cultural Theory.* Routledge.

Erez, M. and Gati, E. (2004). A dynamic, multi-level model of culture: From the micro level of the individual to the macro level of a global culture. *Applied Psychology,* Volume 53(4), 583–598.

Flin, R., Mearns, K., O'Connor, P. and Bryden, R. (2000). Safety climate: Identifying the common features. *Safety Science,* Volume 34, 177–192.

Gelfand, M. J., Nishii, L. H. and Raver, J. L. (2006). On the nature and importance of cultural tightness-looseness. *Journal of Applied Psychology*, Volume 91(6), 1225.

Guldenmund, F. (2000). The nature of safety culture: A review of theory and research. *Safety Science*, Volume 34, 215–257.

Guldenmund, F. (2007). The use of questionnaires in safety culture research-an evaluation. *Safety Science*, Volume 45, 723–743.

Guldenmund, F. (2017). *Are Safety Culture Assessments Really Necessary?* Paper presented at the International Conference on Applied Human Factors and Ergonomics.

Gupta, V., Hanges, P. J. and Dorfman, P. (2002). Cultural clusters: Methodology and findings. *Journal of World Business*, Volume 37, 11–15.

Helmreich, R. L. (1999). *Building Safety on the Three Cultures of Aviation*. Paper presented at the Proceedings of the IATA Human Factors Seminar.

Hofstede, G. (1983). The cultural relativity of organizational practices and theories. *Journal of International Business Studies*, Volume 14(2), 75–89.

Hofstede, G. (2001). *Culture's Consequences: Comparing Values, Behaviours, Institutions and Organizations Across Nations*. Thousand Oaks, CA: Sage Publications, Incorporated.

Hofstede, G. (2002). Dimensions do not exist: A reply to Brendan McSweeney. *Human Relations*, Volume 55, 1355–1361.

Hofstede, G. (2003). *Culture's Consequences: Comparing Values, Behaviours, Institutions and Organizations Across Nations*. Newbury Park, CA: Sage Publications.

Hofstede, G., Hofstede, G. J., and Minkov, M. (2010). *Cultures and Organizations: Software of the Mind, Revised and Expanded* (3rd ed.). New York: McGraw-Hill.

Hollnagel, E. (2014). Is safety a subject for science? *Safety Science*, Volume 67, 21–24.

Hsu, S., Lee, C., Wu, M., and Takano, K. (2008). A cross-cultural study of organizational factors on safety: Japanese vs. Taiwanese oil refinery plants. *Accident Analysis and Prevention*, Volume 40, 24–34.

Janssens, M., Brett, J. M. and Smith, F. J. (1995). Confirmatory cross-cultural research: Testing the viability of a corporation-wide safety policy. *Academy of Management Journal*, Volume 38(2), 364–382.

Jesuino, J. C. (2002). Latin Europe cluster: From south to north. *Journal of World Business*, Volume 37, 81–89.

Leaver, M. P. and Reader, T. (2019). Safety culture in financial trading: An analysis of trading misconduct investigations. *Journal of Business Ethics*, Volume 154(2), 461–481.

Leung, K., Bhagat, R., Buchan, N. R., Erez, M., and Gibson, C. B. (2010). Beyond national culture and culture-centricism: A reply to Gould and Grein (2009). *Journal of International Business Studies*, Volume 42, 177–181.

Lu, C., Lai, K., Lun, Y. and Cheng, T. (2012). Effects of national culture on human failures in container shipping: The moderating role of Confucian dynamism. *Accident Analysis and Prevention*, Volume 49, 457–469.

Manzey, D. and Marold, J. (2009). Occupational accidents and safety: The challenge of globalization. *Safety Science*, Volume 47, 723–726.

McSweeney, B. (2002). Hofstede's model of national cultural differences and their consequences: A triumph of faith-a failure of analysis. *Human Relations*, Volume 55(1), 89–118.

Mearns, K., Kirwan, B., Reader, T., Jackson, J., Kennedy, R. and Gordon, R. (2013). Development of a methodology for understanding and enhancing safety culture in Air Traffic Management. *Safety Science*, Volume 53, 123–133.

Mearns, K., Whitaker, S. M. and Flin, R. (2003). Safety climate, safety management practice and safety performance in offshore environments. *Safety Science,* Volume 41(8), 641–680.

Mearns, K. and Yule, S. (2008). The role of national culture in determining safety performance: Challenges for the global oil and gas industry. *Safety Science,* Volume 47, 777–785.

Mearns, K. and Yule, S. (2009). The role of national culture in determining safety performance: Challenges for the global oil and gas industry. *Safety Science,* Volume 47, 777–785.

Merritt, A. and Helmreich, R. (1996). Human factors on the flight deck: The influences of national culture. *Journal of Cross-Cultural Psychology,* Volume 27, 5–24.

Noort, M. C., Reader, T., Shorrock, S. and Kirwan, B. (2016). The relationship between national culture and safety culture: Implications for international safety culture assessments. *Journal of Occupational and Organizational Psychology,* Volume 89(3), 515–538.

Nordlöf, H., Wiitavaara, B., Winblad, U., Wijk, K. and Westerling, R. (2015). Safety culture and reasons for risk-taking at a large steel-manufacturing company: Investigating the worker perspective. *Safety Science,* Volume 73, 126–135.

Paté-Cornell, M. E. (1993). Learning from the piper alpha accident: A postmortem analysis of technical and organizational factors. *Risk Analysis,* Volume 13(2), 215–232.

Pidgeon, N. (1998). Safety culture: Key theoretical issues. *Work and Stress,* Volume 12, 202–216.

Reader, T., Noort, M. C., Kirwan, B. and Shorrock, S. (2015). Safety Sans frontieres: An international safety culture model. *Risk Analysis,* Volume 35, 770–789.

Reader, T. and O'Connor, P. (2014). The deepwater horizon explosion: Non-technical skills, safety culture, and system complexity. *Journal of Risk Research,* Volume 17, 405–424.

Reason, J. (1997). *Managing the Risks of Organisational Accidents.* Aldershot: Ashgate.

Reason, J. (2000). Safety paradoxes and safety culture. *Injury Control and Safety Promotion,* Volume 7(1), 3–14.

Reiman, T. and Rollenhagen, C. (2014). Does the concept of safety culture help or hinder systems thinking in safety? *Accident Analysis and Prevention,* Volume 68, 5–15.

Schein, E. (1992). *Organizational Culture and Leadership.* San Francisco: Jossey-Bass Publishers.

Schneider, B., Ehrhart, M. G., and Macey, W. H. (2013). Organizational climate and culture. *Annual Review of Psychology,* Volume 64, 361–388.

Schwartz, S. H. (1992). Universals in the content and structure of values: Theoretical advances and empirical tests in 20 countries. *Advances in Experimental Social Psychology,* Volume 25, 1–65.

Sivak, M., Soler, J., Tränkle, U. and Spagnhol, J. M. (1989). Cross-cultural differences in driver risk-perception. *Accident Analysis and Prevention,* Volume 21, 355–362.

Soeters, J. and Boer, P. (2000). Culture and flight safety in military aviation. *The International Journal of Aviation Psychology,* Volume 10, 111–113.

Spangenberg, S., Baarts, C., Dyreborg, J., Jensen, L., Kines, P. and Mikkelsen, K. (2003). Factors contributing to the differences in work related injury rates between Danish and Swedish construction workers. *Safety Science,* Volume 41, 517–530.

Steinzor, R. (2011). Lessons from the North Sea: Should "safety cases" come to America? *Boston College Environmental Affairs Law Review,* Volume 38, 417–444.

Strauch, B. (2015). Can we examine safety culture in accident investigations, or should we? *Safety Science,* Volume 77, 102–111.

Taras, V., Kirkman, B. L. and Steel, P. (2010). Examining the impact of culture's consequences: A three-decade, multilevel, meta-analytic review of Hofstede's cultural value dimensions. *Journal of Applied Psychology,* Volume 95(3), 405.

Tear, M., Reader, T., Shorrock, S. and Kirwan, B. (In Press). Safety culture and power: Interactions between perceptions of safety culture, organisational hierarchy, and national culture.

Vincent, C. (2011). *Patient safety* (2nd ed.) London: John Wiley & Sons.

Vinken, H., Soeters, J. and Ester, P. (2004). *Comparing cultures: Dimensions of Culture in a Comparative Perspective.* Leiden: the Netherlands Brill.

Waarts, E. and Van Everdingen, Y. (2005). The influence of national culture on the adoption status of innovations: An empirical study of firms across Europe. *European Management Journal,* Volume 23, 601–610.

Weber, E. U. and Hsee, C. (1998). Cross-cultural differences in risk perception, but cross-cultural similarities in attitudes towards perceived risk. *Management Science,* Volume 44, 1205–1217.

Williamson, D. (2002). Forward from a critique of Hofstede's model of national culture. *Human Relations,* Volume 55, 1373–1395.

Zohar, D. (2010). Thirty years of safety climate research: Reflections and future directions. *Accident Analysis and Prevention,* Volume 42, 1517–1522.

3

Governance for Safety in Inter-Organisational Project Networks

Nadezhda Gotcheva, Kirsi Aaltonen and Jaakko Kujala

CONTENTS

3.1 Governance in Project Networks..42
3.2 Key Elements of Governance in Inter-Organisational Project
    Networks..43
    3.2.1 Goal Setting...44
    3.2.2 Rewarding..45
    3.2.3 Monitoring...46
    3.2.4 Coordination..47
    3.2.5 Roles and Decision-Making 47
    3.2.6 Capability Building 48
3.3 Concluding Remarks and Future Research Directions.........................49
Acknowledgements...50
References..51

Introduction: Bridging Project Governance and Safety Science

We observe a significant rise in the frequency, complexity and magnitude of global projects in areas such as energy production, transportation, aerospace or infrastructure construction. Such projects often pertain to potentially high risks to the environment or society, and safety is fundamentally a prerequisite for their existence (Oedewald and Reiman, 2007). These entities bring together a large number of stakeholders, such as investors, contractors, subcontractors, local interest groups, government organisations, local communities, political decision makers and environmental groups (Flyvbjerg et al., 2003). The diversity and dynamics of these stakeholders adds to the project complexity, making the management or governance of safety in project networks particularly challenging (Oedewald and Gotcheva, 2015). The *concept of governance* refers to the notion that project networks cannot be fully controlled by a single organisation; governance in project networks is influenced by various factors such as network structure, previous relationships between actors, as well as the potential for future collaboration. Governance differs from project management in that it integrates practical knowledge from economics, law, sociology and management to steer interaction processes and structures in inter-organisational networks to nurture further collaboration. Our research complements the traditional approach on managing project networks, which has been emphasising the role of the focal actor; it has been dominated by the planning and control-oriented approach.

The focal finding of project management literature has been that the key challenges in complex projects are *organisational* isswn character, rather than technical. Organisational complexity has been found to have an amplifying effect on the escalation of (often non-linear) consequences, leading to dysfunctional performance (Fleming and Zyglidopoulos, 2008). The need to understand the links between project governance and safety issues became evident in the context of the challenges, experienced in the process of constructing the new nuclear build project Olkiluoto 3 (OL3) in Finland, which is the first European Pressurised Water Reactor (EPR) plant unit in the world. Construction began in 2005, and the unit itself was supposed to start commercial operation in 2009. However, it is still not operational as of 2018. At the beginning of the project, the Radiation and Nuclear Safety Authority in Finland (STUK) identified a number of non-conformances, largely associated with project management (STUK, 2006). Thereafter, this ongoing project has faced many challenges throughout its design and construction phases such as communication problems due to a long supply chain, unclear roles and responsibilities and shortcomings in quality assurance related to an insufficient understanding of the requirements expected of subcontractors, most of whom had limited experience in the nuclear industry (STUK, 2011). Oedewald et al. (2011) further investigated the topical issue of managing safety in complex subcontractor networks in the case of the OL3 project. The management model of this project was, to a large degree, a traditional top-down hierarchy (Oedewald and Gotcheva, 2015), which has proved inappropriate for timely identification and dealing with these organisational challenges.

Although there are many examples of projects in the safety-critical domain that have ended in disaster, the specifics of each project's context and governance approach have rarely been explicated during accident investigations. For instance, in the offshore industry, the Petrobras P-36 oil drilling platform sank in 2001 – before even being operated – due to underlying issues relating to the poor design of key safety-critical parts; component failure and project management issues evident in focus on cost-cutting and insufficient communication (Wander, 2008). Likewise, the loss of the space shuttles *Challenger* and *Columbia* have been largely associated with a performance culture, gradually infused with cost efficiency and focus on meeting production goals and deadlines (Vaughan, 1996; Starbuck and Farjoun, 2005). Pinto (2014) indicated that the 'normalisation of deviance' (Vaughan, 1996) represents a significant problem in project organisations, and examined the role of organisational learning and corporate governance in identifying and minimising this negative impact.

On the other hand, well-designed and implemented governance mechanisms are instrumental for ensuring appropriate project planning and delivery, including a good safety performance. Some success stories of complex projects have been, for example, Heathrow Terminal 5 in the UK, and the Tampere Lakeside Tunnel project in Finland. Both of these projects have been implemented with an integrated project delivery form, which emphasises collaboration between project key stakeholders (Brady and Davies, 2010; Hietajärvi et al., 2017). This approach ensured that an integrated management system was in place for the overall project network, while motivating the key project actors to work towards joint project goals, including strict safety-related goals.

An additional challenge in a project context is that the safety of the project's end product is influenced by decisions made during different project lifecycle stages, particularly during the planning, design or construction phases. Different lifecycle stages of a given project each have distinctive goals and features, challenged by their specific safety culture (Gotcheva and Oedewald, 2015). The initiation phase may be negatively affected by unclear definitions or differing opinions regarding the objectives of the project (Ward and Chapman, 1995). There is a need to understand how the involvement of diverse and multiple stakeholders, often at different geographical locations, could have an impact on safety, as actors may interpret differently the requirements and related responsibilities (Gotcheva et al., 2016; Gotcheva and Oedewald, 2015). Design has been identified as a significant contributor to recent major accidents, accounting for about 50% of accidents and incidents in the aircraft and nuclear industries (Kinnersley and Roelen, 2007). To support safe design, engineers and project managers need to have similar levels of decision-making power to allow for competing goals to be openly discussed (Hayes, 2015). As the project proceeds to the implementation phase, it is challenging to facilitate a shared understanding of safety due to a multitude of temporary actors. Cagno et al. (2002) highlighted the importance of coordination and good project management during the commissioning phase, when major accident safety risks are significantly higher, and there are multiple organisational and technical interfaces.

Despite the growing number of inter-organisational projects in diverse safety-critical sectors, safety research has focussed primarily on safety management in permanent and single organisational settings, often on operations, with some exceptions (e.g. Oedewald and Gotcheva, 2015). The main theoretical premises in safety science, such as accident causation models or different safety models (safety culture, high-reliability organisations and resilience engineering) have not been explicitly based on studying projects (Le Coze, 2016). Some scholars have recognised inter-organisational aspects as critical – yet insufficiently understood – in safety research. While delineating future research needs in the field of high-reliability organisations, Karlene Roberts acknowledged the need to know more about how errors occur at the *'interstices'*, such as *'shift changes, relationships between hospital pharmacies and wards, the relationship of organisations with their contractors, the relationships among geographically separate parts of organisations'* (Bourrier, 2005, p. 95). The project context provides timely and unique opportunities for advancing our understanding of how safety is ensured in such challenging settings.

3.1 Governance in Project Networks

An inter-organisational project network can be viewed as an intentionally created, temporarily dynamic network of organisations that combine the resources, capabilities and knowledge of all participating actors to contribute to a common goal and fulfil the needs of the owner (Järvensivu and Möller, 2009; Ruuska et al., 2011). Some of the project's parties might have little experience and understanding of the specific national and international regulatory requirements, standards and industry practices in general. In this context, ensuring that the safety and quality requirements are adequately understood and fulfilled by every party is a challenging task. Many of the project's participants (e.g. contractors or subcontractors) also work in other, non-safety critical industries, so they cannot necessarily be expected to share the same values, knowledge and working methods in support of the overall safety goals of the project.

The *project governance* challenge is to align and coordinate numerous interrelated project roles and activities, characterised by an ambiguity of cause–effect relationships, in addition to a marked difficulty to understand and control the behaviour of the various project network actors (Brady and Davies, 2010; Ahola et al., 2014). In a project context, two distinct streams of governance research can be identified: project governance as *external* to any specific project, and project governance as *internal* to a specific project (Ahola et al., 2014). As our research focus is on the internal coordination of activities in a given project, in order to avoid confusion, we define governance in project networks as the coordination, adaptation and safeguarding mechanisms enabling multiple organisational actors in project networks to work towards a range of shared goals. We suggest that the concept of governance in project networks is useful for understanding coordination and decision-making in inter-organisational project networks. Governance-related research on safety in inter-organisational projects is relatively limited, and a better understanding is needed of how to design governance systems that take into account safety objectives, especially in situations where a diverse range of stakeholders have somewhat conflicting objectives and differing views about the importance of safety (or how it could best be achieved and sustained). It is nevertheless challenging to effectively define safety objectives in terms of major accident hazards for a project when potential issues may only manifest themselves years into the future, long after the project's organisation has been disbanded.

The organisational structures of complex projects are often characterised as hierarchical contract organisations (Morris and Hough, 1987). The purpose of a project's contract organisation is to create a cooperative system, and this is accomplished by achieving common objectives by properly incentivising the contractors to work with contract pricing terms. Ruuska et al. (2011) analysed the supply networks of the Olkiluoto 3 and Flamanville 3 nuclear power plant projects, and found that they support Miller and Lessard (2000) by suggesting that there should be a shift away from viewing multi-firm projects as hierarchical contract organisations, to viewing them as supply networks characterised by a set of complex, networked relationships. Furthermore, Ruuska et al. (2011) suggest that there should be a shift in emphasis away from the predominant modes of governance, market and hierarchy, towards novel governance approaches emphasising network-level mechanisms, such as self-regulation within the project's themselves.

Governance in project networks can be based on a shared effort between participants, or else it can be exercised by a single network member (a lead organisation) or by an external network administrator organisation (Provan and Kenis, 2008). Despite recent efforts to take a broader look at governance in project networks (Ahola et al., 2014; Kujala et al., 2015) and network forms of governance, in general (Jones et al., 1997), going beyond a single-hubfirm-driven approach is still a less discussed perspective in the previous literature on project networks and strategic business networks.

3.2 Key Elements of Governance in Inter-Organisational Project Networks

This framework synthesis key elements of governance in inter-organisational project networks and is based on the integration of two systematic literature reviews. These followed a structured approach for analysing and categorising the literature in the fields of project management and safety science. Based on the analysis of project management literature, we identified six key elements of governance in inter-organisational project networks and their corresponding mechanisms (Kujala et al., 2016). These key elements were used for the subsequent systematic literature review, which focussed on verifying whether these dimensions were found in the safety literature (Starck, 2016). These reviews, along with the identified taxonomy of governance dimensions and mechanisms, laid the foundation of the governance framework in inter-organisational project networks (Figure 3.1).

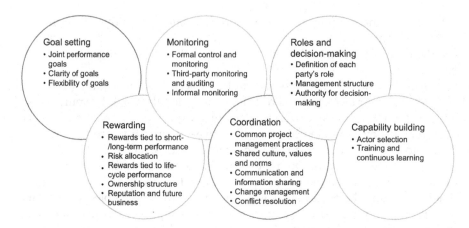

FIGURE 3.1
Key elements and mechanisms associated with governance in inter-organisational project networks.

Governance mechanisms influence how work is *coordinated and organised in project networks*, and ultimately determine the organisations' *commitment and capability* to work towards achieving their project goals. The capability to understand safety and hazards, as well as the marked commitment to act in a manner that ensures safety in daily activities, are both seen as being critical for a good safety performance (Reiman and Pietikäinen, 2010; Reiman et al., 2012). Governance mechanisms may also influence commitment to safety in several ways: by enabling project actors to jointly create project safety-related goals, having effective monitoring practices to ensure that these goals are achieved, as well as providing some sort of reward for the meeting of goals. Similarly, project governance mechanisms such as effective communication and information sharing, or the selection of actors based on their competence, influences the ability of actors to meet project safety goals.

3.2.1 Goal Setting

Goal setting seeks to create shared performance goals for the project that are properly understood by all project actors. In general, project goals include both short-term goals focussing on the project implementation process, and long-term goals relating to the use and benefit gain from the end result (Nisar, 2013). The impact of joint performance goals on safety is a relevant issue, especially in relation to the different positions and roles of stakeholders. In safety literature, performance indicators for high-hazard industries have been studied; in particular, the need to define leading indicators – measures that intend to predict patterns or trends of events (Hopkins, 2009; Reiman and Pietikäinen, 2012). In safety-critical projects, potential goal conflicts between schedule, cost, scope and safety can all influence the capability of project actors to understand the links between short-term actions and long-term consequences.

The existing research acknowledges the importance of the early involvement of all project actors in the process of establishing joint goals in an inter-organisational project network (Guo et al., 2014; Davies et al., 2014). This is needed to ensure an optimal understanding of needs, as well as the proper interpretation of safety requirements, despite any individual differences in the actors' experiences and backgrounds. Setting shared goals with regard to safety; for example, how safety culture is measured and continuously improved, includes the establishment of shared principles and expectations among the project actors. As complex projects always involve changes that cannot be anticipated, project goals and the contracting process itself must strive to be sufficiently flexible in order to respond to unforeseen risks and opportunities (Davies et al., 2014). Additionally, the continuous improvement of safety calls for the flexibility of goals (as the ways in which safety can be achieved is always context dependent; there are multiple ways it can be achieved and maintained in different organisations).

3.2.2 Rewarding

Rewarding refers to aligning actors' goals with project goals by means of incentives. These may include monetary incentives linked to joint performance goals (Davies et al., 2014), conditional future payments, work prompting a lifecycle approach (van den Hurk and Verhoest, 2015; Nisar, 2013), as well as reputational scoring systems (Guo et al., 2014). However, the effectiveness and challenges of using bonuses (and other rewards) has been discussed in safety literature as well. For instance, Maslen and Hopkins (2014) reminded us that in recent offshore industry disasters the incentives were dysfunctional, and actually worked against process safety. Drawing on the literature of human motivation and incentives, they studied how senior managers are motivated by incentives in their daily decisions in hazardous industries. The main insight from this work was that incentives do influence priorities and behaviours because they tap into basic human needs, such as the need for approval, and the need to be recognised when making a valuable contribution. The authors conclude that safety, in terms of major accident prevention, should be incentivised at least in the same way as financial and business performance. However, these incentives would need to be carefully designed.

McDermott et al. (2017) investigated senior executives long-term incentive plans and safety objectives in the construction sector in Australia, and identified that these plans were exclusively associated with financial indicators; that is, senior executives were not incentivised to align their long-term decision-making with long-term safety objectives. Furthermore, using financial incentives to boost productivity in safety-critical industries may trigger unsafe acts (Sawacha et al., 1999). For instance, collective bonus schemes were found to inhibit reporting of safety concerns for fear of losing one's bonus (Collinson, 1999). Incentives may discourage some actors to report safety deficiencies if they believe that it could cause immediate additional costs (or negatively affect future business opportunities). Incentives such as the determination of who carries additional costs related to safety-related risks may be used to discourage such behaviour. For example, in alliance contracts risks are shared by all project participants. The determination of who carries the cost of risk is especially relevant if a supplier's risk-bearing capacity is low, as they are more likely to leave errors unreported (Osmundsen et al., 2010). In that sense, creating a sense of ownership through contractual means among participating organisations is also important (Guo et al., 2014).

3.2.3 Monitoring

Monitoring seeks to ensure that all actors behave as expected, thereby enabling the use of performance-based incentives. To be useful, project milestones and performance targets must not only be realistic, they must be monitored (Nisar, 2013). Abednego and Ogunlana (2006) highlighted the importance of continuous project control and monitoring in pursuit of common goals in order to satisfy all interests.

In safety-critical organisations, monitoring includes system activities, boundaries and potential sources of risk in the system (Reiman et al., 2015). Recent safety literature on monitoring and audit also warns of the perils of an over-focus on compliance, particularly in multi-tier contracting chains (e.g. McDermott and Hayes, 2018). Governance mechanisms could be structured to provide balance in this regard; managers must not only attend in general to the behaviours and interactions that characterise a group or organisation – such as violations of rules and procedures – they must organise the monitoring of technology and physical processes, which are critical for system safety. Osmundsen et al. (2008) noticed that there is an increased use of key performance indicators and incentive schemes in the Norwegian petroleum industry, and also observed that there is lack of effective incentives for safety since the main emphasis is on more tangible, easier to monitor performance dimensions. Hence, it is challenging to create good performance indicators for monitoring safety (and use them for incentives) because safety includes a lot of 'soft' aspects, which are difficult to measure.

As acknowledged by Weick (1998), monitoring system activities and boundaries depends on organisational decisions about what information is relevant for safety (and what can be ignored). In the safety literature, Burns (2006) emphasised the importance of proactive monitoring in the petrochemical industry, as done in three stages: deviation detection, problem prediction (by determining what effect the deviation will have in future) and taking compensatory action. Based on a self-regulation framework, Griffin and Hu (2013) indicated that monitoring done by leaders in a learning, supportive environment could improve safety; the emphasis is on the leader's aware-ness of unsafe behaviour and errors, as well as their ability to encourage their employees to learn and challenge the current safety system in place.

3.2.4 Coordination

Coordination is required to align the behaviour of each actor so that they can effectively work together. Formally, project contracts and plans define the processes for implementing project work, as well as specifying tools and practices for carrying out the work. Informal types of coordination such as shared values and behavioural norms can have a significant impact on how project actors work together in order to ensure safety. These behavioural norms may also be included as part of the contract document, which may include values, norms and expectations for a project (Caldwell et al., 2009). The challenge posed by these informal coordination mechanisms is that cultural and organisational boundaries can hinder the creation of shared behavioural norms for a project (Bresnen and Marshall, 2000; Dossick and Neff, 2011).

Reason and Hobbs (2003) indicated that 'a breakdown in coordination is one of the most common circumstances leading to an incident. In many cases, coordination breaks down when people make unspoken assumptions about a job, without actually communicating with each other to confirm the situation'. Another important aspect of coordination is to ensure that all par-ties have the information they need to complete their work in collaboration with other project parties. Information sharing is ensured by practices such as regular meetings or the co-location of project teams. Brady and Davies (2014) described a single model environment, which ensured that all parties could access a central repository for all digital data, as one of the key success factors in the Heathrow T5 project.

3.2.5 Roles and Decision-Making

Roles and decision-making refers to giving actors the necessary informa-tion to understand the effect of decisions on overall performance, thereby enabling the actors to make appropriate decisions. Roles of different proj-ect parties in safety-critical industries are oftentimes mandated by law. For example, the licensee in a nuclear industry project is ultimately responsible for safety. In case of turnkey delivery nuclear industry projects, the plant supplier is usually responsible for the design, construction, installation and commissioning of the plant. Although the formal roles and responsibilities of each party are defined contractually (Lu et al., 2015), effective governance incorporates suitable project management structures and decentralised decision-making principles (Abednego and Ogunlana, 2006; Nisar, 2013; Pitsis et al., 2014). For example, creating a leadership team that is legally and spatially separate from the parent organisations may help in coping with unknowable future events and avoiding disputes (Sanderson, 2012), while clearly defined management team can support and assume responsibility for daily execution of the project (van Marrewijk and Smits, 2015).

Effective governance requires also that decision-making power is appropriately distributed between actors (Eriksson, 2010; Nisar, 2013). According to Ruuska et al. (2011), companies' responsibilities should be allocated on the basis of their competence, capabilities and risk-bearing capacity, while Abednego and Ogunlana (2006) highlighted the importance of equality and active participation in making the 'right decisions at the right time'. In projects emphasising relational governance, democratic decision-making mechanisms are needed to build trust and confidence in relationships (Nisar, 2013). As implied by polycentric governance (Woods and Branlat, 2011): on one hand, project network actors have sufficient autonomy and decision-making power within their domain of competence, while on the other hand, they have responsibilities for achieving their shared project goals. In terms of safety performance, polycentric governance approach assigns an active role even for small subcontractors at the periphery of the project network. In this process, knowledge sharing also involves sharing of cultural information about safety (Oedewald and Gotcheva, 2015).

3.2.6 Capability Building

Capability building ensures that project actors have adequate abilities and potential to meet performance expectations. The aim is to ensure that appropriate skills and expertise are identified and tied to the project at an early stage, and that sufficient attention is devoted to resourcing the project team (Nisar, 2013). The use of suitable competitive tendering processes and selection criteria play an important role in capability building (van den Hurk and Verhoest, 2015). Beyond procurement, systematic training and continuous learning (with regard to safety and its management during projects, and from project to project to get operational lessons into the project process) can both enhance capability (Davies et al., 2014; Guo et al., 2014; Ruuska et al., 2011). Practical means to develop capabilities in a temporary project context can include, retrospectives, systematic lessons and learned sessions, joint workshops, joint trainings and education, capability and competence mapping as well as related staffing policy, the building of common ground and collective capabilities through agreeing on joint processes and reviews, and so on.

From a safety perspective, capability building for the continuous improvement of safety is also highly relevant. This can take the form of preventive action, corrective action and reviews. Combining project and safety research domains also points to the issue of learning in such dynamic environments. As indicated by Le Coze (2013) is his reflection about learning from accidents, interests between different actors may lead to difficulties in reporting near misses or incidents. Capability building in safety-critical domains also entails increased mindfulness to make sense of situational requirements, and to heighten awareness of when it is appropriate to cross boundaries (e.g. to receive information or use available resources) or to adjust or interpret rules according to situational requirements (Reiman et al., 2015). Actors involved in safety-critical projects need sufficient understanding and knowledge on the nature of hazards, possible new hazards and the need for strict rules and instructions on how to fulfil various requirements (Oedewald and Gotcheva, 2015). Good relationships and trust are essential for developing a situational self-organising or adjustment capability in inter-organisational project networks (again requiring effective governance to steer positive interactions between project actors).

3.3 Concluding Remarks and Future Research Directions

This chapter advances the understanding of what constitutes governance for safety in the context of inter-organisational project networks, comprising multiple autonomous organisations and which interests need to be aligned in order to achieve a joint system-level goal. Our focus is on project network governance elements to coordinate, adapt and safeguard work between autonomous project actors. Governance does not stay the same throughout the project – it can be seen as 'impermanent' as it needs to be continuously tailored and adapted. The notion of *impermanence* (Weick, 2001) is seen as inherent in organisations, insofar as the process of organising is ongoing and adaptive rather than stable and fixed; continuously occurring 'between smoke and crystal' (Atlan, 1979). While 'smoke' relates to unpredictability in terms of variety and complexity, 'crystal' refers to stability in terms of repetition and habit in organisations (Taylor and Van Every, 2000). The underlying assumption of the governance in project networks framework is that project actors will collectively act as 'system architects', creating the governance structure in the project network through their decision-making processes, capabilities, negotiations and interactions. This approach complements the view that emphasises the role of the 'strong owner' (Sallinen et al., 2008; Winch and Leiringer, 2016) in setting the tone and culture for the entire project network.

In order to bridge the gap between project governance and safety science, in this chapter, we have described the key elements associated with governance in safety-critical inter-organisational project networks. Further conceptual development is needed to inform empirical research and build detailed propositions on actual decision-making and formation processes of project governance in inter-organisational settings (van Marrewijk and Smits, 2015). Our research also highlights the need for a deeper understanding of how the structural – and other – characteristics of a given project network affect safety performance culture and governance for safety. A single project network may accommodate diverse governance regimes, raising interesting questions about how this could affect safety at the project network level and how the different governance elements should be tailored to support safety across the boundaries of the different parts of a project network.

The variety of connections in complex project networks has been underestimated – for instance, vertical and horizontal interactions co-exist. The polycentric governance approach is central for complex nuclear projects in order to acknowledge actors on the periphery of the network (e.g. small subcontractor companies) as active partners in developing a shared project culture (Ruuska et al., 2011). From that perspective, horizontal interactions between project partners could be seen as equal in importance to bilateral hierarchical connections (Oedewald and Gotcheva, 2015). It is therefore important to understand how governance elements could be utilised to constructively affect the ability of the project owner/licensee (and other project actors) to ensure that the project's safety goals are met during its whole lifecycle.

The governance for safety framework developed here also invites a contingency perspective on governance of safety-critical projects, which requires tailoring the governance elements to a project's specific context and characteristics. Future research could seek to identify the contingency factors most relevant to safety. The role of temporality is also interesting in terms of governing safety in different points of time (and different phases of the project), since safety is continuously co-created by project actors' daily decisions and activities.

Inter-organisational project networks within contemporary safety-critical domains are typically global and multicultural, involving a large number of participants from different occupational and institutional backgrounds. Future research should explore the implications of this diversity for safety performance, and the shared understanding of safety. Given that governance mechanisms differ across companies in the project network (as well as across national backgrounds), it becomes increasingly important to understand the role of governance in safety-critical projects, such as nuclear industry new builds or decommissioning projects. Another potentially fruitful area for further research concerns the role of regulation, as local regulations significantly influence the development and implementation of both governance systems and safety management systems.

Acknowledgements

This chapter extends on previous research carried out by Kujala et al. (2016), presented at the European Academy of Management (EURAM) conference, 1–4 June 2016, Paris, France, and Starck (2016). The work summarised in this chapter was conducted as part of a research project, Management

Principles and Safety Culture in Complex Projects (MAPS). The authors wish to acknowledge the funding provided by SAFIR2018, The Finnish Research Programme on Nuclear Power Plant Safety (2015–2018), University of Oulu and VTT Technical Research Centre of Finland.

References

Abednego, M.P. and Ogunlana, S.O. (2006). Good project governance for proper risk allocation in public-private partnerships in Indonesia. *International Journal of Project Management*, Volume 24, 622–634.

Ahola, T., Ruuska, I., Artto, K. and Kujala, J. (2014). What is project governance and what are its origins? *International Journal of Project Management*, Volume 32, 1321–1332.

Atlan, H. (1979). *Entre le cristal et la fume [Between crystal and smoke]*. Paris: Editions du Seuil.

Bourrier, M. (2005). An interview with Karlene Roberts. *European Management Journal*, Volume 23(1), 93–97.

Brady, T. and Davies, A. (2010). Learning to deliver a mega-project: The case of Heathrow terminal 5. In Caldwell, Nigel and Howard, Mickey (eds.), *Procuring Complex Performance: Studies of Innovation in Product-Service Management*. New York, USA, Routledge, pp. 174–198.

Brady, T. and Davies, A. (2014). Managing structural and dynamic complexity: A tale of two projects. *Project Management Journal*, Volume 45, 21–38.

Bresnen, M. and Marshall, N. (2000). Motivation, commitment and the use of incentives in partnerships and alliances. *Construction Management and Economics*, Volume 18, 587–598.

Burns, C.M. (2006). Towards proactive monitoring in the petrochemical industry. *Safety Science*, Volume 44, 27–36.

Cagno, E., Caron, F. and Mancini, M. (2002). Risk analysis in plant commissioning: The Multilevel Hazop. *Reliability Engineering and System Safety*, Volume 77(3), 309–323.

Caldwell, N., Roehrich, J. and Davies, A. (2009). Procuring complex performance in construction: London Heathrow Terminal 5 and a Private Finance Initiative hospital. *Journal of Purchasing and Supply Management*, Volume 15(3), 178–186.

Collinson, D.L. (1999). Surviving the rigs: Safety and surveillance on North Sea oil installations. *Organisation Studies*, Volume 20(4), 579–600.

Davies, A., Macaulay, S., Debarro, T. and Thurston, M. (2014). Making innovation happen in a megaproject: London's crossrail suburban railway system. *Project Management Journal*, Volume 45(6), 25–37.

Dossick, C.S. and Neff, G. (2011). Messy talk and clean technology: Communication, problem-solving and collaboration using building information modelling. *The Engineering Project Organisation Journal*, Volume 1(2), 83–93.

Eriksson, P.E. (2010). Improving construction supply chain collaboration and performance: A lean construction pilot project. *Supply Chain Management: An International Journal*, Volume 15(5), 394–403.

Fleming, P. and Zyglidopoulos, S. (2008). The escalation of deception in organisations. *Journal of Business Ethics*, Volume 81, 837–850.

Flyvbjerg, B., Bruzelius, N. and Rothengatter, W. (2003). *Megaprojects and Risk: An Anatomy of Ambition*. Cambridge.

Gotcheva, N. and Oedewald, P. (2015). SafePhase: Safety culture challenges in design, construction, installation and commissioning phases of large nuclear power projects, February 2015:10, ISSN: 2000-0456, Swedish Radiation Safety Authority (SSM).

Gotcheva, N., Oedewald, P., Wahlström, M., Macchi, L., Osvander, A.-L. and Alm, H. (2016). Cultural features of design and shared learning for safety: A Nordic nuclear industry perspective. *Safety Science, Special Issue: Learn and Train for Safety*, Volume 81, 90–98.

Griffin, M.A. and Hu, X. (2013). How leaders differentially motivate safety compliance and safety participation: The role of monitoring, inspiring, and learning. *Safety Science*, Volume 60, 196–202.

Guo, F., Chang-Richards, Y., Wilkinson, S. and Li, T. (2014). Effects of project governance structures on the management of risks in major infrastructure projects: A comparative analysis. *International Journal of Project Management*, Volume 32, 815–826.

Hayes, J. (2015). Investigating design office dynamics that support safe design. *Safety Science*, Volume 78, 25–34.

Hietajärvi, A.-M., Aaltonen, K. and Haapasalo, H. (2017). Managing integration in infrastructure alliance projects: Dynamics of integration mechanisms. *International Journal of Managing Projects in Business*, Volume 10(1), 5–31.

Hopkins, A. (2009). Thinking about process safety indicators. *Safety Science*, Volume 47, 460–465.

Järvensivu, T. and Möller, K. (2009). Metatheory of network management: A contingency perspective. *Industrial Marketing Management*, Volume 38(6), 654–661.

Jones, C., Hesterly, W. and Borgatti, S. (1997). A general theory of network governance: Exchange conditions and social mechanisms. *The Academy of Management Review*, Volume 22(4), 911–945.

Kinnersley, S. and Roelen, S. (2007). The contribution of design to accidents. *Safety Science*, Volume 45(1–2), 31–60.

Kujala, J., Aaltonen, K., Gotcheva, N. and Pekuri, A. (2016). Key dimensions of project network governance and implications for safety in nuclear industry projects, *European Academy of Management (EURAM)*, 1–4 June 2016, Paris, France.

Le Coze, J.C. (2013). What have we learned about learning from accidents? Post-disasters reflections. *Safety Science*, Volume 51(1), 441–453.

Le Coze, J.C. (2016). Vive la diversité! High reliability organisation (HRO) and resilience engineering (RE). *Safety Science*.

Lu, P., Guo, S., Qian, L., He, P. and Xu, X. (2015). The effectiveness of contractual and relational governances in construction projects in China. *International Journal of Project Management*, Volume 33, 212–222.

Maslen, S. and Hopkins, A. (2014). Do incentives work? A qualitative study of managers' motivations in hazardous industries. *Safety Science*, Volume 70, 419–428.

McDermott, V. and Hayes, J. (2018). Risk shifting and disorganisation in multi-tier contracting chains: The implications for public safety, *Safety Science*, Volume 6, 263–272.

McDermott, V., Zhang, R.P., Hopkins, A. and Hayes, J. (2017). Constructing safety: Investigating senior executive long-term incentive plans and safety objectives in the construction sector. *Construction Management and Economics*, Volume 36(5), 276–290.

Miller, R. and Lessard, D. (2000). *The Strategic Management of Large Engineering Projects: Shaping Institutions, Risks, and Governance*. Cambridge, MA: MIT Press.

Morris, P.W.G. and Hough, G.H. (1987). *The Anatomy of Major Projects: A Study of the Reality of Project Management*. Chichester, UK: John Wiley and Sons.

Nisar, T.M. (2013). Implementation constraints in social enterprise and community public private partnerships. *International Journal of Project Management*, Volume 31, 638–651.

Oedewald, P. and Gotcheva, N. (2015). Safety culture and subcontractor network governance in a complex safety critical project. *Reliability Engineering and System Safety*, Volume 141(September), 106–114.

Oedewald, P., Gotcheva, N., Reiman, T., Pietikäinen, E. and Macchi, L. (2011). Managing safety in subcontractor networks: The case of Olkiluoto 3 nuclear power plant construction project. *4th Resilience Engineering International Symposium*, Sophia-Antipolis, France, 8–10 June.

Oedewald, P., Gotcheva, N., Viitanen, K. and Wahlström, M. (2015). Developing safety culture and organisational resilience in nuclear industry throughout the different lifecycle phases, "Managing safety culture throughout the lifecycle of nuclear plants" (MANSCU), 2011–2014, *Final project report. VTT Technology*.

Oedewald, P. and Reiman, T. (2007). Special characteristics of safety critical organisations. *Work psychological perspective, Espoo, VTT Publications 633*.

Osmundsen, P., Aven, T. and Tomasgard, A. (2010). On incentives for assurance of petroleum supply. *Reliability Engineering and System Safety*, Volume 95(2), 143–148.

Osmundsen, P., Aven, T. and Vinnem, J.E. (2008). Safety, economic incentives and insurance in the Norwegian petroleum industry. *Reliability Engineering and System Safety*, Volume 93(1), pp. 137–143.

Pinto, J. (2014). Project management, governance, and the normalisation of deviance. *International Journal of Project Management*, Volume 32(3), 376–387.

Pitsis, T.S., Sankaran, S., Gudergan, S. and Clegg, S.R. (2014). Governing projects under complexity: Theory and practice in project management. *International Journal of Project Management*, Volume 32, 1285–1290.

Provan, K. and Kenis, P. (2008). Modes of network governance: Structure, management, and effectiveness. *Journal of Public Administration Research and Theory*, Volume 18(2), 229–252.

Reason, J. and Hobbs, A. (2003). *Managing Maintenance Error. A Practical Guide*. Hampshire: Ashgate.

Reiman, T. and Pietikäinen, E. (2012). Leading indicators of system safety - Monitoring and driving the organisational safety potential. *Safety Science*, Volume 50(10), 1993–2000.

Reiman, T., Rollenhagen, C., Pietikäinen, E. and Heikkilä, J. (2015). Principles of adaptive management in complex safety-critical organisations. *Safety Science*, Volume 71(Part B), 80–92.

Ruuska, I., Ahola, T., Artto, K., Locatelli, G. and Mancini, M. (2011). A new governance approach for multi-firm projects: Lessons from Olkiluoto 3 and Flamanville 3 nuclear power plant projects. *International Journal of Project Management*, Volume 29, 647–660.

Sallinen, L., Ruuska, I. and Ahola, T. (2008). How governmental stakeholders influence large projects: The case of nuclear power plant projects. *International Journal of Managing Projects in Business*, Volume 6(1), 51–68.

Sanderson, J. (2012). Risk, uncertainty and governance in megaprojects: A critical discussion of alternative explanations. *International Journal of Project Management*, Volume 30, 432–443.

Sawacha, E., Naoum, S. and Fong, D. (1999). Factors affecting safety performance on construction sites. *International Journal of Project Management*, Volume 17(5), 309–315.

Starbuck, W.H. and Farjoun, M. (Eds.) (2005). *Organisation at the Limit: Lessons from the Columbia Disaster*. Malden, MA: Blackwell Publishing.

Starck, M. (2016). Key dimensions of project governance and their relation to nuclear safety- An explorative study of nuclear industry projects, Master's thesis, Department of Management and Organisation, Hanken School of Economics, Helsinki.

STUK. (2006). Management of safety requirements in subcontracting during the Olkiluoto 3 nuclear power plant construction phase. Investigation Report 1/06, Nuclear reactor regulation translation, 1.9.2006.

STUK. (2011). Investigation of the procurement and supply of the emergency diesel generators (EDG) and related auxiliary systems and equipment for the Olkiluoto 3 nuclear power plant unit. The Radiation and Nuclear Safety Authority.

Taylor, J.R. and Van Every, E.J. (2000). *The emergent organisation: Communication as its site and surface*. Mahwah, NJ: Erlbaum.

Van den Hurk, M. and Verhoest, K. (2015). The governance of public-private partnerships in sports infrastructure: Interfering complexities in Belgium. *International Journal of Project Management*, Volume 33, 201–211.

van Marrewijk, A. and Smits, K. (2016). Cultural practices of governance in the Panama Canal Expansion Megaproject. *International Journal of Project Management*, Volume 34(3), 533–544.

Vaughan, D. (1996). *The Challenger Launch Decision*. Chicago: University of Chicago Press.

Wander, S. (2008). System failure case studies: That sinking feeling. *NASA Safety Center*, Volume 2(8), 1–4.

Ward, S.C. and Chapman, C.B. (1995). Risk-management perspective on the project life cycle. *International Journal of Project Management*, Volume 13(3), 145–149.

Weick, K.E. (1998). Foresights of failure: an appreciation of Barry Turner. *Journal of Contingencies and Crisis Management*, Volume 6, 72–75.

Weick, K.E. (2001). *Making Sense of the Organisation: The Impermanent Organisation*. Oxford: Blackwell Publishing.

Winch, G. and Leiringer, R. (2016). Owner project capabilities for infrastructure development: A review and development of the "strong owner" concept. *International Journal of Project Management*, Volume 34(2), 271–281.

Woods, D. and Branlat, M. (2011). How human adaptive systems balance fundamental trade-offs: Implications for polycentric governance architectures. *4th Resilience Engineering Symposium*, Sophia-Antipolis, France, 8–10 June.

4

Coping with Globalisation: Robust Regulation and Safety in High-Risk Industries

Ole Andreas Engen and Preben Hempel Lindøe

CONTENTS

4.1 What Is Regulation?..56
    4.1.1 A General Concept..56
    4.1.2 Different 'System Logics' 57
4.2 Risk Regulating Regimes..59
    4.2.1 A General Framework...59
    4.2.2 Hierarchy of Norms...60
    4.2.3 Function-Based Regulation 61
4.3 Balancing Regulation and Innovation 63
    4.3.1 Regulation and Technological Development.................63
    4.3.2 Dynamics of Exploitation and Exploration..................64
    4.3.3 Combining Enforcement and Management by Rule.................64
    4.3.4 Flexibility in Handling Rules and Norms in a Global
        Context...66
4.4 Lessons Learned 66
    4.4.1 Different Modes of Risk Regulation................................66
    4.4.2 Trust and Distrust...68
4.5 Concluding Remarks..70
Notes 72
References...73

Introduction

Globalisation and new technologies introduce new risks; they challenge national risk-regulating regimes. The French safety researcher, Jean Christophe Le Coze argues that globalisation has 'reconfigured the landscape and operating constraints of high-risk systems'. Le Coze underlines how such trends challenge national risk-regulating regimes, and how they require a discussion of these regimes linked to the broader pattern of globalisation (Le Coze, 2017). It is undoubtedly true that different patterns of globalisation, along with outsourcing, digitalisation and standardisation, challenge the flexible leeway that nations enjoy when forming and shaping their own regulatory means and ends – that is rule-making, enforcement and compliance strategy. Global trends pose occupational, technological and environmental risks. They also increase the involvement of a broad range of stakeholders and create new relations across national borders. Regulatory regimes in a global era are a multi-layered phenomenon on both national and international levels. However, national laws, national political objectives and local industrial contexts are fundamental variables in regulatory regimes, crucial for understanding the relationship between regulatory instruments and the safety level as it actually is (Engen et al., 2017; Lindøe et al., 2013; Lindøe et al., 2014; Paul et al., 2017; Renn, 2008).

This chapter discusses how national risk-regulating regimes cope with the increased complexities and uncertainties following in the wake of global challenges, such as new technologies, outsourcing and standardisation. Do the different national regulatory styles still have comparative advantages in securing a robust safety level in their respective national industrial contexts? What are the challenges, and what are the possibilities? Moreover, how does globalisation affect the interrelationship between national political goals, economic requirements and the specific regulatory regimes?

We follow up – and attempt to answer – these questions conceptually and empirically. The chapter begins with an outline of the concepts of regulation and risk-regulation regimes, before going on to discuss how these concepts constitute different governance strategies. We then continue with the issues of innovation and regulation, asking whether the new trends and trajectories require new conceptual approaches. Naturally, this requires a discussion of new perspectives concerning regulation, inspection practices and new scientific pillars in risk analysis. We then elaborate on the challenges connected to risk regulation (and managing risk) by connecting it to the dilemmas of exploration and exploitation. This prompts a discussion of the various modes of regulation, as well as how increased stakeholders' involvement in handling risk implies a complex set of power relations and a fine-tuned balance of trust and mistrust. Finally, we raise the issue of regulatory responses to the variety of hazards in the workforce, infrastructure and ecosystems.

4.1 What Is Regulation?

4.1.1 A General Concept

The area of regulation has considerably matured over the past decade. Because of its interdisciplinary character, the concept of regulation has many definitions – both overlapping and contradictory. Accordingly, Baldwin et al. (2012) suggest three main concepts of regulation:

- The promulgation of an authoritative set of rules, accompanied by some mechanism ... for monitoring and promoting compliance of these rules
- All efforts of state agencies to steer the economy
- All mechanisms of social control

The first concept mainly refers to the regulation as it is applied in industrial contexts, the second applies to market regulations and the last applies to all kinds of social mechanisms aiming to customise the behaviour of individuals, organisations and institutions. It also includes governmentality – how power relations 'auto-correct' individuals according to the dominant logic of governing (Power, 2007; Sabel et al., 2017).

Koop and Lodges' (2017) review of the concepts of regulation categorises the different definitions into two basic camps: an essence-based definition or a pattern-based definition. The essence-based definition provides that regulation is 'intentional intervention in the activities of a target population', while the pattern-based definition provides that regulation is an intentional intervention in a target population where the intervention is typically direct – involving binding standard setting, monitoring and sanctioning – and exercised by public-sector actors on the economic activities of private-sector activities.

Both definitions are relevant in the context of risk regulation, while the essence-based definition includes the broad area of risk governance literature. The pattern-based definition touches on, but does not include, Black's definition – where regulation is 'information gathering' and 'behaviour modification' on private and public actors (Black, 2011). The last clarification requires a clear organisational separation between regulator and regulatee. Accordingly, regulation in the context of high-risk industries relates to the pattern-based definition supplemented with Black's focus on information gathering and an organisational separation of the regulator and the regulatee. In this context, both mechanisms promoting compliance and studies analysing the 'auto-correction of individuals' are relevant. Furthermore, we add the adverbial 'robust' to the term 'regulation'. Here, 'robust regulation' refers to the strategy to provide the regulatory regime with the capability to handle shifting and challenging environmental conditions. The main conditions for such capabilities are flexibility, stakeholder involvement, legitimacy and trust; these are conditions that are often positively interrelated – for example, stakeholder involvement and trust.

4.1.2 Different 'System Logics'

High-risk industries are complex systems of political issues, containing a variety of stakeholders with different agendas, and different administrative and legal practices. One core element in this 'system' is a *policy-driven process* with legally binding laws as an outcome. On the other hand, economic actors and stakeholders constitute the industry with the main purpose of value-creation. Another core element in this 'system' is, therefore, a process driven by knowledge and new technology with 'best practices' and industrial standards as an outcome. In Figure 4.1, we present how these two different 'system logics' interact.

The left-hand side of the figure describes the control mechanism driven by legislation, and the administrative and regulatory control agencies based on values, preferences and a common understanding of risk acceptance in society. This means that legally binding rules (laws and regulations) are laid down in place for the government, citizens and economic enterprises, and so on. In an open and democratic society, many actors with preferences and interests influence this process. On the right-hand side of the figure, the control-mechanism focusses on financial risks, ensuring quality and the pace of technological innovation, but also incorporating occupational health, technical integrity and the environment.

Hence, risk assessment and risk governance aspire to be a part of the technological and organisational processes within value creation where research institutions, laboratories, universities, and so on, are all key producers of knowledge. Along with professional and experience-based knowledge, this

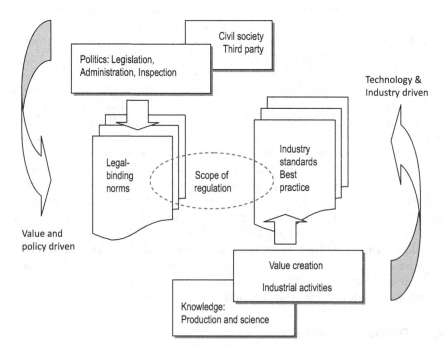

FIGURE 4.1
Different system logics: societal values and policy versus technology and value creation.

will amalgamate with best practices in businesses. Besides constituting increased uncertainty and complexity, such processes induce ambiguity. A certain amount of leeway and flexibility of the function-based regulatory regime allows actors to play different roles and push technological and organisational limits. The trade-off between economics and safety will challenge inspection practices, targeting the boundaries of accepted risk behaviour. Throughout the history of high-risk regulation, economic and environmental challenges have created several notable accidents: the Toulouse disaster, Deepwater Horizon and Fukushima Daiichi serve as three of the latest relevant examples (Hopkins, 2012).

4.2 Risk Regulating Regimes

4.2.1 A General Framework

Risk regulation and risk management are the result of a modernisation process whereby social production of wealth systematically accompanies technical, medical and social risk. In the 1970s and 1980s, major disasters revealed high-hazard levels in certain industries: the chemical production industry (Seveso in 1976 and Bhopal in 1984), the nuclear power industry (Chernobyl in 1986 and Fukushima in 2011) and the offshore petroleum industry (Alexander Kielland in 1980, Piper Alpha in 1988 and Deepwater Horizon in 2010). These major accidents mobilised public opinion, increased debates on safety and prompted self-reflection within the industries themselves. During the following decades, new organising principles and methods of assessing, managing and regulating risk were developed. In this endeavour, many actors took part: companies and unions, politicians who framed economic and social regulations and allocated resources, scientific communities, stakeholders within civil society, non-governmental organisations (NGOs), scientific communities, and so on. They developed new concepts of regulation and safety management, wherein part of the regulatory process was delegated from the authorities to industrial stakeholders.

Risk regulation can thus be defined as a binding command, excluding commands contained within standards developed by nongovernmental bodies (such as professional and industry trade associations which have sanctioning authority). These binding commands include those found in statutes, rules issued by government agencies that administer statutes, as well as legal principles arising from court decisions. The report from the Transportation Research Board in the United States uses this definition. Recognising that the many different terms in use are often imprecise, inconsistent or even overlapping, the report introduces a conceptual framework describing four design types – micro-means, micro-ends, macro-means and macro-ends – as

TABLE 4.1

Mapping Common Regulation Descriptions

	Means	**Ends**
Micro	Micro-Means • Prescriptive regulation • Design standards • Technology-based regulation • Specification standards	Micro-Ends • Performance-based regulation • Output-based regulation • Market-based regulation
Macro	Macro-Means • Management-based regulation • Performance-based regulation • System regulation • Goal-based regulation • Safety-case regulation • Enforced self-regulation	Macro-Ends • Tort and ex post liability • General duty provisions • Outcome-based regulation

Source: Adapted from Coglianese (2010).

presented in Table 4.1. They argue that this framework captures more precisely and accurately the underlying differences that most professionals have in mind when discussing regulatory designs.[1]

4.2.2 Hierarchy of Norms

A regulatory regime is formally organised through hierarchies of 'norms' or 'rules'. A hierarchy of 'rules' ranges from legally binding laws and regulations to voluntary technical standards, best practices and professional judgements. Table 4.2 presents the hierarchy of rules with two main categories and sub-categories. The first category, 'legally binding rules', includes commands found in laws and regulations, statutes, rules issued by government agencies that administer statutes and legal principles arising from court decisions. The second category includes all kinds of non-legally binding rules, such as guidelines to the regulation, on-statutory instruments and global industrial standards, as well as company-specific standards. In between these two main categories, we have put the term 'legal standard' – referring to norms

TABLE 4.2

Hierarchy of Norms

Hierarchy of rules	**Sub-categories**	
Laws and regulation as legally binding rules	• Laws • Regulations authorised by the laws	
'Legal standards' or 'guiding principles' as a linking pin		
Non-legally binding rules	• Guidelines to the regulation • Non-statutory instruments	• Recognised 'global' industry standards • Company-specific requirements and guidelines

or rules that, by themselves, do not belong to the domain of law as an 'undetermined legal concept'. Examples of the practical use of legal standards or 'guiding principles' in risk and safety contexts are the BACT and ALARP-principle – 'Best Available Control Technology' and 'As Low as Reasonably Possible', respectively.

The practical implementation of legal standards takes place within the particular professional community in accordance with their prevalent use, professional judgements, values and attitudes. The content of the legal requirement embedded in the law will change along with social (and technological) changes.[2] Using such legal standards implies that the regulatory activities involve professionals in the formulation of safe practices, and thereby commit companies to comply with applicable legislation. While detailed and prescribed rules will lag behind fast-developing technology, legal standards enable a more easily updated regulatory practice, especially for achieving appropriate regulation of complex organisational developments and fast-changing technology (Marchant et al., 2013). Legal standards refer to norms and practices besides the law, which changes over time – for example, consequences of new technology, organisational procedures or historical and social contexts. Legal standards tie the unchanging word of law to the ever-changing implementation of norms and ideas embedded in that law. The use of legal standards aims to achieve an appropriate level of regulation within the context of these highly dynamic industries. It is as an expression of respect for the importance of expert knowledge to ensure safety and quality in key areas of society in changing circumstances.

4.2.3 Function-Based Regulation

Furthermore, modes of risk regulation consist of rule compliance versus risk management (Hopkins, 2011), with a distinction between 'hard regulation' based on 'command-and-control' with prescriptive rules, and 'soft regulation' with concepts coined as 'self-regulation' (Short and Toffel, 2010) and 'meta-regulation' (Gilad, 1997). Industrial activities based on 'command-and-control' and the state as an external controller with performance-based rules and functional requirements have different logics of action. Adjustment from one system towards the other represents a change in goal setting, use of legally binding norms, industrial standards and professional and legal competence. The transfer or delegation of safety responsibilities provides new and cooperative approaches in the implementation of regulatory regimes. An important question, however, is whether or not these different modes of regulation represent dichotomies, or whether they are complementary. According to Sinclair, strict dichotomies do not exist in reality. In practice, risk regulation regimes will combine roles and responsibilities in a public–private partnership with a top-down approach composed of legally binding norms, as well as a bottom-up approach composed of industrial standards and best practices (Sinclair, 1997).

Principles of enforced self-regulation (functional regulations) rely on an industry's capability to manage its risks according to accepted norms and standards. However, such processes are vulnerable due to the comprehensive, frequent and multifaceted patterns of interaction among government, operators/suppliers and labour unions. Functional risk regulation requires a balance of power and mutual trust among the intervening actors. Therefore, function-based regulation needs some form of discretionary criteria such as 'legal standards' in order to link functional requirements in the law to industrial standards.

High-risk regulatory regimes have been on the forefront in developing function-, purpose- and goal-based regulations. Such regimes rest on the assumption that the involved parties have a common interest in maintaining the system, and that conflicts of interest will naturally be solved without threatening the foundation of trust between the relevant parties. How much power each of the agents possesses will vary depending on a range of variables. The new era of free trade and globalisation causes constant reorganisation in industries trying to seize opportunities and increase competitiveness. Therefore, today's high-risk industries undergo major changes due to downsizing and merges, which inevitably affects and challenges their safety levels.

A consistent application of function-based regulations requires a comprehensive and systematic review of how the various provisions are understood, as well as how the appropriate standards should be used to meet the requirements. Procedures must stipulate relationships between laws, regulations and technical/professional standards. For regulatory authorities and inspectors, this can be a demanding and comprehensive system to try and keep up to date, as it requires that standards keep pace with developments and new knowledge (Bieder and Bourier, 2013). There is an inherent tension between following comprehensive guidelines and best practices, and the desire to require an industry to continually innovate and implement any new expertise and scientific knowledge that might improve safety. Risk regulation with stakeholder involvement and internal control may be described as a balance of power between state control and industrial degrees of freedom. A function-based regulatory system from such a perspective is flexible, adaptable and expedient towards the globalisation process. On the other hand – and underlined by Le Coze (2017) – globalisation induces externalisation, which implies networked structures of safety-critical systems, 'multiplying interfaces between organisations while creating complex interaction issues across a range of various entities'. Instead of self-regulation, self-control based on high liability can easily end in the pulverisation of responsibility (Le Coze, 2017).

4.3 Balancing Regulation and Innovation

4.3.1 Regulation and Technological Development

Combining regulatory approaches with perspectives on global innovation and technological development in high-risk industries, sociotechnical theory may serve as a lens to identify influencing factors such as prices, cost-efficient technological development and environmental change. Above, we argued that risk regulation and technology are in a dialectical relationship. Arguments for a closer relationship between technology and regulatory practices are also underlined in the definitions of technology systems that comprise regulatory elements. According to Bijker, Hughes and Pinch (1987), technological systems include technical devices, organisational routines and procedures, legislative artefacts, as well as scientific and other knowledge elements such as skills, rules of thumb and norms for the handling of technology. These elements are not distinctive or autonomous factors but form a 'seamless web' that constitutes technological pathways.

Innovation and new technologies produce solutions to bottlenecks. New technology is required in order to reduce cost, reduce emissions and organise production in new ways. On the other hand, it also induces uncertainty and cerates unforeseen events due to inexperience in organisations and a lack of absorptive capacity among the actors. Moreover, new technology carries, almost without exception, increased complexity. From a regulatory point of view, technology represents the answers and instruments to comply with international environmental regulations. However, it also requires constant monitoring from the regulator – that is, adjusting the laws, regulations, legal standards, and so forth – and active renewal of expertise to keep pace with development. Accordingly, stakeholder involvement will accelerate in accordance with the increased need for knowledge associated with new risks.

Different dimensions characterise high-risk industries. One dimension is the value chain of operators, contractors, suppliers, distributors, customers and end users. Interwoven with these actors are research communities, environmental and consumer interest groups, the media, and so on. Another dimension is the hierarchal levels ranging from local workplaces, departments and installations/plants, to companies and industrial groups. Political institutions and governmental agencies that surround the firms and industries create political, economic and legal frameworks. Interactions between the different system levels constitute a flow of information upward and mechanisms of steering and control downward, and further include the dynamics between design, modification and daily operation (Rasmussen, 1997; Leveson, 2004).

4.3.2 Dynamics of Exploitation and Exploration

How do organisations adapt to dynamic environments, changing markets, regulatory frames and new technology? Basic organisational learning theory combines two distinctive elements. The 'first-order learning' element concerns providing existing services and products better or more safely by managing and operating in accordance with established rules. A 'second-order learning' element implies changing the rules, and thereby developing and providing new and improved services and products (Argyris and Schon, 1996). Organisational learning combining first- and second-order learning is further developed through the concepts of *exploitation* and *exploration*. (March, 1991). Exploitation entails the efficient use of existing competence, which required a company to survive in the short term, while exploration requires managing new opportunities and flexibility to create a strategic direction to ensure long-term survival. Organisational learning implies a management process in which knowledge is transferred through different stages, from tacit to documented knowledge, such as routines or standardised operating procedures. There is, however, a limit on how tacit and explicit knowledge (with rich experience) from communities of practice can be codified into written formalised procedures, manuals and industrial standards (Gerhardi and Nicolini, 1999; Weick et al., 1999). Handling the innovation dilemma requires the ability to balance exploitation and exploration, which is a challenge for both industry in developing technological policy, as well as for regulators (Noteboom, 2000).

4.3.3 Combining Enforcement and Management by Rule

To balance different stakeholder interests and the innovation dilemma within a regulatory framework implies a typology of a 'top-down bottom-up approach'. Industries (and companies) represent arenas that focus on adapting to the market with a need for 'self-regulating' or internal control. Ideally, the companies act responsibly, assessing their actual risk, implementing safety barriers and adapting management systems, as well as negotiating with the state and individual firms to establish regulations particularised to each firm (Aires and Braithwaite, 1992). As shown in Figure 4.2, there is an 'interface' or overlap of the two approaches, with an array of possible combinations between the two systems. In practice, a regulatory regime has elements of both modes of regulation with a mixture of responsibilities and roles in a public–private partnership. For example, In the United Kingdom, government departments apply regulatory standards through primary and secondary legislation, enforcing them by regulatory bureaucracies governing the private sector (Baldwin and Cave, 2012).

In Figure 4.2, the modes of risk regulation and risk management discussed so far have been linked together as three 'control loops'. Reading the figure from the bottom-up, the companies and industries can be seen as innovative

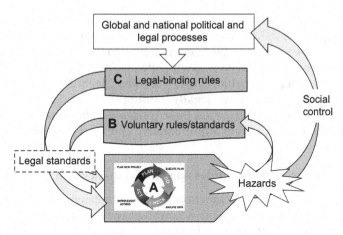

FIGURE 4.2
Enforcement by rules and enterprise risk management.

production systems accompanied with hazards, incidents and accidents that represent threats to employees, the environment and third parties. However, an integrated element of a resilient production system is the development and maintenance of activities, processes, best-practices procedures and enterprise risk management (ERM). Such processes are compatible with programs of 'quality management', 'continuous improvement' and 'quality circles' Plan-Do-Check-Act, and so on.[3] Such a system of improvement and learning is labelled (A) in the figure. These processes reflect the 'system logic' of technology and knowledge-based development as presented on the right-hand side in Figure 4.1.

As pointed out above, a company is part of a value chain of customers, suppliers, distributors, customers and end users, interwoven within a network of stakeholders, interest groups, media, and so on. Customers, suppliers, industrial associations and professional groups may give important and necessary inputs to 'best practices' and operation standards within the company. This is a process of 'voluntary enforcement' by procedures and rules, which may be codified and documented as industrial standards.[4]

Hazards are exposed to the environment; incident and accidents will be exposed as public issues, involving different forms of social control. In some cases, civil society actors, citizens or non-governmental organisations may exert pressure on the industry to voluntarily improve its practices. This could take place within the existing legal framework and accepted societal norms without intervention from regulatory agencies. In both cases, a 'learning loop' of improvement by voluntary rules and standards takes place with influence or pressure from within the value chain or from external stakeholders. The letter (B) denotes this loop in Figure 4.2.

Concerns about this approach include uncertainty about compliance requirements, given that each industrial activity has a unique mix of characteristics and ambiguity about governmental intervention and enforcement. In other cases, risk raises issues of public engagement, political debate and legislative processes. Public and political concerns and debates can occur on a national or global level. As an outcome, new laws and regulations aimed at protecting individuals, the workforce, technical infrastructure and the environment emerge as legally binding rules, as denoted by (C) in Figure 4.2. If we return to Figure 4.1, these processes reflect the 'system logic' of political issues, agendas and administrative and legal practices.

4.3.4 Flexibility in Handling Rules and Norms in a Global Context

Feedback loops with possible sanctions act as enforcement mechanisms so that the industry adjusts its practice according to new norms, rules and standards. As indicated in the figure, the use of 'legal standards' could be the key in developing a mechanism for coordination and adjusting legally binding and voluntary rules and standards. Accordingly, the concept of 'legal standards' may serve as a mechanism to link national requirements with an increasing globalisation movement.

Using legal standards requires that the regulatory activities involve local professionals in the formulation of safe practices, thereby committing companies into complying with applicable legislation. While detailed and prescribed rules will lag behind fast-developing technology, legal standards enable a more easily updated regulatory practice, especially for achieving the appropriate regulation of complex organisational global developments and fast-changing technology (Marchant et al., 2013). However, this assumes governance structure where safety issues are reasonably balanced towards financial goals, as well as where the regulators have sufficient competence and resources with which to develop regulatory frameworks and possess a good overview of the complexity and functionality of the high-risk industry in question.

4.4 Lessons Learned

4.4.1 Different Modes of Risk Regulation

In the anthology 'Risk Governance in Offshore Oil and Gas Operation', Andrew Hale finds the most striking difference between regimes is the way in which the rules are formulated (Hale, 2014). Rules exist on a spectrum where goals are at one end, and specific actions or states to be carried out or achieved are at the other (Hale and Swuste, 1998). Goals state the outcome to be achieved, sometimes in concrete, measurable terms, but more often in generic terms, so as to achieve as low a level of risk as reasonably practicable'

(ALARP). The company is then left to decide exactly how to operationalise and justify its choice to the regulator. On the other hand, action/state rules specify exactly what must be done in terms of behaviour – for example, wearing specified personal protective equipment or using a specific scaffolding or pressure-release system. In between, there is a category of rules that specify the processes (risk management) to be carried out to arrive at the specific, locally appropriate risk control specifications. Such rules outline how to translate goals into actions. This section seeks to discuss possible regulatory responses to global pressures that have been introduced and, when relevant, apply examples from empirical studies, particularly from Norway, the European Union and the United States.

On the right-hand side of Figure 4.3, the state acts as a bureaucracy with a regulatory focus on upholding established rules. The outcome of such a regime is coined 'command-and-control'. Baldwin and Cave argue that regulators who operate command-and-control techniques are sometimes equipped with rule-making powers, as is often the case in the U.S. command-and-control regulations developed by the state in its capacity to dictate legally binding norms, enforcement; sanctions to protect citizens, institutions and infrastructure against hazards and risks. An inspection with detailed checklists usually follows the prescribed rules. Accordingly, deviations from the rules is subject to enforcement and sanctions. Some criticisms of this instrumental and procedural approach include: the risk of ignoring human and organisational factors; an inability to keep pace with rapid technological change, as well as systemic failures when this approach is implemented at higher institutional levels (Baldwin and Cave, 2012).

As pointed out above, the workforce and its unions, oil and gas customers, the public, the regulator, as well as various interest and pressure groups all have different understandings, perceptions and opinions about the risk

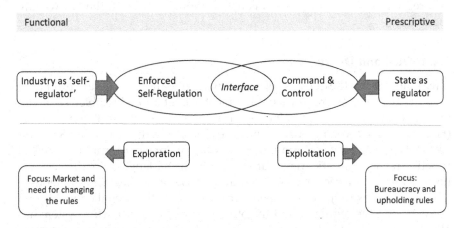

FIGURE 4.3
Different modes of risk regulations.

issues at stake. Hale (2014) finds the Norwegian Petroleum offshore regime to be the most explicit and articulated with the *tripartite structure* as one of the key foundations. The role of the unions is most powerful when compared to the more limited tripartism of the United Kingdom, or its almost complete absence in the U.S. offshore industry in the Gulf of Mexico. Consequently, the potential role of unions and workplace involvement in safety and health is perceived differently in the regimes. In the Norwegian social democracy, the tripartite system is a constructive and creative way to organise and balance the interests of the main stakeholders. The strength of the unions can be crucial in keeping a balance of the three parties, and tripartism allows shifting alliances when developing the regime. Safety representatives may see themselves in alliance with the regulator when safety is threatened, but the unions may ally themselves with employers' cases where too-intensive enforcement threatens the industry's future. Hale summarises the tripartite system by using a metaphor of the stability of a three-legged stool, where a stool with three equal-length legs compares favourably to one with only two legs or three uneven legs.

The balance of frames is an important issue in regulatory regimes. The Norwegian offshore regime has developed with technical and organisational problem solving and systems thinking as a dominant 'engineering frame'. Within the tripartite system, the industry and trade unions have both made additional practical input. However, the lawyers' frame of prime cause seeker, clarity of requirements and judicial review is not dominant, even where the system itself creates problems of jurisdiction (Kaasen 2014). The U.S. regime has the opposite bias, while the United Kingdom takes a position in between with the 'safety case' emphasising the engineering frame. However, according to Paterson (2014), a weakness is failing to ensure the practical input in a workplace involvement. The difference in frames has a parallel in the dimension from the dominance of conflict in the U.S. regime to the dominance of collaboration and trust in the Norwegian regime.

4.4.2 Trust and Distrust

If we follow up on the question of coping with globalisation, we are prompted to ask which of these regimes are the most robust and adaptable? Hale tentatively concluded that the Norwegian regime meets most of such criteria (Hale, 2014). Its development began in the 1980s, with the Alexander Kielland accident as driving force, and moved towards its present form. Its principal functional characteristics have survived several challenges, mobilised stakeholders and prompted discourse to retain those principles. Simultaneously, it has shown flexibility and resilience in adapting to those challenges. A joint effort among the regulator and the other parties in the tripartite system has allowed them to develop the regime as a learning organisational system (Korneef, Foor and van der Meulen, 2000).

The United States, with prescriptive rules, detailed inspections and the threat of sanctions and lawsuits, seems to undermine trust and increase the adversarial relationship between the industry and regulators. Conversely, one of the building blocks of the Norwegian regime is a tripartite governance based on mutual trust. The function-based regulations presuppose that the regulators trust the oil companies when it comes to organisational capacity (as well as trusting that they have the necessary skills and competence in implementing enforced self-regulation). For instance, companies have to trust subcontractors to do the drilling job correctly.

The Norwegian regime provides an illustration of the role of trust within the tripartite system. In the balance of power between the authorities, the industry and unions, the legal framework seldom provides a clear-cut threshold for acceptable risk and regulatory compliance. Kringen (2014) points to reputational concerns and a compliance-friendly regulatory incentive structure as the reason for a lack of sharp confrontations over legal issues, and very few lawsuits. Enforcement strategies with legal alternatives regarding industrial response strategies rely on functional trust for best results. The regulator may also escalate enforcement and sanctions along the trust-distrust dimension. Rosness and Forseth (2014) use a narrative of the tripartite collaboration with periods of erosion, conflicts, negotiations, joint action and subsequent revitalisation, to emphasise that all stakeholders as a whole have to work to ensure trust and reputation.

Within the two-folded 'system logics' presented in Chapter 1, the regulator must trust the companies and their ability to develop robust and resilient systems and procedures for safe work, both for themselves and for their suppliers. Further down the value chain, oil companies must trust contractors, subcontractors, and employees to perform their work according to a set of agreed-to rules and standards of quality and safety. Trade unions have to trust that companies and merchants have established the right supervisory and control chains.

In a study of safety culture on an oil platform, Tharaldsen (2011) discusses how trust and distrust can occur in functional and dysfunctional features through a variety of possible combinations, as presented in Table 4.3. Based on reasonable precaution, trust may be functional (1), but too much

TABLE 4.3

Functional and Dysfunctional Trust and Distrust

	Trust		
Functional	1	2	Dysfunctional
	Trust based on reasonable precautions	Naïve and blind trust	
	3	4	
	Distrust, based on realistic precautions	Distrust based on detailed surveillance and control	
	Distrust		

confidence may lead to a dysfunctional relationship when there is naivety and blindness (2). On the other side, distrust may be functional when the relationship contains realistic precautions (3), while dysfunctional distrust may occur in rigid control strategies (4).

In the process of value creation, professionals are interrelated, and dependency can be denoted as *trust chains* (Grimen 2009). A high degree of trust promotes the flow and quality of information, reducing transaction costs (Hardin, 2002). Confidence and trust among actors creates positive expectations about others' intentions and behaviour in the chain – and it reduces complexity – yet it also introduces vulnerability. However, as Table 4.3 illustrates, too much confidence may turn into naivety, while too much distrust may end up in rigid control strategies. Therefore, *control chains* are a compensating mechanism representing an institutionalised form of 'distrust' that reduces faults, errors and omissions that may occur in trust chains. Trust chains and control chains are mutually dependent in professional organisations. The successful completion of producing goods and services depends on the individual department and professional expertise. The Deepwater Horizon case is a good illustration of dysfunctional trust and control chains between the main actors BP, Transocean, Halliburton/Sperry Sun and Cameron.[5]

Hale (2014) points to trust as a key factor in comparing regimes. An important issue in deciding which of the regimes described in this book should serve as the template for other industries in a specific country is one of trust. Balancing trust and distrust seems to be one of the most important factors promoting a robust risk-regulation regime.

As shown in the Norwegian regime, long-term collaboration within the tripartite system, along with an institutional framework in place, promotes a fruitful balance between confidence givers and trust beneficiaries alike. On the other hand, conflicting interests among the involved actors can ultimately have an eroding effect. If the context and conditions for a fruitful balance of trust and distrust is lacking, a regime based on distrust with prescriptive rules and 'command-and-control' seems to be the most reasonable option.

4.5 Concluding Remarks

As mentioned above, 'robust regulation' refers to the strategy to provide the regulatory regime with the capability to handle shifting and challenging environmental conditions. The main conditions for such capabilities are flexibility, stakeholder involvement, legitimacy and trust. These are conditions that are largely positively interrelated – for example, stakeholder involvement and trust. Increasing complexity of high-risk industries, along with the interconnectedness of technical and social systems, has given rise to the need for a broader, more integrated understanding of safety. The globalisation of risk regulation represents challenges of combining different 'system logics'. On the one hand, regulation is a policy-driven process with legally binding laws as an outcome. On the other hand, the industry is a 'system' driven by knowledge and developing technology. The former tends to produce a 'top-down' strategy of regulation and enforcement, while the latter produces a 'bottom-up' strategy of self-regulation. This is a challenge embedded within the dilemma of innovation and change within industries in general. A mechanism to balance 'top-down' and 'bottom-up' strategies with risk regulation has to be a reflexive learning system, combining multiple learning and control loops, which is critical of its own performance, structure and functioning.

Balancing trust and distrust seems to be one of the most important factors for promoting a robust risk regulation regime. Collaboration within a tripartite system, along with an institutional framework in place, promotes a fruitful balance between confidence givers and trust beneficiaries. On the other hand, a range of conflicting interests among the involved actors ultimately has an eroding effect. If the context and conditions for the fruitful balance of trust and distrust are lacking, a regime based on distrust with prescriptive rules and 'command-and-control' seems to be the most reasonable option. Transformations and trajectories challenging the offshore industry impose regulatory regimes and a response to occupational health and safety (OHS), technical safety and integrity of platforms and installations, as well as an external environment with pollution to sea and shores. Learning from major accidents over a number of years reinforces the need for cumulative learning to improve robust offshore regulatory regimes.

The transfer of operators, management and technology between globally operating firms may produce new risks regarding failures, misuse and accidents. A systematic assessment of the societal, economic and industrial framing for such transfers is necessary in order to determine if a regulatory approach is optimal in a specific context. A strategy of enforced self-regulation with a 'bottom-up' strategy of developing industrial standards could be effective when it is likely that consensus can be achieved among actors with long-term relations. In contrast, if there is a continuing need or desire for regulators to provide stable norms and controls in an unambiguous way, a 'top-down' strategy with prescriptive rules, regulations and sanctions would be better. The political, administrative and legal structures play a major role in explaining differences in the regimes.

It becomes obvious that the tripartite system of cooperation and balancing\interests among authority, industry and the workforce/unions may inhibit flexibility to adjust to external requirements. One core element in the tripartite system is the trust developed between the regulator and the regulated industry, as well as their recognition of the legitimate role of the workforce as a basis for industrial relations. This enables the regulator to implement a strategy in which some of the major responsibilities for control are shifted over to the industry, and there is a presumption that there is both willingness and capability to collaborate on the continuous development of accepted norms and standards. In this context, such abilities are seen as a competitive advantage related to new technologies and requirements that follow in the wake of globalisation.

A risk-informed, trust- and dialogue-based, as well as a functional-based authority enforcement of self-regulation means that regulators set the general requirements the industry must meet. The strengths of such regulatory regimes are a strong stakeholder involvement, as well as an adopted capability building among the industry and regulatory body. The tripartite system is tailor-made to ensure large stakeholder involvement in safety discussions. However, there are vulnerabilities in such a function- and trust-based regime, where, for instance, global political and economic issues can easily weaken the trust between parties and undermine cooperation. Similarly, the win–win principles between safety and the economy during economically difficult times is challenging to maintain. When continuously confronted with global challenges, small economies may have to adjust to safety regulations and standardisations, which are less adaptable to national styles and traditions; they will be forced to change and harmonise their regulatory regimes. The question is thus not *if*, but *how* local functions and trust-based regulatory regimes survive in a global *post-regulatory* era.

Notes

1. Designing Safety Regulations for High-Hazards Industries. Transportation Research Board. Special Report 324. 2017.
2. Translated from E. Boe, Introduction to law (Universitetsforlaget 2010), 278. In particular, when illustrating the Norwegian Working Environment Act, Boe emphasises 'a standard of welfare at all times consistent with the level of technological and social development of society'.
3. The so-called PDCA circle is also known as the "Deming circle" after the American statistician Edward Deming (1900–1993).
4. These are ISO standards, such as ISO 14000 Environmental Management, ISO 31000 Risk Management or OHSAS 18001 Occupational Health and Safety Management Systems. Within the framework of our discussion, standards of safety management, occupational health and safety, technical safety and environmental protection are the most relevant.
5. For a discussion of the involved parties, their relationship and critical roles in the accidents, see U.S. Chemical Safety Board, Investigation Report, Volumes 1 and 2, 6/5 2014, http://www.csb.gov/macondo-blowout-and-explosion/.

References

Aires, I. and Braithwaite, J. (1992). *Responsive Regulation*. Oxford.

Argyris, C. and Schön, D. (1996). *Organizational Learning II. Theory, Method and Practise*. Addison-Wesley.

Baldwin, R. and Cave, M. (2012). *Understanding Regulation*. Oxford University Press.

Bieder, C. and Bourrier, M. (2013). *Trapping Safety into Rules. How Desirable or Avoidable Is Proseduralization?* Ashgate.

Bijker, W., Pinch, T., and Hughes, T. (1987). *The Social Construction of Technological Systems*. MIT Press.

Black, J. (2011). "The rise, fall and fate of principles-based regulation", in Alexander, K. and Moloney, N. (eds.) *Law Reform and Financial Markets*, pp. 3–34. Elgar Financial Law Series. Cheltenham, UK: Elgar.

Boe, E. (2010). *Introduction to Law*. Oslo: Universitetsforlaget.

Engen, O. A. and Lindøe, P. H. (2017). "The Nordic model of offshore oil regulation: Managing crises through a proactive regulator", in Balleisen, Edward J. et al. (eds.) *Policy Shock*, pp. 181–203. Cambridge: Cambridge University Press.

Engen, O. A., Lindøe, P. H., and Hansen, K. (2017). "Power, trust and robustness – The politicisation of HSE in the Norwegian petroleum regime", *Policy and Practice in Health and Safety*.

Gherardi, S. and Nicolini, D. (2000). "The organisational learning of safety in communities of practice", *Journal of Management Inquiry* 7–18.

Gilad, S. (2010). "It runs in the family: Meta-regulation and its siblings", *Regulation and Governance* 485–506.

Grimen, H. (2009). *Hva er tillit? (What is trust?)*. Universitetsforlaget.

Hale, A. (2014). "Advancing robust regulation", in Lindøe, Baram and Renn (eds.) *Risk Governance of Offshore Oil and Gas Operations*, pp. 403–424. Cambridge.

Hale, A. and Swuste, P. (1998). "Safety rules: Procedural freedom or action constraint?" *Safety Science* 163–177.

Hardin, R. (2002). *Trust and Trustworthiness*. Russel Sage Foundation.

Hopkins, A. (2011). "Risk management and rule-compliance: Decision-making in hazardous industries". *Safety Science* 110–120.

Hopkins, A. (2012). *Disastrous Decisions: The Human and Organisational Causes of the Gulf of Mexico Blowout*. CCH.

Kaasen, K. (2014). "Safety regulation on the Norwegian continental shelf", in Lindøe, Baram and Renn (eds.) *Risk Governance of Offshore Oil and Gas Operations*, pp. 103–131. Cambridge.

Koop, C. and Lodge, M. (2017). "What is regulation. An interdisciplinary concept analysis". *Regulation and Governance*, Volume 11, 95–109.

Korneef, Foor, Meulen, and Meile van der (2000). Computer Safety, Reliability, and Security, 19th International Conference, SAFECOMP 2000, Rotterdam, The Netherlands.

LeCoze, J. C. (2017). "Globalisation and high risk systems". *Policy and Practise in Health and Safety*, Volume 15(1), 57–81.

Leveson, N. (2004). "A new accident model for engineering safer systems". *Safety Science* 237–270.

Lindøe, P. H. (2017). "Risk regulation and resilience in offshore oil and gas operations", in A. Herwig and M. Simoncini (eds.) *Law and the Management of Disasters*, pp. 105–123. Routledge.

Lindøe, P. H., Baram, M., and Renn, O. (eds.) (2014). *Risk Governance of Offshore Oil and Gas Operations*. New York: Cambridge University Press.

Lindøe, P. L., Baram, M., and Paterson, J. (2013). "Robust offshore risk regulation – an assessment of US, UK and Norwegian approaches", in Marchant, Abbott and Allenby (eds.) *Innovative Governance Models for Emerging Technologies*, pp. 235–253. London: Edward Elgar.

March, J. (1991). "Exploration and exploitation in organisational learning". *Organisation Science* 71–87.

Marchant, G., Abbot, K., and Allenby, B. (eds.) (2013). *Innovative Governance Models for Emerging Technologies*. Elgar Publishing.

Nooteboom, B. (2000). *Learning and Innovation in Organisations and Economics*. Oxford University Press.

Paul, R., Molders, M., Bora, A., and Munte, P. (2017). *Society, Regulation and Governance. New Modes for Shaping Social Change*. Cheltenham, UK: Elgar Publishing.

Paterson, J. (2014). "Safety regulation on the UK continental shelf", in Lindøe, Baram and Renn (eds.) *Risk Governance of Offshore Oil and Gas Operations*, pp. 132–153. Cambridge.

Power, M. (2007). *Organising Uncertainty*. Oxford: Oxford University Press.

Rasmussen, J. (1997). "Risk management in a dynamic society: A modelling problem". *Safety Science* 183–213.

Renn, O. (2008). *Risk Governance, Coping with Uncertainty in a Complex World*. London: Earthscan.

Sabel, C., Herrigal, G., and Kristensen, P. H. (2017). "Regulation under uncertainty: The coevolution of industry and regulation". *Regulation and Governance*. John Wiley.

Short, J. M. and Toffel, M. W. (2010). "Making self-regulation more than merely symbolic: The critical role of the legal environment". *Administrative Science Quarterly* 361–396.

Sinclair, D. (1997). "Self-regulation versus command and control? Beyond false dichotomies". *Law and Policy* 529–559.

Tharaldsen, J. (2011). *In safety we trust – Safety, risk and trust in the offshore petroleum industry*. PhD thesis, unpublished, University of Stavanger.

Weick, K. E., Sutcliffe, K. M., and Obstfeld, D. (1999). "Organisation for high reliability: Process of collective mindfullness", in Sutton, R. S. and Staw, B. M. (eds.) *Research in Organisational Behaviour*, Volume 1, pp. 81–123. Jai Press.

5

On Ignorance and Apocalypse: A Brief Introduction to 'Epistemic Accidents'

John Downer

CONTENTS

5.1   The Limits of Interrogation 77
5.2   Epistemic Accidents 79
5.3   New Perspectives: NAT and EAT 80
References 84

Introduction: A Close Call?

In the late 1990s, a German biotech company altered a common bacterium – *Klebsiella planticola (or K. planticola)*,* which plays a role in decomposition – by adding a gene that would make it convert plant material into alcohol. The newly modified bacterium, which they called 'SDF20', was intended for farmers. It would allow them to take the waste they would usually burn and convert it into ethanol they could sell, as well as creating a nutrient-rich 'sludge', which they could use as fertiliser.

An organism that simultaneously reduced burning and produced two valuable resources seemed like it would be a win-win, but SDF20's appeal sharply waned after a study conducted at the University of Oregon (Holmes et al., 1999). The Oregon researchers noticed that the leftover sludge contained live SDF20 and decided to explore the bacterium's effects on the ecosystem. In a series of experiments, they introduced the sludge to soil samples containing wheat plants. Within two weeks, the plants had died. The live SDF20 had interacted with micro-organisms in the soil (its 'biota') in ways that were ultimately fatal to the crops it supported.

The study and its implications remained relatively obscure and uncontroversial until 2001, when the lead researcher, Elaine Ingham, invoked it while testifying about the potential hazards of GMOs to the New Zealand

---

* *Klebsiella planticola* has since been reclassified as *Raoultella planticola*.

Royal Commission on Genetic Engineering. She claimed that the U.S. Environmental Protection Agency (EPA) had missed the danger that SDF20 posed because its initial tests were conducted with sterile (biota-less) soils, and, as a result, had approved it for field trials, which were only cancelled in the wake of her study. Had the field trials gone ahead, she suggested that SDF20 might easily have spread worldwide: displacing the ubiquitous natural *K.planticola*, damaging soils and decimating crops around the globe (Porterfield, 2016; Walter et al., 2001).

Ingham's testimony quickly became embroiled in the acrimonious politics around GMOs. Environmental advocates invoked it as evidence of a narrowly averted apocalypse; painting dire portraits of a runaway, laboratory-born organism racing across the planet, leaving wastelands and famines in its wake (e.g. Robbins, 2002; Brockway, 2010). Proponents of GMOs responded robustly. In a rebuttal to Ingham's original testimony, a group of scientists denied that the EPA had formally approved field tests; highlighted an erroneous citation in her paper, and claimed her testimony went beyond the published evidence. Others built on this testimony to discredit Ingham herself: casting the erroneous citation as a falsified source, and a routine departmental review as an investigation of her integrity (Walter et al., 2001; Krebs, 2001). Such efforts were well-funded and rhetorically effective. Ingham retracted her claim about formal EPA approval, corrected the erroneous citation and allowed that her fears went beyond her direct observations. None of these concessions amounted to an admission of wrongdoing or substantial error – scientists routinely speak to matters beyond their direct observations – yet the optics were bad, and the alarm is now remembered by many as an example of environmentalist hysteria.

As the dust settles on the controversy it should be clear that Ingham's fears were far from ungrounded. Her research strongly indicated that should it have spread SDF20 might have endangered plant life. And given the ubiquity of natural *K.planticola*, it was reasonable to imagine that it might have spread uncontrollably if it was released into the wild, as other organisms have in the past. It might have been true, moreover, that the EPA had not issued formal approvals for field tests prior to her study, yet it is easy to imagine that it could have done so, and would have done so, in time. At the time of the incident, the agency's standard GMO assessment practices paid little account to the wider ecosystem; framed, as they were, by assumptions better suited to chemicals than to living organisms. Its initial lab tests with sterile soil would not have highlighted the dangers arising from SDF20's interaction with soil biota. Nor would other tests, mandated by the Toxic Substances Control Act or the Federal Insecticide, Fungicide, Rodenticide Act. Reasonable people can disagree about how close the world came to a disaster with SDF20, but it would be foolish to dismiss Ingham's concerns as baseless alarmism.

Somewhat troublingly, it is at least *plausible* that the U.S. Environmental Protection Agency, in conducting its routine safety assessments, might have unwittingly unleashed the biological apocalypse. Perhaps even more trou- blingly, it is difficult to imagine who would have been at fault in this sce- nario. The scientists involved would have been following proper protocols, performing all the mandated tests prior to testing SDF20 in the wild. And while it is true that those tests, with their sterile soil, were unrepresentative of the real world in consequential ways, it is also true that the true represen- tativeness of any test (much like that of any experiment, simulation or model) is always an inherently imperfect and unknowable property (see e.g. Pinch, 1993; Mackenzie, 1996; Downer, 2007).

5.1 The Limits of Interrogation

The value of tests – whether they be of a bacterium, a reactor or even a jet engine – lies in their ability to *represent* the 'real world', while simultaneously *differing* from it in key respects. In some cases, this means stripping away cer- tain facets of the world that might otherwise occlude the phenomena from being investigated. Tests designed to compare the tensile strength of metals under pressure, for example, are maximally reductive. They seek to isolate those two properties – tensile strength and pressure – by strictly controlling every other significant variable (moisture, purity, temperature, manufactur- ing imperfections, etc.) that might skew the result and keep it from being comparable across samples. In other cases, the phenomenon under investiga- tion relates directly to performance in the real world, and here tests must be maximally representative. In these circumstances, testers seek to reproduce every facet of the world that might affect the result, while simultaneously retaining the differences that give the test its value. Most 'safety' tests fit into this category. They seek to recreate the world in every way except for those that might create actual hazards – flight tests are conducted without passengers for example. In every case, however, the quality of a test is a func- tion of testers' ability to identify and control significant variables: either to include or exclude them, depending on the context. A test for measuring the strength of a metal is meaningless if it doesn't control the metal's exposure to the world; and a test for measuring the safety of a GMO is meaningless if it doesn't control the world's exposure to GMOs.

Therefore, identifying and controlling significant variables is key to ensur- ing a test's validity, but how are experts to ensure that they have identified and controlled every significant variable? Philosophers sometimes call this the 'problem of relevance'. It is implicated in every technological test, scien- tific experiment, theoretical model and computer simulation. And it has no satisfying solution.

The essence of the problem is that the number of factors that might be significant to a test are potentially infinite, while the number of factors that testers can control is inherently finite.* Hence, every test, by its nature, might be unrepresentative in a significant way that the tester has not considered. This has complex ramifications; not least because it implies that there is a fundamental circularity to our interrogations. We use tests (experiments, models, etc.) to examine the validity of our theories about the world (whether they be about the safety of a GMO or the strength of a metal). Yet every test also embodies its own theories about the world, in the sense that it reflects assumptions about the factors it must represent and how well it has represented them. But if testing theories must always mean relying on other theories (i.e. about the representativeness of the test itself), then there can be no real certainty. For every finding can be questioned by disputing the bases on which it was observed. A test that 'proves' the safety of a bacterium, in other words, could always be an imperfect test.†

Dilemmas like those outlined above are at the crux of modern epistemology. For hundreds of years, great philosophical minds wrestled with the 'scientific method', dissecting the logics of induction and deduction in an effort to formulate an ineluctable route to definitive facts. None succeeded. Instead, the great 20th century breakthroughs came from those who embraced the impossibility of the task. In different ways, thinkers such as Kripke (1982), Kuhn (1996 [1962]), Feyerabend (1975), Quine (1951) and Bloor (1976), many of them building on Wittgenstein (2001 [1953]), demonstrated the impossibility of the 'perfect proof'. Collectively, they taught scholars to embrace fundamental indeterminacy; encouraging them to recognise that evidence is always contingent on prior assumptions, and that facts are always open to revision.

In their wake, a generation of STS (Science and Technology Studies) scholars have built on these 'finitist' insights by exploring their material ramifications in contemporary science and engineering. Sociologists such as Collins (1985), Latour (1987) and Mackenzie (1990) have illustrated how the 'problem of relevance' creates meaningful uncertainties in contemporary scientific and technological knowledge claims. Others have explored the implications of this in a wide range of contexts, from the courtroom (e.g. Lynch, 1998) to the halls of government (e.g. Jasanoff, 1990). Few, however, have closely and systematically examined how an epistemological understanding of uncertainty might contribute to our understanding of accidents.

This neglect is unfortunate, for the inherent uncertainties of tests and models would seem to have direct bearing on the question of why systems fail. Tests are essential to both achieving and measuring technological safety. They underpin the foundational premises (such as measurements of tensile

---

* Hence, the term 'finitism', which describes the philosophy that builds on this insight.
† Formulated slightly differently, in terms of the 'underdetermination' of theory by evidence, this is a foundational dilemma in the philosophy of knowledge, sometimes referred to as 'the Duhem-Quine problem' or 'confirmation holism' (see Quine, 1951).

strength) on which experts design technologies to perform safely; and they simultaneously underpin expert assessments of those technologies' safety performance. If tests of the design are unrepresentative, then assumptions about its behaviour could be wrong in ways that have catastrophic potential. And if tests of that behaviour are similarly unrepresentative (perhaps for the same reason), then that catastrophic potential might be invisible. Therefore, insofar as the representativeness of a test cannot be known for certain, a system might fail because an assumption implicit in its design proves to be erroneous, even though there were logically consistent grounds for experts to hold that assumption before (although not after) the event. I have proposed to call such failures 'Epistemic Accidents' (Downer, 2011).

5.2 Epistemic Accidents

We need not look far to find evidence that epistemological dilemmas play a significant role in accidents. Accident investigations routinely identify causes (or causal factors) in the form of logically held (but nevertheless erroneous) assumptions about the design of a system or its assessment. Take, for example, the SDF20 scare outlined above, where the EPA's safety tests were framed by a misleading assumption about the (ir)relevance of soil biota. A more detailed example can be found in Downer (2011), which explores the fate of *Aloha Airlines Flight 243*: a Boeing 737 that suffered a massive mid-flight fuselage failure over Hawaii in 1988. This failure stemmed from a misunderstanding of how aluminium fatigues in highly specific conditions (relating to imperfections in an airplane's manufacturing process, a specific rivet configuration and the saltwater-infused operating environment of the Pacific Islands). The airplane's designers had misunderstood this fatigue behaviour because they had never experimented with these exact conditions, and the safety tests they performed for the airplane's regulators did not reveal the misunderstanding for the same reason (Downer, 2011).

If we look further afield to some of the more iconic accidents of the last few decades, we again see evidence of epistemological finitism wreaking havoc with the best-laid plans of experts. A good case could be made, for instance, that the 2011 Fukushima accident was an Epistemic Accident, at least in part. In the most fundamental sense, the reactors failed because the plant's flood defences were overwhelmed by the tsunami that struck them. Those defences performed as designed, however, the problem was that their design was inadequate, and it was inadequate because it was premised on an erroneous belief about the maximum possible tsunami. This erroneous belief about tsunamis, in turn, stemmed an erroneous belief about the maximum size of potential earthquakes off the coast – a belief, which nevertheless reflected the best seismology available at the time the plant was designed.* And the error was not revealed in testing because the same earthquake assumptions that informed the design also informed the models through which regulators assessed the flood defences. At the time the plant was constructed, no engineering analysis could have identified the circumstances of its demise.

A similar case might be made for the 1986 *Challenger* disaster. In a far-reaching (1996) book on the tragedy, Diane Vaughn lays much of the blame for the disaster on a tendency among NASA engineers to normalise 'deviances' in the behaviour of a key component (the 'o-rings' sealing the booster rockets) over time. Yet, her account is rich enough to support a rival interpretation: specifically, that the meaning of 'deviance' in a complex technical system – where it would be normal for performance to imperfectly match predictions in myriad ways – could not have been transparent to engineers in advance of the failure.

Accidents are complex phenomena, and any specific example of an Epistemic Accident will be contestable, yet the principle stands on its own logic. If we accept the premise that some accidents result from erroneous beliefs about the functioning of a system. And if we further accept that even the most rigorous and well-founded beliefs about the functioning of systems necessarily contain uncertainties, then it follows that Epistemic Accidents must exist.

5.3 New Perspectives: NAT and EAT

What is the value of understanding accidents in these terms? There are several ways to address this question. One straightforward but significant answer is that it shows why some failures are inherently *unpredictable* and, therefore, *unavoidable*. To recognise the existence of Epistemic Accidents is to acknowledge that no amount of organisational restructuring, application of intelligence or hard work will ever 'solve' the problem of accidents.

Recognising this limitation is not the same as saying that the prevalence of accidents cannot be greatly reduced, or that attempts to reduce them are not valuable and important, but it does have far-reaching implications in certain contexts. Most notably, perhaps, in the context of debates around critical systems, where exceptional hazards are often offset by extreme claims to reliability. Safety debates in most technological spheres accept that entirely failure-free operations are an unreasonable aspiration. In spheres where

---

* There is a case to be made that Fukushima's design did not reflect the best seismology available at the time of the accident (Nöggerath et al., 2011). Insofar as we expect operators to reassess, redesign and rebuild nuclear plants in light of (contested) changes in the upstream science on which those plants are premised, therefore, then Fukushima's claim to being an Epistemic Accident is arguably diminished.

failures would be most consequential, however, this common-sense under-standing has long been suspended. Reactors, perhaps most notably, would not be politically tenable unless the possibility of catastrophic accidents was excluded entirely from decision-making processes (Downer, 2016; Rip, 1986).* Here – where accidents can incur costs (direct and indirect) of over a trillion dollars, together with unknown, but potentially vast, long-term health risks – polities embrace an ideal of technological mastery that is highly at odds with basic epistemology.

However, the argument that accidents are inherently unavoidable is nei-ther new, nor unique. Most notably, it has been asserted by Perrow (1999 [1984]), and other proponents of his 'Normal Accident Theory' (NAT) (e.g. Sagan, 1993). The argument outlined above does not seek to undermine Perrow's thesis in any way, but rather to compliment it. Epistemic Accidents are not the same as Normal Accidents, and examined closely the two have meaningfully different ramifications. Indeed, the act of exploring these dif-ferences serves as a useful way of examining the implications of Epistemic Accidents. It is to this that we now turn.

NAT has many facets and deserves more unpacking than can be afforded here (see Le-Coze, 2015; Downer, 2015, 2011). At its core, however, it is a sim-ple but profound insight: that accidents caused by highly improbable 'bil-lion-to-one' confluences of otherwise minor events (which no expert could anticipate in advance) are statistically probable in systems where there are a great many opportunities for them to occur. (Perrow, for instance, points to the extraordinary chain of events that led to the meltdown at Three Mile Island). If Epistemic Accidents are an emergent property of epistemological limitations, we might say, then Normal Accidents are an emergent property of structure and probability.

Perrow's thesis offers an elegant explanation for why some accidents (albeit only a small minority by his accounting)† are fundamentally unavoidable. The kinds of events that come together to make them are too innocuous in themselves to serve as meaningful warning signs; and the specific sequences that make those events catastrophic are too improbable and variable to ever register as a threat in advance. At the same time, however, there are accidents that would not register as Normal Accidents, but which can nevertheless be understood as unavoidable from the perspective of epistemology. Herein lies one specific argument for thinking epistemologically as well as probabilisti-cally. To wit:

> i. *Epistemic Accidents broaden the spectrum of potentially unavoidable failures.*

---

* Or so it is believed by most nuclear authorities. A similar argument might be made for nuclear deterrence networks, although the case here is more complicated (see Sagan, 1993).
† Perrow (who more often refers to such accidents as 'system accidents') uses the word 'normal' to connote inevitability rather than commonness.

By Perrow's calculus, accidents that arise from a single point-of-failure are not 'Normal', and ought to be foreseeable and avoidable (Perrow, 1999, pp. 70–71).* Yet, Epistemic Accidents can arise in this way: from a single understanding that proves to be erroneous (about the relevance of sterile soil, for instance, or the fatigue properties of aluminium), rather than from an unanticipated confluence of otherwise anticipated events. This does not imply that all non-Normal Accidents are unavoidable, but it does imply that there are unavoidable failures that NAT would not recognise.

Once an understanding has been identified as erroneous, however, then there is no reason for it to be unavoidable again. And herein lies a second distinctive feature of Epistemic Accidents relative to Normal Accidents…

   ii. Epistemic Accidents imply a distinct relationship to institutional learning and technological achievement.

One important property of Normal Accidents is that they rarely challenge common understandings of the component-level events that cause them. The specific circumstances that come together to instigate a Normal Accident – failed warning lights, stuck valves, human error, and so on – are usually unremarkable in themselves. Normal Accidents are only remarkable by virtue of the unique manner in which these circumstances combine. The investigators of TMI were not surprised to find that valves sometimes stick, or that warning lights sometimes fail: they were surprised to find such an improbable combination of these failures.

This odd combination of accident-level uniqueness and component-level banality is important because it leads to a second property of Normal Accidents, this one is highly consequential: the fact that they are not very *edifying*. The banality of the events that line-up to create a Normal Accident means that they offer few insights to their investigators, who are unlikely to learn anything new about valves, warning lights or human behaviour. Learning from Normal Accidents is difficult, because they are 'one-of-a-kind' events. For although it is logical to anticipate a fatal 'one-in-a-billion' coincidence in systems that allow for billions of such coincidences, it is not logical to expect the same exact coincidence twice. The exact circumstances that led to TMI might never occur again, even if the exact same plant ran for another ten thousand years. Acting to prevent the recurrence of a sequence that might never reoccur is not useful as the next fatal coincidence is vanishingly unlikely to involve the exact same sequence. For these reasons, Normal Accidents usually only teach one important lesson, and it is always the same lesson: that unpredictable confluences of seemingly trivial events can instigate disasters.

---

* He calls them 'Component Failure Accidents'.

Epistemic Accidents are very different in these respects. Unlike Normal Accidents, they are caused by specific, consequential events, rather than by random, one-in-a-billion coincidences of events. And, almost by definition, they challenge conventional understandings of those events. They arise from misunderstandings about the nature of a system's functioning, and if those misunderstandings are not remedied then the same failures are liable to reoccur. EPA testing regimens that continued to mandate sterile soil, for example, would pose a hazard to soil biota every time the agency tested a new bacterium.

For the same reasons, however, Epistemic Accidents can be edifying. The EPA's close encounter with SDF20 offered insights that allowed it to change its tests. And the fact that other bacteria could pose similar risks in the future meant that those changes could contribute meaningfully to the safety of its work. In other words, the fact that Epistemic Accidents reveal shortcomings in our knowledge – ways in which our tests and models are unrepresentative – means that experts can leverage hindsight to improve systems and practices over time. If a future bacterium ever destroys the world's soils because regulators failed to test its effects on soil biota, then the resulting catastrophe would not be an Epistemic Accident.

This property of Epistemic Accidents is important because it offers a unique perspective on innovation and safety. Elsewhere (Downer, 2017), for example, I have argued that one of the ways that modern civil aviation has achieved such remarkable levels of safety is by leveraging its vast well of service experience. Over tens of millions of flights its experts have slowly honed their understandings by: (a) fastidiously deconstructing accidents for epistemological insights about their designs and assessment practices and (b) adhering to a common, very specific, jetliner paradigm ensuring that any hard-earned insights continue to be relevant. This is to say that decades of service with very similar airplanes has allowed aviation engineers to slowly weed-out the significant uncertainties in both their designs and the tests through which they verify those designs.

Understanding technological achievement this way highlights the limits of technological ambition and helps outline effective routes to achieving ultra-high levels of safety in complex systems. At the same time, however, it emphasises the costs of such achievements. By directly linking the impressive safety of modern civil aviation to the industry's long and painful (but ultimately instructive) history, it highlights the difficulty of replicating that success without that history. The modern governmentality of many critical technologies is premised on an understanding that experts can achieve and assess ultra-high levels of performance on the basis of tests and models alone. This is deeply misleading. The fact that we can build safe jetliners does not mean that we can build safe reactors, and when dealing with new hazardous technologies – be they bombs or bacteria – no amount of scrutiny or oversight will ever fully shield us from accidental catastrophe. It is important that we recognise this.

References

Bloor, D. (1976) *Knowledge and Social Imagery*. Routledge and Kegan Paul, London.

Brockway, R. (2010) "How a Biotech Company Almost Killed the World (With Booze)." cracked.com April 3. 2010. online: http://www.cracked.com/article_18503_how-biotech-company-almost-killed-world-with-booze.html (accessed 1 June 2017).

Collins, H. (1985) *Changing Order*. SAGE, London.

Downer, J. (2007) "When the Chick Hits the Fan: Representativeness and Reproducibility in Technological Testing." *Social Studies of Science,* Volume 37(1): 7–26.

Downer, J. (2011) "'737-Cabriolet': The Limits of Knowledge and the Sociology of Inevitable Failure." *American Journal of Sociology,* Volume 117(3): 725–762.

Downer, J. (2015) "The Unknowable ceilings of Safety: Three Ways that Nuclear Accidents Escape the Calculus of Risk Assessments," in Taebi, B. and Roeser, S. (eds.) *The Ethics of Nuclear Energy: Risk, Justice and Democracy in the Post-Fukushima Era*. Cambridge University Press, Cambridge, UK, 35–52.

Downer, J. (2016) "Resilience in Retrospect: Interpreting Fukushima's Disappearing Consequences," in Herwig, A. and Simoncini, M. (eds.) *Law and the Management of Disasters: The Challenge of Resilience*. Routledge, London, 42–60.

Downer, J. (2017) "The Aviation Paradox: Why we can 'Know' Jetliners but not Reactors." *Minerva,* Volume 55(2): 229–242.

Feyerabend, P. (1975) *Against Method: Outline of an Anarchistic Theory of Knowledge*. Redwood Burn, London.

Holmes, M. T., Ingham, E. R., Doyle, J. D. and Hendricks, C. W. (1999) "Effects of Klebsiella Planticola SDF20 on Soil Biota and Wheat Growth in Sandy Soil." *Applied Soil Ecology,* Volume 11(1): 67–78.

Jasanoff, S. (1990) *The Fifth Branch: Science Advisors as Policymakers*. Harvard University Press, Cambridge, MA.

Krebs, A.V. (2001) "Commentary: Searching for a Fair Resolution Concerning Controversial Story on Possible Effects of Klebsiella P on the Environment." In *The Agribusiness Examiner*. Issue #119 online: http://gmwatch.org/en/news/archive/2001/8951-full-story-of-the-dr-elaine-ingham-controversy-over-kleb-siella-p (accessed 1 June 2017).

Kripke, S. (1982) *Wittgenstein on Rules and Private Language*. Harvard University Press, Cambridge, MA.

Kuhn, T. (1996 [1962]) *The Structure of Scientific Revolutions* (3rd Edition). The University of Chicago Press, Chicago.

Latour, B. (1987) *Science in Action: How to Follow Scientists and Engineers through Society*. Harvard University Press, Cambridge, MA.

Le-Coze, J. C. (2015) "1984–2014. Normal Accidents. Was Charles Perrow Right for the Wrong Reasons?" *Journal of Contingencies and Crisis Management,* Volume 23(4): 275–86.

Lynch, M. (1998) "The Discursive Production of Uncertainty: The OJ Simpson "Dream Team" and the Sociology of Knowledge Machine." *Social Studies of Science,* Volume 28(5/6): 829–869.

MacKenzie, D. (1990) *Inventing Accuracy: A Historical Sociology of Nuclear Weapon Guidance*. MIT Press, Cambridge, MA.

MacKenzie, D. (1996) "How Do We Know the Properties of Artifacts? Applying the Sociology of Knowledge to Technology," in Fox, R. (ed.) *Technological Change*. Amsterdam, Harwood, 249–251.

Nöggerath, J., Geller, R. and Gusiakov, V. (2011) "Fukushima: The myth of safety, the reality of geoscience." *Bulletin of the Atomic Scientists*, Volume 67(5): 37–46.

Perrow, C. (1999 [1984]) *Normal Accidents: Living With High-Risk Technologies* (2nd Edition). Princeton, New York.

Pinch, T. (1993) "'Testing – One, Two, Three … Testing!': Toward a Sociology of Testing." *Science, Technology and Human Values*, Volume 18(1): 25–41.

Porterfield, A. (2016) Did you hear about the GMO that almost destroyed all life? Genetic Literacy Project. 6 September. online: https://geneticliteracyproject. org/2016/09/06/hear-gmo-almost-destroyed-life/ (accessed 1 June 2017).

Quine, W. v. O. (1951) "Two Dogmas of Empiricism." *The Philosophical Review*, Volume 60: 20–43.

Rip, A. (1986) "The Mutual Dependence of Risk Research and Political Context." *Science and Technology Studies*, Volume 4(3/4): 3–15.

Robbins, J. (2002) "A Biological Apocalypse Averted." *Earth Island Journal*. Winter 2002. online: http://www.earthisland.org/journal/index.php/eij/article/a_biological_ apocalypse_averted/ (accessed 3 July 2017).

Sagan, S. (1993) *The Limits of Safety*. Princeton University Press, Princeton.

Vaughan, D. (1996) *The Challenger Launch Decision*. University of Chicago Press, Chicago.

Walter, C., Berridge, M. and Tribe, D. (2001) "Genetically engineered Klebsiella planticola: A threat to terrestrial plant life? Rebuttal of written and verbal evidence presented by Dr Elaine R. Ingham." Presentation to the Royal Commission on Genetic Engineering, 1 February 2001. Press Release: New Zealand Life Sciences Network. online: http://www.scoop.co.nz/stories/print.html?path= SC0102/S00070.htm (accessed 1 June 2017).

Wittgenstein, L. (2001 [1953]) *Philosophical Investigations*. Blackwell Publishing, London.

6

Revisiting the Issue of Power in Safety Research

Stian Antonsen and Petter Almklov

CONTENTS

6.1 Safety Culture and the Issue of Power..88
6.2 Advances in Understanding the Role of Power in Safety.....................91
    6.2.1 The Two-Faced Relationship between Power and Safety..........91
    6.2.2 Power in Discourse, Models and Concepts.................................92
    6.2.3 Power in Boundary Processes...93
    6.2.4 Power and Safety Regulation..94
    6.2.5 National Cultures and Power Distance.....................................95
6.3 What Do Power-Oriented Studies Contribute to the Study
    of Safety?..96
    6.3.1 A Critical Perspective on Theory and Practice...........................96
    6.3.2 Changing the Workplace or Changing the Worker? 96
6.4 Future Challenges Involving Power and Safety.....................................97
    6.4.1 Power in Safety Science...97
    6.4.2 Digital Surveillance and Accountability....................................98
    6.4.3 Concentration of Power and Risk...99
6.5 Conclusion 99
References...100

Introduction

The importance of including power, differentiation and status differences in safety research has been stressed in several studies. As early as 1984, Gephart called for a more power-oriented approach to the study of disaster causation. Criticising Turner's (1978) accident model, he argued that:

> *divergent views of reality emerge in the written and verbal statements of government, industry, and public critics involved in the disasters. These views of reality compete for acceptance as the dominant reality.*

> (Gephart, 1984, p. 205)

In the years after Gephart's publication, several other researchers of safety and disasters have made similar points. t'Hart (1993) argued that symbols, rituals and power were the 'lost dimensions' of crisis management, and proposed a more power-critical approach, while Collinson (1999) studied the politics of accident reporting in the United Kingdom's oil and gas sector through a Foucauldian approach. As the wave of safety culture research gained momentum in the late 1990s and early 2000s, the call for more power-sensitive research was reiterated through the writings of Pidgeon and O'Leary (2000), Richter and Koch (2004), Haukelid (2008), Silbey (2009) and Antonsen (2009).

In the last decade or so, several studies have been published, contributing to an increased understanding of the role of power in organisational safety. In this chapter, we review some of the more recent literature, discuss the contributions of power-oriented approaches in safety science, and delineate some of the future challenges for power-oriented safety research.

6.1 Safety Culture and the Issue of Power

Since one of the authors of this chapter has previously written about power issues in safety culture research, we will use this article as a point of departure, although the topic has been addressed decades before this article (as evident in the references above).

In 2009, Antonsen published an article called 'Safety Culture and the Issue of Power' in *Safety Science*. The motivation behind the article was a growing worry that the concept of safety culture led researchers and practitioners to employ an overly integrative and harmonic model of organisational life, while underplaying the role of differentiation, power relations and conflict.

The article started out by quoting Charles Perrow's comments on the usefulness of a cultural approach to safety, stating that: 'we miss a great deal when we substitute power with culture' (Perrow, 1999, p. 380). While agreeing with Perrow that understanding organisational life without including power relationships would at best be superficial and simplified, the option of choosing either culture or power as a perspective for understanding safety in organisations was not very appealing. The article was therefore devoted to exploring the intersections between culture and power.

The conception of culture underlying the article is based on a view of culture as a socially constructed phenomenon, continually produced and reproduced through the interaction of individuals within a context of existing social structures. Placing the 'origin' of organisational culture (a topic rarely addressed in the literature on safety culture) within social interaction means that culture is not an entity that can be modified according to the goals of managers (or others). Also, if culture grows out of interaction, then the patterns created will be differentiated: different social units consisting of people with different backgrounds doing different tasks, are likely to produce different norms and hold different things as important. Thus, a large organisation is bound to comprise different subcultures. Moreover, these subcultures might not always share common goals and interests. This is one of the theoretical intersections where the concepts of culture and power become entwined, and where it is difficult to understand one without understanding the other.

The conception of power used in the article was based on Lukes' (1974) theoretical framework. Lukes synthesises different forms of power into a multifaceted framework consisting of three dimensions. The first dimension is based on the Weberian conception where 'A has power over B to the extent that he can get B to do something that B would not otherwise do' (Dahl, 1957, cited in Lukes, 1974, p. 11). The sources of power may be many, including formal position, information and expertise, control over rewards, resources and sanctions, alliances and networks or even personal charisma (Pfeffer, 1992). Either individually or combined, these different sources of power can increase a person or group's ability to realise their interests. While the first dimension refers to overt and visible use of power in the decision-making processes, the second dimension turns the attention to less visible forms of power that do not only play out in decision-making processes. A decision is a choice between action alternatives, but as Bachrach and Baratz (1962) have shown, the ability to influence which alternatives are actually on the table is equally as important. This is the power of 'non-decisions' – the struggle over which alternatives are to be regarded as viable options in a given decision-making process. Here, power is seen as being more covert, and playing out in a range of agenda-setting processes preceding decision-making.

In the first two dimensions, power is used in intentional acts directed at achieving desired results. However, not all forms of power can be traced back to individual and intentional actions. Social systems tend to have a bias reflecting the values and worldviews of a few actors, but without this bias being the result of any individual's deliberate acts. This is Lukes' third dimension, seen as a 'supreme form of power', where compliance is secured by 'controlling their thoughts and desires' (Lukes, 2005, p. 27). From this perspective, power becomes a cultural phenomenon; a subtle system of values, not necessarily recognised by those who are subjected to it. Here, power is seen more as a series of structures rather than possessions, in accordance with a more Marxist line of thinking. Lukes' framework of power is multidimensional, meaning that all three dimensions are needed in order to understand what power is, and how it works.

Antonsen (2009) illustrated the links between power- and culture-oriented approaches to safety with an analysis of the *Challenger* disaster, which will not be repeated here. The main conclusion of the discussion was that there was indeed reason to be sceptical about purely integrative approaches to safety culture. Organisational cultures are never politically neutral – they are influenced, but not determined by, the interests that the organisations serve; furthermore, they reflect the worldviews of dominant groups and individuals (corresponding to the third dimension of power in Lukes' framework). In this respect, Perrow is right in his statement that we would miss a great deal of organisational life if we were to substitute power with culture, as quoted above. This position resonates with Scholte's criticism of Geertz's (1973, p. 5) concept of culture, which states that culture consists of webs of significance that man 'himself has spun', where Scholte underlines that 'very few do the actual spinning, while the majority is simply caught' (Scholte, 1984, cited in Keesing, 1987, p. 162).

However, this does not mean that we should turn it the other way around and substitute culture with power. First, organisational members are not 'cultural dopes' (Giddens, 1979, p. 71), that passively respond to the cultural patterns around them. Culture is also a bottom-up process, where workers at all levels construct patterns of, and patterns for, meaning. Worker resistance to change and the formation of strong collectives shows that resources of power do not necessarily follow the lines of organisational charts (Lysgaard, 2001). Second, the position that the real issue of safety matters is power, is merely a form of reductionism, boiling the complexity of organisational life down to one single factor. Accepting the fact that conflict exists is not the same as accepting that integration does not. The most important parts of organisational life are found in the intersection between power and culture. For instance, power is not always expressed and used in overt forms, but it is expressed through, and embedded in, symbols. If an observer of a large company had no knowledge whatsoever of what symbolises power within an organisation – for instance, the size and location of its offices, or the way in which its employees dress – he or she would likely miss out on the subtle markers of power that an insider would know immediately. Therefore, in order to understand power, you will sometimes need to understand culture.

If we accept the proposition that culture and power are entwined, then several lessons for safety culture research could be extracted. The first is that when studying what is shared, we must understand what is *not* shared, and by whom it is shared and not shared. The second lesson concerns the role of whistle-blowers – the ones voicing concerns over unsafe conditions. The leverage for (and treatment of) whistle-blowers is central to safety culture, but it is also strongly related to power. The cultural acceptance of somebody being a messenger of bad news, and often bypassing the chain of command, cannot be understood without taking power structures into account. The third lesson lies in a power-oriented view of behaviour-based safety approaches (BBS), which have been used as a basis for safety culture campaigns. BBS represents something of a Lutheran turn with regard to safety improvement, where the responsibility for one's own health and safety is individualised. Employees are not addressed as a group, but as individuals. Furthermore, the cultural variant of BBS aims to influence employees' hearts and minds, as well as imposing sanctions for errors and 'bad behaviour'.

This is a form of symbolic power aiming to influence the frames of reference through which organisational members interpret risk, centred around a notion of there being 'one best way' of performing the work in question. The ethical dimension of such tacit influence needs to be considered, as it balances on a very fine line the difference between motivation and manipulation.

6.2 Advances in Understanding the Role of Power in Safety

Antonsen's (2009) article was primarily a conceptual discussion of the links between power-oriented (and cultural) approaches to safety within organisations. While discussing the theoretical aspects of cultural differentiation, the subtle and systemic forms of power, national culture and regulation as contextual factors for organisational culture, the article did not present empirical analyses of these issues. Fortunately, this has since been carried out by other researchers during the last decade. In the following sections, we will give a brief review and discussion of some of the recent contributions to power-oriented safety research.

6.2.1 The Two-Faced Relationship between Power and Safety

Dekker and Nyce (2014) provide a theoretical discussion of the role of power and politics in organisational accidents. Their core argument is that the relationship between power and safety goes both ways – there is 'safety in power' and 'power in safety'. This expression, 'safety in power' refers to the fact that some actors are more powerful than others in defining what has gone wrong during an accident's investigation. The explanation of accidents, including causation and the attribution of error and blame, is not an objective process. There will always be competing views of reality regarding what has caused a given accident, particularly when it comes to organisational factors. Dekker and Nyce's point is that 'holding the pen' while the many narratives of accident causation are being written can potentially become a source of safety for some within the organisation (as well as the wider regulatory regime)…

> There is a lot of safety in having the power to say what happened, to be able to identify risk, to write history, to assign cause and consequences after an accident. Legitimating both the actions and the interests of elite groups is what determines whose view of events becomes the accepted or dominant one.

> (Dekker and Nyce, 2014, p. 46)

'Finding' the causes of accidents is a process of social construction, where interests, scapegoating and power relations most definitely matter. Here, we would like to add some reflections on Dekker and Nyce's argument. The description, or rather construction, of organisational factors contributing to accidents does not stay within the practitioners' field. They also travel into safety science, as researchers use accident investigations as a source of knowledge and data. If we accept the premise that accident investigations reflect group or individual interests, then the use of investigations in safety research means that the same bias is imported into models of accident causation. Thus, an implication of the 'safety in power' perspective is that there is a link between power relations in high risk industries, and the knowledge created within safety science. The other way around, the perspective of safety science returns to the practitioners' field as a form of discursive power, as shown by Almklov et al. (2014).

Furthermore, Dekker and Nyce have also stressed that there is 'power in safety'. Closely related to the first dimension, this term refers to the ubiquity of power in the basic structures of safety work, for example, procedures, discourses and relationships within and across organisations. Power is everywhere in safety, in both overt and 'capillary' forms as described by Foucault (1980). From this perspective, the 'bureaucratisation of safety' (Dekker, 2014, 2018) and the 'trapping of safety into rules' (Bieder and Bourrier, 2013) both become structures of power that should be studied and understood as such.

Dekker and Nyce delineate a research agenda to enable safety science to move beyond the mere acknowledgement of the power of elite groups to 'consider how the vocabularies, institutions and methods that characterise this domain as a science have emerged and continue to be legitimised'. Essentially, they call for studies to be made into the link between power and knowledge within safety science.

6.2.2 Power in Discourse, Models and Concepts

Several researchers have taken up on the challenge from Dekker and Nyce's (2014) research agenda. Henriqson et al. (2014) is one example of this, describing the constitution and effects of safety culture as a knowledge object in the discourse of accident prevention. They label their study 'Foucauldian' as they study both the historical conditions for the birth of the safety culture concept (the 'archaeology') and its effects as a form of knowledge (the 'genealogy'). The authors provide a rich account of the way in which the concept of safety culture emerges as a combination of pre-existing knowledge and historical events in need of explanation, as well as a critical discussion of discursive effects that include normative homogeneity, disciplinary power, new forms of bureaucratisation and a 'governmentality', linking individual behaviour to organisational norms. While many of these effects have been mentioned in the literature before, their approach allows for the analysis of safety as part of a broader organisational context where safety is not an output variable, but an 'evolving, political, negotiated order in organisations' (Henriqson et al., 2014, p. 474).

Another Foucauldian line of research is the language-oriented study of safety-related communication (e.g. Rasmussen, 2011, 2013; Amernic and Craig, 2017). In a close-up analysis of the dialogue between managers and safety representatives around the implementation of a BBS campaign, Rasmussen (2011) shows how power manifests itself in a concrete communication process concerning safety. The case studied is a mundane work situation where the dialogue about safety includes a subtle direction of conduct and identity, as well as an individualisation of the responsibility for safety. Rasmussen's study is a novel way of studying 'power in safety', where power is placed at the very centre of the analysis, and safety is seen more as the discursive field rather than a dependent variable for study.

Safety communication between managers and employees has a frontstage and a backstage: the frontstage being a series of formally prepared speeches; the backstage as an ongoing dialogue. Amernic and Craig (2017) studied the frontstage safety rhetoric in their interpretative analysis of 19 speeches delivered by BP's CEO before the Deepwater Horizon disaster. They illustrate how top management's 'happy talk' on safety sometimes seriously under-communicates the conflicting objectives of safety and cost efficiency, while at the same time aiming to create consensus around efficiency as a core value. Such speeches should not be overinterpreted as they are directed at audiences besides the company's employees (e.g. shareholders, regulators and the media). Nevertheless, messages from CEOs influence organisational attention – what is regarded as important, and which actions can affect employees' careers. Therefore, such analyses have a place in the study of power in safety.

6.2.3 Power in Boundary Processes

Sometimes, the study of power in organisations seems like a game between anonymous elite groups, and 'the rest'. On a micro- or meso-level, the boundaries and power struggles between groups becomes relevant without one of them necessarily representing a hegemonic elite. Almklov et al. (in press) studied the so-called 'boundary processes' between different public actors involved with societal safety and security in Norway during the aftermath of the 2011 terrorist attacks in Oslo. 'Boundary processes' refers here to the meetings between cultural units and the way in which different frames of reference, areas of expertise, language and method conventions all influence the way ill-structured matters of safety and security are dealt with. Several different boundary processes are described, of which we will now provide one example here.

The efforts to build safer societies depends on collaboration between agencies responsible for safety (e.g. assessing risks related to natural hazards and industrial accidents in Seveso sites) and agencies responsible for security (e.g. assessing risks related to terrorist attacks and information security). These two domains have different competence, language and methods when it comes to risk assessment and emergency preparedness, as well as different regulations when it comes to the sharing of information. In a context where financial resources are limited and related to prestige and political attention, they will find themselves in competition at precisely the same time that the need for their cooperation is increasing. No matter who 'wins' the battles, these power struggles need to be understood in order to improve the way that the Norwegian public sector as a whole is able to deal with problems that cross sectoral boundaries. Moreover, understanding the narratives, stereotypes and cultural processes operating on these boundaries is key for anyone interested in understanding the culture of these organisations.

6.2.4 Power and Safety Regulation

Power is also a topic for research on safety regulation. King and Hayes (2017) presented an important paper on risk regulation within the Australian pipeline industry, studying how legislation and regulatory practices are influenced by mechanisms of power, in addition to researching how knowledge relationships between the regulator and the regulated industry operate. They employed a Foucauldian perspective where power is not 'possessed' by a single actor or any group of actors, but rather exercised through an apparatus of institutional mechanisms and knowledge structures. Hence, power is not a characteristic of actors (i.e. someone being powerful) but a characteristic of the relationships between actors. Furthermore, it is a concept where power and knowledge are so closely knit together that '[i]t is not possible for power to be exercised without knowledge, it is impossible for knowledge not to engender power' (Foucault, 1980, p. 98).

King and Hayes described a form of 'regulatory capture' that occurs when regulators operate so closely with an industry that they fail to enforce the law against powerful companies in the industry. Kings and Hayes expanded the existing understanding of regulatory capture. Instead of viewing the capture from a classical perspective of power where power is assigned to powerful industry actors, they highlighted the production and negotiation of the underlying logic of regulatory practices – the ways of thinking, knowing and doing that influences the actions and decisions of both the regulator and the industry. They concluded the following:

> As regulators rely on knowledge developed by industry and their attempts to establish good relationships with industry, regulators have unwittingly put themselves into a captured and unequal position through the process of attaining and establishing relevant knowledge. The relational effect is that regulators have become less effective and less powerful in the process of risk regulation.
>
> (King and Hayes, 2017, p. 11)

In this analysis, power is not good or bad per se. It has both positive and negative consequences, although the authors strongly indicate that the regulatory capture negatively influences the independence of authorities and the effectiveness of safety regulations. From a Norwegian context, Rosness and Forseth (2013) also discuss power dimensions of regulation. Here, however, the tripartite collaboration unions, regulator and management, seems to generate a dynamic and shifting power balance, which, according to the authors, may be constructive for safety. Regardless of the effects, the dynamics of regulation constitute a subtle form of power inscribed in the foundational structures of safety, which is an important topic for safety research.

6.2.5 National Cultures and Power Distance

In the last decade or so, there has been an increase in the number of studies addressing the possible variations across national contexts in management conventions and cultural characteristics, entailing differences in the behaviour and decision-making that is most relevant for safety. Such differences are also relevant for those studying the relationship between power and safety. The investigation into the Fukushima disaster provides an example of this:

> What must be admitted – very painfully – is that this was a disaster "Made in Japan." Its fundamental causes are to be found in the ingrained conventions of Japanese culture: our reflexive obedience; our reluctance to question authority; our devotion to 'sticking with the program'; our groupism; and our insularity. Had other Japanese been in the shoes of those who bear responsibility for this accident, the result may well have been the same.
>
> (Fukushima Nuclear Accident Independent Investigation Commission, 2012, p. 9)

The quotation points to traits of Japanese culture that corresponds to Hofstede's (1984) cultural dimensions of power distance and individualism. Power distance refers to 'perceptions of the superior's style of *decision-making* and of colleagues' fear to disagree with superiors, and with the type of decision-making which subordinates prefer in their boss' (Hofstede, 1980, p. 65). Individualism refers to 'the relationship between the individual and the collectivity which prevails in a given society' (Hofstede, 1984, p. 148) and thus the leverage for individually motivated action that is regarded as culturally accepted. Helmreich and Merrit's (1998) massive survey study was among the first to use Hofstede's framework within safety research. They found that pilots from cultures with high-power distance were more likely to obey orders and adhere to procedures, while pilots classified as belonging to individualist cultures seemed to be more flexible in their use of safety procedures. Mearns and Yule (2009) followed a similar line of research in the oil and gas industry. They also found power distance as being potentially significant for behaviour (although their data also showed that management's commitment to safety was easily as important a determinant as national culture was). Reniers and Gidron (2013) studied the correlation between

Hofstede's dimensions and fatal work injuries in 11 European countries. They found that countries with a high-power distance score had a higher rate of fatal work injuries, while countries with a high individualism score had fewer injuries. The interpretation of such data can be challenging, which we will return to in Section 6.4.2.

6.3 What Do Power-Oriented Studies Contribute to the Study of Safety?

Having reviewed some lines of research connecting different forms of power to safety, it is time to ask why this research is important, and how power-oriented research can continue to contribute to understanding and improving safety in the future.

6.3.1 A Critical Perspective on Theory and Practice

One of the most obvious contributions is that power-oriented studies provide critical perspectives on the established truths of safety science and practice. They enable us to ask fundamental questions about existing models, logics and practices regarding safety: where and from whom do the established and taken-for-granted knowledge and safety practices come from? How does this taken-for-granted knowledge influence the way people think about safety, the questions we ask, the way we do research and the way safety interventions are designed? The list of possible questions could be much longer, but they are all related to a foundational form of learning. For science to develop, there is the constant need for a critique of the existing paradigms; power-oriented perspectives can have this function when it comes to safety in organisations.

An important part of this is to be sensitive to the historical context and conditions for the production of knowledge, as King and Hayes (2017) convincingly did in the study described earlier in this chapter. By paying more attention to the 'birth' of knowledge and concepts, we are immediately reminded that they do not come about by discovery, but through a construction process which is embedded in interest structures.

6.3.2 Changing the Workplace or Changing the Worker?

Whether safety is to be seen as a property of the workplace or a property of the individual is a fundamental dilemma within safety science (Rasmussen, 2013). We find traces of it everywhere, but discussions about human error in accident causation are probably where the dilemma is the most pronounced: is human error a (root) cause of accidents, or is it a consequence of error-producing work situations (Reason, 1997)? It is also seen in discussions about safety improvement, such as Rollenhagen (2010), asking if focus on safety culture can become an excuse for not rethinking technological design. This dilemma is important both from a 'safety in power' and a 'power in safety' perspective, to use the distinction from Dekker and Nyce (2014) – assigning the label of 'human error' can be both a coercive and discursive form of power, while at the same time protecting others from blame and liability.

The dilemma between safety as an individual or collective phenomenon has been a troublesome issue for research on safety culture in particular. The interpretation of the results from Reniers and Gidron's (2013) study provides us with an example here. Using Hofstede's dimensions of culture, they find that the countries within the 'individualism' category have better safety levels that their 'collectivist' counterparts. This could be interpreted as supporting the behaviour-based safety doctrine. As Reniers and Gidron (2013, p. 80) conclude: '[T]he more workers are expected to look after themselves, they are more safe'. Does this mean that managers are 'off the hook' and that what they really need to do is to leave workers alone and let them take care of themselves? This is by no means the only possible interpretation of the correlation between individualism and accident rates. It could also be that the correlation indicates that organisations with empowered workers have better technology and work organisation due to better bottom-up communication regarding hazards, human error or unsafe technologies and working conditions. The individual leverage can thus be a basis for improving the *collective* conditions for safety. It could also be related to a cultural acceptance of 'bending' the rules in order to fit situational characteristics, instead of choosing compliance for the sake of compliance.

Therefore, we see the need to reiterate, once again, that organisational safety is not only created at the sharp end of organisations. Of course, operative behaviour matters, but safety science should be wary of reinventing the 'bad apple' theory on safety in disguise, viewing operative personnel as a bunch of latent bad apples in need of constant monitoring and adjustment to keep them on the right path.

6.4 Future Challenges Involving Power and Safety

6.4.1 Power in Safety Science

Both the advances and the future challenges described thus far represent challenges to the theoretical frameworks of safety science. Many of the existing perspectives (e.g. HRO, resilience engineering) are system perspectives where organisational traits and system functions are placed at the centre, while the subjects are referred to the background. The absence of subjects in the theoretical frameworks runs the risk of continuing to underplay the power dimension of organisational life, despite the warnings from Gephart (1984), Perrow (1984) and others. There is a need for empirical studies taking into account the role of discursive and coercive power in HROs, resilient organisations and normal operations.

Another challenge for future research on power in safety science lies in the way power relations play out in different cultural contexts. The quotation above, stating that the Fukushima disaster was 'Made in Japan', points to cultural differences in the distribution of power in organisational hierarchies. Many of the established theoretical frameworks are based on empirical research from European and North American contexts. This gives the theoretical canon of safety science an ethnocentric flavour – which is in itself a form of power – needing to be challenged in order to test its relevance across a range of cultural contexts. As globalisation turns major organisations into nexuses for people with different cultural backgrounds, concepts like power distance become relevant for processes within organisations, not only as macro comparisons between nations and regions.

6.4.2 Digital Surveillance and Accountability

Despite the advances made in the understanding of power within safety science, there are ongoing changes in the world that have yet to be taken into account, which probably make a power-oriented approach more important than ever. The digitalisation of work and production systems play a key part in this.

Work can now be automatically monitored in new ways. As we discussed in this book (Almklov and Antonsen, this volume, Chapter 1), digital traces of work can provide the basis for new forms of surveillance and accountability. In addition to the increase in management control, an equally important development can be summed up in Zuboff's observation that we are moving towards a situation where we are monitored by a generalised 'big other', rather than a single 'big brother'. The generation of digital data about behaviour creates a 'surveillance capitalism', producing new markets of behavioural prediction and influence (Zuboff, 2015). Zuboff uses the term 'the big other' to denote the new form of surveillance growing out of the Internet of Things, as well as the marked increase in behavioural data following out from it. The digitalisation of work processes is where the 'big other' leaves the consumption sphere and comes to work. When companies have digital management systems, create digital solutions for conducting risk assessments and make use of more digital tools in carrying out safety-critical work, this will create new data about how work is actually done. This, in turn, creates new forms of accountability, as there might be better, more detailed data available for the consideration of errors and liability. The power dimension of this proliferation of data is, of course, far from settled.

This development sets the stage for a panopticon-like surveillance of work. The increased surveillance capacity might mean that we expect our actions to be monitored in some way, even when they are in fact not being monitored. Moreover, the digital side of power means that it might be increasingly hard to know 'who is watching', both from the view of the individual employees, and the organisations trying to protect their critical systems from intrusion and attack.

6.4.3 Concentration of Power and Risk

When considering digitalisation in relation to power and risk, one cannot avoid noticing the ongoing concentration of power in the hands of a few top executives in companies like Facebook, Google, Amazon and Apple. Their power lies in their ownership of big data about individual citizens, and the way these data can be used to influence public opinion, as well as consumer and voting behaviour. The 2018 Facebook scandal, where it was revealed that the personal data of 87 million Facebook users had been shared with a political consultancy company aiming to influence the American election in 2016, illustrates that a digital risk society is emerging.*

As of yet, though, the links between this form of risk and industrial safety are mostly indirect. However, the possible convergence between risks related to privacy, cybersecurity and industrial safety does need to be taken seriously. As industrial systems become more connected, we might be creating levels of interorganisational and technological complexity and coupling that far exceeds the worries expressed by Charles Perrow in *Normal Accidents*. The concentration of power that lies in the ability of a few to subtly influence the behaviour and opinions of many is thus also related to a concentration of risk. In interconnected systems, the externalities of errors of (mis)management grow exponentially, creating a massive challenge for both regulation and regulatory oversight that should also be taken as a joint challenge for the research communities within safety, reliability and security.

6.5 Conclusion

The knowledge about power in organisational safety has progressed considerably over the last decade. However, it is still an immature part of safety

---

* While we expect the power dimensions of digitalisation to largely be a shift towards centralisation where elites are the main winners, there is also a democratic potential in data. One example can be found in the UK Transparency Agenda, enabling citizens to scrutinise the state in new ways via digital data (Ruppert, 2015). Within corporations, systems of detailed digital accountability can be more than management tools, providing possibilities for unions and individual employees to lift problematic issues to the agenda.

science. Our review of recent studies on power illustrates that the field has taken a Foucauldian turn, where power is ever present as a systemic feature. While this research has brought new theoretical perspectives and innovative empirical studies into safety research, it is important not to lose sight of the more overt forms of power. High-risk organisations operating in a commercial environment will always be characterised by goal complexity, balancing long-term safety investments with short-term profit. In this context, there will always be managers and top executives making profit-maximising decisions which will more or less directly influence safety. This overt form of power fades into the background in discourse-oriented studies. This is somewhat paradoxical, considering the important role assigned to management's prioritisation of safety matters, for instance, in the literature on safety culture and the works of Perrow. The study of power in safety science would be best served by not focussing *either* on overt conflict and intentional decisions, *or* systemic power. This is in line with Lukes's (2005) model of power as a multidimensional concept, as well as the multi-layered research agenda on the relationship between power and safety that emerged through Le Coze's (2015) re-reading of Charles Perrow's authorship. The main strength of a multidimensional approach is that it allows for addressing the systemic power structures that make the overt use of power in decisions appear legitimate and meaningful in decision-making processes.

Furthermore, we see a need for critical research regarding the disruptive innovation processes that are currently ongoing under the heading of digitalisation. Digitalisation enables an unprecedented level of standardisation and surveillance, which is sometimes claimed to improve safety by means of reducing the potential for human error. While this might possibly be true, the research community should nevertheless take on the role of devil's advocate and question whose interests these innovations truly serve; which perceptions of safety gain prominence and which voices are (not) heard. Furthermore, the concentration of both power and risk emerging around digital technology is a major challenge for future research.

References

Almklov, P. G., Antonsen, S., Bye, R. J. and Øren, A. (in press). Collaborating with the others: Cultural differentiation as an interactional process. *Safety Science*.

Almklov, P. G., Rosness, R. and Størkersen, K. (2014). When safety science meets the practitioners: Does safety science contribute to marginalisation of practical knowledge? *Safety Science*, Volume 67(Supplement C), 25–36.

Amernic, J. and Craig, R. (2017). CEO speeches and safety culture: British Petroleum before the Deepwater Horizon disaster. *Critical Perspectives on Accounting*, Volume 47(Supplement C), 61–80.

Antonsen, S. (2009). Safety culture and the issue of power. *Safety Science*, Volume 47(2), 183–191.

Bachrach, P. and Baratz, M. S. (1962). Two faces of power. *American Political Science Review*, Volume 56, 947–952.

Bieder, C. and Bourrier, M. (2013). *Trapping safety into rules: How desirable or avoidable is proceduralisation?* Farnham: Ashgate.

Collinson, D. L. (1999). Surviving the rigs': Safety and surveillance on North Sea oil installations. *Organisation Studies*, Volume 20(4), 579–600.

Dahl, R. A. (1957). The concept of power. *Behavioral Science*, Volume 2, 201–215.

Dekker, S. (2018). *The safety anarchist: Relying on human expertise and innovation, reducing bureaucracy and compliance*. London: Routledge.

Dekker, S. W. A. (2014). The bureaucratisation of safety. *Safety Science*, Volume 70 (Supplement C), 348–357.

Dekker, S. W. A. and Nyce, J. M. (2014). There is safety in power, or power in safety. *Safety Science*, Volume 67, 44–49.

Foucault, M. and Gordon, C. (1980). *Power/knowledge: Selected interviews and other writings 1972–1977*. Brighton: Harvester Press.

Geertz, C. (1973). *The interpretation of cultures: Selected essays*. New York: Basic Books.

Gephart, R. P. (1984). Making sense of organisationally based environmental disasters. *Journal of Management*, Volume 10, 205–225.

Giddens, A. (1979). *Central problems in social theory*. London: Macmillan.

Haukelid, K. (2008). Theories of (safety) culture revisited—An anthropological approach. *Safety Science*, Volume 46(3), 413–426.

Helmreich, R. L. and Merritt, A. C. (1998). *Culture at work in aviation and medicine: National, organisational, and professional influences*. Aldershot: Ashgate.

Henriqson, É., Schuler, B., van Winsen, R. and Dekker, S. W. A. (2014). The constitution and effects of safety culture as an object in the discourse of accident prevention: A foucauldian approach. *Safety Science*, Volume 70, 465–476.

Hofstede, G. (1984). *Culture's consequences*. Newbury Park: Sage.

Keesing, R. M. (1987). Anthropology as interpretive quest. *Current Anthropology*, Volume 28(2), 161–176.

King, D. K. and Hayes, J. (2017). The effects of power relationships: Knowledge, practice and a new form of regulatory capture. *Journal of Risk Research*, 1–13.

Kuhn, T. (1962). *The structure of scientific revolutions*. Chicago: University of Chicago Press.

Le Coze, J.-C. (2015). 1984–2014. Normal accidents. Was Charles Perrow right for the wrong reasons? *Journal of Contingencies and Crisis Management*, Volume 23, 275–286.

Lukes, S. (1974). *Power: A radical view*. Macmillan: London.

Lukes, S. (2005). *Power: A radical view*. Basingstoke: Palgrave Macmillan.

Lysgaard, S. (2001). *Arbeiderkollektivet*. Oslo: Universitetsforlaget.

Mearns, K. and Yule, S. (2009). The role of national culture in determining safety performance: Challenges for the global oil and gas industry. *Safety Science*, Volume 47(6), 777–785.

Perrow, C. (1999). *Normal accidents* (2nd ed.). Princeton: Princeton University Press.

Pfeffer, J. (1992). Managing with power. Harvard Business School Press: Boston, MA.

Pidgeon, N. and O'Leary, M. (2000). Man-made disasters: Why technology and organisations (sometimes) fail. *Safety Science*, Volume 34, 15–30.

Rasmussen, J. (2011). Enabling selves to conduct themselves safely: Safety committee discourse as governmentality in practice. *Human Relations,* Volume 64(3), 459–478.

Rasmussen, J. (2013). Governing the workplace or the worker? Evolving dilemmas in chemical professionals' discourse on occupational health and safety. *Discourse and Communication,* Volume 7(1), 75–94.

Reason, J. (1997). *Managing the risks of organisational accidents.* Aldershot: Ashgate.

Reniers, G. and Gidron, Y. (2013). Do cultural dimensions predict prevalence of fatal work injuries in Europe? *Safety Science,* Volume 58, 76–80.

Richter, A. and Koch, C. (2004). Integration, differentiation and ambiguity in safety cultures. *Safety Science,* Volume 42, 703–722.

Rollenhagen, C. (2010). Can focus on safety culture become an excuse for not rethinking design of technology? *Safety Science,* Volume 48(2), 268–278.

Rosness, R., Blakstad, H. C. and Forseth, U. (2011). *Exploring power perspectives on robust regulation.* Trondheim: SINTEF.

Rosness, R. and Forseth, U. (2013). Boxing and dancing. Tripartite collaboration as an integral part of a regulatory regime. In P. H. Lindøe, M. Baram and O. Renn (Eds.), *Risk governance of offshore oil and gas operations* (pp. 309–339). Cambridge: Cambridge University Press.

Ruppert, E. (2015). Doing the transparent state: Open government data as performance indicators. In R. Rottenburg, S. Merry, S. Park, and J. Mugler (Eds.), *The world of indicators: The making of governmental knowledge through quantification* (pp. 127–150). Cambridge: Cambridge University Press.

Scholte, B. (1984). Comment on: The thick and the thin: On the interpretive theoretical program of Clifford Geertz, by P. Shankman. *Current Anthropology,* Volume 25, 261–270.

Silbey, S. S. (2009). Taming prometheus: Talk about safety and culture. *Annual Review of Sociology,* Volume 35(1), 341–369.

t'Hart, P. (1993). Symbols, rituals and power: The lost dimensions of crisis management. *Journal of Contingencies and Crisis Management,* Volume 1(1), 36–50.

Turner, B. (1978). *Man-made disasters.* London: Wykenham Science Press.

Zuboff, S. (2015). Big other: Surveillance capitalism and the prospects of an information civilisation. *Journal of Information Technology,* Volume 30, 75–89.

7

Sensework

Torgeir Haavik

CONTENTS

7.1 A Very Brief Introduction to ANT.. 106
7.2 Safety Research and ANT – Symmetries 107
7.3 Safety Research and ANT – Controversies 108
7.4 Sensework 109
    7.4.1 Sensework in Offshore Operations 110
    7.4.2 Sensework in Surgical Teams... 111
    7.4.3 Sensework within and beyond the Cockpit 112
7.5 A Research Agenda for Sensework 112
References.. 113

Introduction

Sensework (Haavik 2016a,b, 2014b) is a theoretical perspective that addresses sociotechnical work in high-risk domains. Sensework denotes the collective and cross-disciplinary work of professionals and technologies, putting together pieces of information to give meaning to, and work through familiar and unfamiliar situations. It combines perspectives from safety research with Actor Network Theory (ANT), a subfield of Science and Technology Studies (STS) (for an overview of ANT and STS, see Latour, 2005 and Sismondo, 2010) to formulate both a theory and research agenda for safety in sociotechnical systems.

The sensework perspective draws on sources of inspiration from different strands within safety research, and lends support from theoretical concepts, methodologies and vocabularies of ANT to expand on some of the central ideas and calibrate them to the developments of sociotechnical systems that increasingly infiltrate our lives. In particular, the ancestry of sensework can be traced back to Perrow's work on *Normal Accident Theory* (Perrow, 1999, 1984), Weick's work on *Sensemaking* (Weick, 2007, 1995, 1993) and Hollnagel's work on *Resilience Engineering* (2014)*,†,‡ and the ambition is to bring some central insights from these strands together, and capitalise on powerful ideas from ANT to bring forward a sensework perspective.

To get an (non-exhaustive) idea of the kind of sociotechnical systems that are within the scope of sensework, one might think of the sociotechnical systems that Perrow (1999) has been exploring. And surely, the concepts of couplings and complexity have been an indispensable back cloth for the reasoning about these systems; so interwoven are these concepts with the culture of safety research researchers that we are in risk of taking their meaning and richness for granted. In sensework is implicit an attempt to develop meaning and nuances that are already implanted in them so as to keep them potent as the world continues to develop both empirically and theoretically. From Perrow and his colleagues, we have learned important lessons about tight couplings and interactive complexity. In Perrow's own words, tight coupling is a 'mechanical term meaning there is no slack or buffer or give between the two items [in a system]. What happens to one, directly affects what happens in the other' (Perrow, 1984, pp. 89–90). ANT helps to expand our understanding of couplings by adding to transmission the dimension of *transformation*, in order to explain what happens when – as so often is the case in sociotechnical systems – any of the parts are mediators and not merely intermediaries, elaborating on how components' characteristics are as much *results from* interactions, as *determinants of* interactions (Latour, 2005). ANT also offers tools to further explore the interactive complexity of sociotechnical systems, through a *sociology of uncertainty*, elaborated on in Latour's (2005) *Reassembling the Social*, particularly the chapter 'Second source of uncertainty: action is overtaken' (Latour, 2005, pp. 43–62).

From Weick and his colleagues we have learned important lessons about the collective nature of sensemaking. Indeed, this is a perspective that is supported by ANT. However, in ANT, there has been much focus on expanding

---

* Although these strands are often associated with these three individual thinkers, many others have been contributing to developing the strands and their unique position in the field, exemplified by works of Sagan (1995), Weick, Sutcliffe and Obstfeld (2005), Wears, Hollnagel and Braithwaite (2015), Nemeth, Hollnagel and Dekker (2009), Hollnagel, Pariès, Woods and Wreathall (2011), Woods (2015).

† In particular, Resilience Engineering is a perspective with many contributors. Here I choose to mention Hollnagel, since it is in particular the idea and rationale to study and learn from normal work – captured in the concept of *Safety II* – that I want to emphasise.

‡ Sensework is a sociotechnical perspective, implying that the actions of individuals are always considered as taking place in a context that is social and material. With respect to cognition, thus, there are also relations to be found to distributed cognition Hutchins (1995a,b) and inspired strands promoted by e.g. Stanton, Salmon and Walker (2015), Salmon, Walker and Stanton (2015), Salmon, Stanton, Walker and Jenkins (2009).

our notion of the collective – developing theory, methodology and vocabulary to take into account the non-human actors of various collectives* – offering important insights to account for the heterogeneous enterprises of sensework. Sensework has connotations to sensemaking (Weick, Sutcliffe and Obstfeld, 2005; Weick 1995, 1993), expanding on this to address not only social aspects in particular, but also just as much the technological and material ones. Hence, sense connotes not only to a type of meaning that is developed by social collectives, but as much to the indispensable sensors through which phenomena (not readily available to human senses) speak and produce meaning.

From Hollnagel and his colleagues we have learned important lessons about the significance of studying and learning from that which goes right – often expressed by Safety II – in contrast to focussing primarily on failures and accidents – in this context, Safety I. Such symmetry is surely supported by ANT, and at the heart of sensework lies an interest in normal operations and normal work. Safety research being a science that almost by definition must be occupied with outcomes – and given the position that major accidents (organisational accidents, systemic accidents) have, and rightly so – the efforts of resilience engineering to direct the attention of organisations and the public towards those events that are not readily associated with accidents, requires hard work; not only in terms of rhetorical persuasion, but also in terms of empirical persuasion. What ANT offers in this respect, is a long history of studying the pragmatic nature of sociotechnical systems; theorising and conceptualising the mundane in ways that lend support to – and expand theoretically and empirically on – concepts such as efficiency-thoroughness trade-off (ETTO), adaptations and *work as done* drawing particular attention to the co-work of humans and non-humans, and problematising the dividing lines between them (for an entertaining example, see Latour, 1992).

In recent years, we have witnessed a new wave of incorporating sociotechnical perspectives and methods from the social sciences and humanities into safety research, where the social and the technical are treated less by reference to different ontological categories, and more with reference to what it is they *do* in practice. Some of these research contributions are generic and directed towards theory building (Le Coze, 2013b; Haavik, 2014a; Le Coze, 2016, 2013a; Haavik, 2014b), while others are more specifically reporting from case studies in fields such as healthcare (Braithwaite and Plumb, 2015; Haavik, 2016a), aviation (Haavik et al., 2017), petroleum (Haavik, 2016b) and the maritime domain (deVries, 2016). It is common for

---

* The advanced user of STS may look up Latour (2004), while less experienced STS readers will probably get more out of Latour (2005).

these contributions to address the relation itself between the social and the technical as a main topic of exploration, and they do so by drawing on the progressive research traditions of STS in general and ANT in particular.

7.1 A Very Brief Introduction to ANT

ANT is closely related to the larger STS tradition. STS refers to an interdisciplinary research strand occupied with the construction of knowledge and science, the relation between science and politics, as well as the relation between the social and the technical/material. Through the decades from the 1960s – to which we may track the birth of STS as a systematic, academic discipline – contributions from Kuhn (1962), Bloor (1976, 1991), Bijker, Hughes and Pinch (1987) and Latour (1987, 2005) represent a few central contributions to a large body of literature, wherein the relation and the coproduction of the social/political and the technological/material has been addressed in fields such as innovation (Akrich, Callon and Latour, 2002a,b); feminism (Haraway, 2013) and climate research (Ryghaug and Skjølsvold 2010; Hart and Victor, 1993).

ANT is one branch of STS that is particularly associated with Bruno Latour (2005, 1999c, 1993, 1987); but also Michel Callon (1986), John Law (1991) and others. As ANT has developed through the years since the first writings that pointed towards a theory was published (Latour and Woolgar, 1979), some key themes – around which *Reassembling the Social* (Latour, 2005) is organised – have come to stand out as an 'ANT epistemology': (i) we learn more from studying the controversies surrounding the *formation* of categories, than we do by using settled categories to explain outcomes; (ii) when we look for the source of action and outcomes, we must be prepared to trace it through a number of mediators and not apply any absolute stop criteria, only pragmatic ones; (iii) agency is not reserved for humans – objects have agency, too: anything that modifies a state of affairs can be said to act; (iv) facts and truths are states of affairs that are stabilised by networks of humans and non-humans – the denser network, the stronger the fact and (v) good descriptions are accounts that trace networks where all participants are treated as full-blown actors, in contrast to causal explanations drawing on external forces that are not traced or traceable.

This description of ANT may perhaps feel impenetrable for safety scholars not familiar with STS, but there is no need to worry. ANT is occupied with understanding how science, technology, social and political aspects work together in all corners and at all scales of our sociotechnical world – and, in that respect, it should be highly inviting to safety research, particularly for those researchers oriented towards sociotechnical systems.

In the rest of the chapter, we shall see how these ANT perspectives can be made relevant to safety research and add new pages to the collective and accumulative body of knowledge, standing on the shoulders of former generations of authors in the field of safety research.

7.2 Safety Research and ANT – Symmetries

In what was called the *strong programme,** Bloor (1976) argued for *impartiality* and *symmetry* (in addition to causality and reflexivity) in the sociology of knowledge. Impartiality referred to the requirement for the sociology of knowledge to examine successful as well as unsuccessful knowledge claims, while symmetry referred to the requirement to use the same types of explanations for successful and unsuccessful knowledge claims alike. This highlighted the unreasonableness of addressing and explaining phenomena based on the 'final' outcome, and, in practice, had an enormous amount of influence on ANT and its orientation (see Latour, 1999a). Since Bloor's strong programme, symmetry has come to include an equal treatment of other dualisms, such as subjects/objects (Latour, 1992) and mental cognition/sociomaterial action (Latour, 1986).

Other recurring themes in ANT – and STS in general – are those of representation and representational artefacts. Through these themes and others such as situated action, the field of workplace studies (Goodwin and Goodwin, 1996, 1994; Suchman, 2007, 1993, 1995) and distributed cognition (Hutchins, 1995a,b) have also been folded into the repertoire of STS,† or its borderland, and these fields have had a profound impact on our understanding of sociotechnical systems.

On the one hand, there is an interesting parallel between the agenda of Bloor's programme and its influence on ANT in the decades after it, and on the other, the more recent trends in safety research to move away from a dominating failure focus (Hollnagel, 2014) and bring sociotechnical systems to the foreground of analyses by leaning on the perspectives of sociotechnical systems in the social sciences and humanities (Haavik, 2014b;

---

* Not to be confused with the *strong program* of Jens Rasmussen, accounted for by Le Coze (2015).

† The inclusion of distributed cognition into STS is somewhat ambiguous. This ambiguity is educationally elaborated on in Latour's (1996) interesting review of Hutchins' (1995a) book, *Cognition in the Wild.* Latour's delight over the book – leading him to raise his ten years' moratorium one year early – is due to Hutchins' elaboration of cognition as a social process, turning cognitive science inside out. His critical remarks are that the *distribution* is not going *all the way,* and that Hutchins is too hopeful on behalf of developments in psychology, whose ambitions Latour quite rudely refers to as: to 'cram cognition inside an individual mind endowed with consciousness and responsibility'.

Le Coze, 2013b). If we take it seriously, there is a potentially great advantage in bringing the epistemology of ANT into our endeavours in the field of safety within sociotechnical systems. One advantage is a weighty sociological support for the agenda of studying and learning from normal conditions, not primarily failures. We are not only provided with the rationale for this, but also with the methods and theoretical concepts that have been developed within ANT. A second advantage is that we are provided with tools to manage the eternally difficult relation between the social and the technical.*

As we shall see, the sensework perspective – being directly situated in this intersection between this well-explored ANT agenda, and the still-in-the-making safety research agenda that I have referred to above† – represents an opportunity for this underexposed coupling to be developed.

7.3 Safety Research and ANT – Controversies

Apart from the characteristics of ANT which I have already mentioned, the methodology of mapping and exploiting controversies and breakdowns has proven to be very fruitful (Bloor, 1982; Latour, 2005, 1987). The bringing of this methodology into safety research serves two 'points': first, a central idea in ANT is that it is particularly from within the controversies and breakdowns that one gets a really good opportunity to look inside sociotechnical systems, and investigate their ingredients and modes of work. Second, within safety research, controversies and breakdowns (failures) have received much attention, but these kinds of failures have long been associated with accidents and catastrophes. Consequently, the development of theories and vocabularies to account for the normal functioning of sociotechnical work has been limited. The types of controversies and breakdowns that ANT has developed its theories and methods to study are sometimes of a different magnitude – more in the range of the normal variability of systems and work. Think, for example, of the types of breakdowns and successes associated with the sociology of keys (Latour, 1991). However, the ambitions of ANT are far from mundane, and over the last few years we have witnessed an increased interest in issues surrounding the risk of climate change (e.g. Latour 2014).

---

* The problems dealing with this relation in safety research have been reflected through the years in the compartmentalised fashion (humans, technology and organisation) in which sociotechnical systems tend to be addressed.
† Surely, Rasmussen, Hollnagel, Leveson and many others have put sociotechnical systems in the foreground for a long time. This second wave of sociotechnical studies in safety research illustrates the slowness with which science progresses, and, that larger changes, if not paradigmatic (Kuhn, 2012), often require persistent pressure.

By exploring what we may call ANT controversies – that is, the kind of controversies of breakdowns that are not part of safety research's 'failure repertoire', a number of studies have contributed to analysing sociotechnical work in safety-critical systems, such as the petroleum industry, healthcare and aviation, without 'having access' to an accident (Haavik et al., 2017; Haavik, 2016a,b; Haavik, 2014b). This issue of access is important, since many of these domains involve highly reliable organisations where accidents rarely occur. The problem of studying organisations and systems where nothing much appears to happen has given rise to the description of safety as a *dynamic non-event* (Weick, 1987). This description has inspired discussions by Haavik, Antonsen, Rosness and Hale (in press) – with reference to the large body of ANT and STS literature on the nature of work – discussions where spin-offs such as *dynamic events* (see also Hollnagel, 2014) represent outcomes that open themselves up for new reflections. Other examples of controversies that have been used as small cracks through which systems are made available for inspection (and concepts available for critique) include *shared situation awareness* (Haavik, 2011, for a related, but slightly differently contextualised discussion of this topic, see also: Stanton, Salmon and Walker, 2015).

7.4 Sensework

Sensework refers to a type of sociotechnical work within safety-critical operations where groups of professionals in material and technological environments work to put together pieces of digital sensor data – and different sorts of representations – to give meaning to, and work through, both familiar and unfamiliar situations. As a theoretical perspective, sensework operates in the intersection between safety research and ANT. Sensework research seeks to avoid both technological determinism and social constructivism, and is first and foremost marked by a kind of *pragmatism*; the technical and the social interact in unpredictable ways, and the ways in which the different sociotechnical systems work is an empirical question.

A device acquired through many years of empirical studies of collaborative work in high-tech, high-risk industries, is that we never know which actors, processes and effects to take into account before we investigate the actual practices in the field. This is an important reminder accompanying research on *normal work*. When we investigate failures and accidents, we run the risk of 'whiggishly'* explaining why certain outcomes occurred, and from that perspective, failures and accidents can become decoy phenomena, obscuring how it could have been – and can be – different.

---

* A term occurring in Latour (1999a), suited to shed light over the Safety II discourse.

While sensework is still in the making, and resists ready-made descriptions, the following interim statements could serve to circle in on what sensework is about:

- Sensework takes place in environments where multiple professional discourses exist side by side, implying continuous negotiations of matters of concern and ranking of priorities.
- Both humans and non-humans take part in sensework, but who they are and what they do is not static; hence the descriptions of sociotechnical systems are never concluded, and their borders are uncertain and leaking.
- Sensework takes place in regulated environments where formal representations of organisation and work are indispensable artefacts. These representations are resources for sensework but do not themselves describe sensework.
- Sensework relies on (and combines) heterogeneous tools and practices, such as real-time information from sensor instrumentation, formalised models of causal relationships, operational experience, formal procedures and ad hoc workarounds. Sensework is by definition pragmatic.
- The production of sense in sociotechnical systems is a continuous achievement. Decision-making is relevant to sensework, but only as artefacts of formalistic representations that do not themselves describe sensework.

Perhaps a richer description of what sensework *is* would be best provided by showing what sensework *does*. To give an idea, a few examples of sensework research are provided below. It is the hope that many more examples will be added by the research community in the time to come; in that way it will gradually become clearer what sensework *is*.

7.4.1 Sensework in Offshore Operations

This study (Haavik, 2014b) of shore-based teams of petroleum engineers supporting offshore drilling operations in a context of integrated operations (Haavik, 2013; Rosendahl and Hepsø, 2013; Albrechtsen and Besnard, 2013) explored the relation between formalised and non-formalised work practices when faced with drilling problems. Drawing on such ideas and concepts as *work as imagined* and *work as done* (Borys, 2009; Nathanael and Marmaras, 2006; Hollnagel, Woods and Leveson, 2006), plans and situated action (Suchman, 1987), distributed cognition (Hutchins, 1995a,b) and multidisciplinary work (Latour, 1999b), sensework was portrayed as a variety of approaches that brought together theory and practice as shown in Figure 7.1.

FIGURE 7.1
Sensework in Integrated Operations.

Figure 7.1 is descriptive and not normative, and it illustrates the pragmatics of sensework drawing on different kinds of formalised and non-formalised experience that is relevant to the collective. What the figure does not say anything about is the dynamics of enrolling and deploying the resources; the power to enrol and deploy must be explored empirically. Surprisingly often one will find that the organisational charts are not pointing in the right direction. Studies of *reliability professionals* provide some good examples of this (Roe and Schulman, 2008; Pettersen and Schulman, in press).

While the role of digital representation definitely played an important role in working out sensible solutions in this case, the particular study was not rigged to account for this. In a later study of surgical work, however, the role of representations was made central to the study.

7.4.2 Sensework in Surgical Teams

In this study (Haavik, 2016a), we observed neurosurgical teams in sixteen operations, focussing on the role of representations and visualisations – particularly images produced by magnetic resonance (MR), X-ray and ultra-sound technology – in producing workable interpretations of the patients' conditions. Two cases of removing a benign tumour and managing a distended abdomen were made subject to detailed elaboration in the research paper. Two central themes were highlighted. One was referred to as the *sensation of representations*, elaborating on the role of representations when the object of work is not readily available for inspection other than by different types of sensors.* The other theme was the performative nature of decisions – with obvious connotations to both sensemaking (Weick, 1995) and naturalistic

---

* This is surely also the case in offshore petroleum operations and so many other instances of sensework that it may be adequate to simply include this as a central feature of sensework.

decision-making (Zsambok and Klein, 2014) – showing how the course of action could be portrayed by *deferral* and *displacement* with representational support, rather than by decisions. These findings contributed to the formulation of a warning against pitfalls associated with a division of labour between decision-making and execution in geographically distributed organisations; decision-making turned into expertise that is detached from the ongoing operations may, if the heterogeneous and pragmatic nature of sensework is not taken into account and organised for, lead to suboptimal, unsafe decisions. The importance of this scenario is illustrated by the currently high level of interest for organisational arrangements that seek efficiency and expertise by aid of augmented reality and other technological solutions, in combination with a centralised expertise that serves remote locations. Examples are telesurgery, space endeavours, offshore operations in harsh environments (see Haavik, 2016b) and decentralised air traffic management.

7.4.3 Sensework within and beyond the Cockpit

This study (Haavik et al., 2017) studied airmanship as an organisational phenomenon, in a context of increased automation and standardisation. By observing pilots in the cockpit in a number of flights and then interviewing them in the office afterwards, we explored how the traditional meaning of airmanship is changing as aviation becomes an increasingly systemic activity. While pilots of earlier days were highly dependent on individual skills and experience, the safety of commercial passenger flights today results from increasingly tight couplings between individual pilots, computers, organisational regimes, standardised aircrafts and procedures, aircraft manufacturers and international legislation. In this dense actor network, the space of manoeuvre for individual pilots and individual flights has changed, and the variability takes different forms. To account for these changes, we introduced a systemic variant of airmanship: *airlineship*, where de-identification and interchangeability of personnel and aircrafts are both central system parameters. Needless to say, sensework in aviation is also a rapidly changing phenomenon, and such research should definitely be provided by a best-before date.

7.5 A Research Agenda for Sensework

One way to think of sensework is that it is oriented towards exploring uncertainties with respect to actors (human and non-human), actions, relations and network extensions of sociotechnical systems. While safety researchers have many competent allies from the natural sciences to help compute risks and outputs from a given set of system parameters, there is still room for improvement with respect to understanding some of the foundational aspects of sociotechnical systems (see e.g. Downer, 2014). The uncertainty of the calculation of outputs must also be seen in relation to the uncertainty associated with these aspects, whether we address 'traditional' sociotechnical systems of safety research such as aviation, nuclear, marine, petrochemical or military systems or systems that are much more geographically trailblazing, spanning from the local use/global impact – for example, nano-particles travelling up through the food chains – to global processes/ local impacts – for example, global climate change resulting in local hazards such as avalanches and floods.

A research agenda for sensework should therefore particularly invite empirical studies of sociotechnical work where *processes* – not their outcomes – inspire and direct the research. High-tech systems or organisations undergoing technological and/or organisational changes are examples of cases that fit this scope well.

The more empirical the sensework research is, the more likely it is to circumvent the threat of being reversed from description to norm. Sensework is first and foremost a living phenomenon begging for theorisation, not a stable theory or model of a phenomenon; rigid theories and models are inclined to gradually loose accuracy. For sensework, it is the underlying ontology that is stable, while the way it manifests itself is an empirical question, varying with time, place, technology and politics.

Failures and accidents represent occasions where sociotechnical systems are made particularly available for inspection to find out what they are made of, and how they function. However, systems that rarely fail also need to be understood; so the occasions we must look for and use may be breakdowns of a different magnitude. ANT offers theoretical tools, methodologies and vocabularies to analyse and articulate normal work within a context of safety, exploiting these minor breakdowns. Sensework research should lend inspiration from ANT to identify and exploit these breakdowns, which may range from systemic accidents on the one hand, and all the way over to linguistic controversies on the other, in order to account for safety in sociotechnical systems.

References

Akrich, M., M. Callon and B. Latour. (2002a). "The key to success in innovation part I: The art of interessement." *International Journal of Innovation Management*, Volume 6(2): 187–206.

Akrich, M., Michel Callon and B. Latour. (2002b). "The key to success in innovation part II: The art of choosing good spokespersons." *International Journal of Innovation Management*, Volume 6(2): 207–225.

Albrechtsen, E., and D. Besnard, eds. (2013). *Oil and Gas, Technology and Humans: Assessing the Human Factors of Technological Change.* Farnham: Ashgate.

Bijker, Wiebe E., T. P. Hughes and T. Pinch. (1987). *The Social Construction of Technological Systems: New Directions in the Sociology and History of Technology.* Cambridge, MA: MIT Press.

Bloor, D. (1976). "The strong programme in the sociology of knowledge." *Knowledge and Social Imagery*, Volume 2: 3–23.

Bloor, D. (1982). "Durkheim and Mauss revisited: Classification and the sociology of knowledge." *Studies in History and Philosophy of Science Part A*, Volume 13(4): 267–297.

Bloor, D. (1991). *Knowledge and Social Imagery.* University of Chicago Press.

Borys, D. (2009). "Exploring risk-awareness as a cultural approach to safety: Exposing the gap between work as imagined and work as actually performed." *Safety Science Monitor*, Volume 13(2): 1–11.

Braithwaite, J. and J. Plumb. (2015). "Exposing hidden aspects of resilience and brittleness in everyday clinical practices using network theories." In *Resiliente Health Care*, Volume 2. *The Resilience of Everyday Clinical Work*, edited by Robert L. Wears, E. Hollnagel and J. Braithwaite, 115–128. Farnham: Ashgate.

Callon, M. (1986). "Some elements of a sociology of translation: Domestication of the scallops and the fishermen of St Brieuc Bay." In *Power, Action and Belief: A New Sociology of Knowledge*, edited by J. Law, 196–223. London: Routledge.

deVries, L. (2016). "Here be monsters: Investigating sociotechnical interaction in safety-critical work in the maritime domain." Licentiate thesis, Department of Shipping and Marine Technology, Chalmers University of Technology.

Downer, J. (2014). "Disowning Fukushima: Managing the credibility of nuclear reliability assessment in the wake of disaster." *Regulation and Governance*, Volume 8(3): 287–309.

Goodwin, C. (1994). "Professional vision." *American Anthropologist*, Volume 96(3): 606–633.

Goodwin, C and M. H. Goodwin. (1996). "Seeing as a situated activity: Formulating planes." In *Cognition and Communication at Work*, edited by Yrjö Engeström and D. Middleton, 61–95. Cambridge: Cambridge University Press.

Haavik, T. K. (2016a). Representasjonsteknologi, samhandling og sikkerhet i kirurgi. For fagseminar, Røros.

Haavik, T. K., S. Antonsen, R. Rosness and A. Hale. (In press). "HRO and RE: A pragmatic perspective." *Safety Science*. doi: http://dx.doi.org/10.1016/j.ssci.2016.08 .010 (special issue on HRO and RE).

Haavik, T. K., T. Kongsvik, R. J. Bye, J. O. D. Røyrvik and P. G. Almklov. (2017). "Johnny was here: From airmanship to airlineship." *Applied Ergonomics*, Volume 59: 191–202.

Haavik, T. K. (2011). "Chasing shared understanding in drilling operations." *Cognition, Technology and Work*, Volume 13(4): 281–294.

Haavik, T. K. (2013). *New Tools, Old Tasks: Safety Implications of New Technologies and Work Processes for Integrated Operations in the Petroleum Industry.* Farnham, UK: Ashgate.

Haavik, T. K. (2014a). "On the ontology of safety." *Safety Science*, Volume 67: 37–43.

Haavik, T. K. (2014b). "Sensework." *Journal of Computer Supported Cooperative Work*, Volume 23(3): 269–298.

Haavik, T. K. (2016b). "Remoteness and sensework in harsh environments." *Safety Science*, Volume 95: 150–158.

Haraway, D. (2013). *Simians, Cyborgs, and Women: The Reinvention of Nature.* New York, Routledge.

Hart, D. M. and David G. Victor. (1993). "Scientific elites and the making of US policy for climate change research, 1957–74." *Social Studies of Science*, Volume 23(4): 643–680.

Hollnagel, E. (2014). *Safety-I and Safety-II: The Past and Future of Safety Management.* Farnham, Surrey, England: Ashgate.

Hollnagel, E., J. Pariès, D.D. Woods and J. Wreathall. (2011). *Resilience Engineering in Practice: A Guidebook.* Farnham: Ashgate.

Hollnagel, E., D. D. Woods and N. Leveson, eds. (2006). *Resilience Engineering: Concepts and Precepts.* Aldershot, UK: Ashgate.

Hutchins, E. (1995a). *Cognition in the Wild.* Cambridge, MA: MIT Press.

Hutchins, E. (1995b). "How a cockpit remembers its speeds." *Cognitive Science*, Volume 19(3): 265–288.

Kuhn, T. S. (1962). *The Structure of Scientific Revolutions*: University of Chicago Press.

Latour, B. (1986). "Visualisation and cognition: Thinking with eyes and hands." *Knowledge and Society: Studies in the Sociology of Culture Past and Present*, Volume 6: 1–40.

Latour, B. (1987). *Science in Action: How to Follow Scientists and Engineers through Society.* Milton Keynes, UK: Open University Press.

Latour, B. (1991). "Technology is society made durable." In *A Sociology of Monsters: Essays on Power, Technology and Domination*, edited by John Law, 103–131. London, UK: Routledge.

Latour, B. (1992). "Where are the missing masses? The sociology of a few mundane artifacts." In *Shaping Technology/Building Society: Studies in Sociotechnical Change*, edited by Wiebe E. Bijker and John Law, 225–264. Cambridge, MA: MIT Press.

Latour, B. (1993). *We Have Never Been Modern.* New York, USA: Harvester Wheatsheaf.

Latour, B. (1996). "Review: Cognition in the wild." *Mind, Culture and Activity*, Volume 3(1): 54–63.

Latour, B. (1999a). "For David Bloor... and beyond: A reply to David Bloor's 'Anti-Latour'." *Studies in History and Philosophy of Science Part A*, Volume 30(1): 113–129.

Latour, B. (1999b). "Circulating references. Sampling the soil in the Amazon forest." In *Pandora's Hope: Essays on the Reality of Science Studies*, edited by Bruno Latour, 24–79. Cambridge, MA: Harvard University Press.

Latour, B. (1999c). *Pandora's Hope: Essays on the Reality of Science Studies.* Cambridge, MA: Harvard University Press.

Latour, B. (2004). *Politics of Nature: How to Bring the Sciences into Democracy.* Cambridge, MA: Harvard University Press.

Latour, B. (2005). *Reassembling the Social: An Introduction to Actor-Network-Theory, Clarendon Lectures in Management Studies.* Oxford, UK: Oxford University Press.

Latour, B. (2014). "Agency at the time of the anthropocene." *New Literary History*, Volume 45(1): 1–18.

Latour, B and S. Woolgar. (1979). *Laboratory Life: The Social Construction of Scientific Facts.* Beverly Hills, CA: Sage Publications.

Law, J. (1991). *A Sociology of Monsters: Essays on Power, Technology, and Domination.* London, UK: Routledge.

Le Coze, J.-C. (2013a). "New models for new times. An anti-dualist move." *Safety Science*, Volume 59: 200–218.

Le Coze, J.-C. (2013b). "Outlines of a sensitising model for industrial safety assessment." *Safety Science*, Volume 51(1): 187–201.

Le Coze, J.-C. (2015). "Reflecting on Jens Rasmussen's legacy. A strong program for a hard problem." *Safety Science*, Volume 71: 123–141.

Le Coze, J.-C.. (2016). "Vive la diversité! High reliability organisation (HRO) and resilience engineering (RE)." *Safety Science* (this issue). doi: http://dx.doi.org/10.1016 /j.ssci.2016.04.006.

Nathanael, D. and N. Marmaras. (2006). "The interplay between work practices and prescription: A key issue for organisational resilience." In: Paper presented at the 2nd symposium on resilience engineering, Juan-les-Pins, France, November 8–10.

Nemeth, C.P., E. Hollnagel and S. Dekker. (2009). *Resilience Engineering Perspectives: Preparation and Restoration*. Aldershot, UK: Ashgate.

Perrow, C. (1984). *Normal Accidents: Living with High-Risk Technologies*. Princeton, NJ: Princeton University Press.

Perrow, C. (1999). *Normal Accidents: Living with High-Risk Technologies*. Princeton, NJ: Princeton University Press.

Pettersen, K. A. and P. R. Schulman. (In press). "Drift, adaptation, resilience and reliability: Toward an empirical clarification (special issue on HRO and RE)." *Safety Science*. doi: http://dx.doi.org/10.1016/j.ssci.2016.03.004.

Roe, E. and P. R. Schulman. (2008). *High Reliability Management: Operating on the Edge*. Stanford, CA: Stanford University Press.

Rosendahl, T. and V. Hepsø. (2013). *Integrated Operations in the Oil and Gas Industry: Sustainability and Capability Development*. Hershey, PA: IGI Global.

Ryghaug, M. and T. M Skjølsvold. (2010). "The global warming of climate science: Climategate and the construction of scientific facts." *International Studies in the Philosophy of Science*, Volume 24(3): 287–307.

Sagan, S. D. (1995). *Limits of Safety: Organisations, Accidents, and Nuclear Weapons*. Washington, DC: American Association for the Advancement of Science.

Salmon, P. M., N. A. Stanton, G. H. Walker and D. P. Jenkins. (2009). *Distributed Situation Awareness: Theory Measurement and Application to Teamwork*. Farnham: Ashgate.

Salmon, P. M., G.H. Walker and N.A. Stanton. (2015). "Broken components versus broken systems: Why it is systems not people that lose situation awareness." *Cognition, Technology and Work*, Volume 17(2): 179–183.

Sismondo, S. (2010). *An Introduction to Science and Technology Studies*, Volume 1. Chichester: Wiley-Blackwell.

Stanton, N. A., P. M. Salmon and G. H. Walker. (2015). "Let the reader decide a paradigm shift for situation awareness in sociotechnical systems." *Journal of Cognitive Engineering and Decision Making*, Volume 9(1): 44–50.

Suchman, L. (1987). *Plans and Situated Actions: The Problem of Human-Machine Communication*. Cambridge, UK: Cambridge University Press.

Suchman, L. (1993). "Centers of coordination: A case and some themes." In *Discourse, Tools, and Reasoning: Essays on Situated Cognition*, edited by Lauren B. Resnick, 41–62. Berlin: Springer.

Suchman, L. (1995). "Representations of work." *Communications ACM*, Volume 38(9): 33–35.

Suchman, L. (2007). *Human-Machine Reconfigurations: Plans and Situated Actions*. New York, USA: Cambridge University Press.

Wears, R. L., E. Hollnagel and J. Braithwaite. (2015). *Resilient Health Care*, Volume 2: *The Resilience of Everyday Clinical Work*. Farnham: Ashgate.

Weick, K. E. (1987). "Organisational culture as a source of high reliability." *California Management Review*, Volume 29(2): 112–128.

Weick, K. E. (1993). "The collapse of sensemaking in organisations: The Mann Gulch disaster." *Administrative Science Quarterly*, Volume 38(4): 628–652.

Weick, K. E. (1995). *Sensemaking in Organisations*. Thousand Oaks, CA: Sage.

Weick, Karl E. (2007). "The generative properties of richness." *Academy of Management Journal*, Volume 50(1): 14–19.

Weick, K. E., Kathleen M. Sutcliffe, and D. Obstfeld. (2005). "Organising and the process of sensemaking." *Organisation Science*, Volume 16(4): 409–421.

Woods, D. (2015). "Four concepts for resilience and the Implications for the future of resilience engineering." *Reliability Engineering and System Safety*, Volume 141: 5–9.

Zsambok, C. E. and G. Klein. (2014). *Naturalistic Decision Making*. New York: Psychology Press.

8

Drift and the Social Attenuation of Risk

Kenneth Pettersen Gould and Lisbet Fjæran

CONTENTS

8.1 The Social Amplification of the Risk Framework 121
8.2 Attenuation and Three Types of Drift 122
    8.2.1 Type 1: Practical Drift 124
    8.2.2 Type 2: Organisational Drift 125
    8.2.3 Type 3: Societal Drift 126
8.3 Safety Attenuation 127
8.4 Implications for Risk Assessment and Management 127
Acknowledgements 130
References 130

Introduction

Accident and disaster research has shown that the perception and control of catastrophic risks by operators at the sharp end of systems are impacted by managerial responses to risks or risk events far from their everyday workplaces (Perrow, 1984; Turner and Pidgeon, 1997; Rasmussen, 1997; Reason, 1997). Sociologist Diane Vaughan (1997) demonstrated this when she traced how changes in the political and managerial environment surrounding NASA trickled down from the top in a way that weakened risk perception and changed both the structure and the culture of work, so as to normalise the deviant performance of critical technical components. Eventually, this led to the catastrophic loss of the space shuttle *Challenger*.

A growing body of work within safety science has addressed *drift* as a key effect of social and organisational factors producing major accident risk (Pettersen and Schulman, 2016). Based on the theory of self-organisation and autonomy in technological systems (Rasmussen, 1994; Le Coze, 2015), drift in relation to accidents has been described as a gradual behaviour change on the part of individuals at the sharp end of systems; when rules did not match current task demands, individuals adjusted their behaviour accordingly (Snook, 2000). This practical drift, where the baseline performance of

119

a system subtly moves away from its original design, potentially increases the possibility for an incident or accident as performance moves too far from the expected baseline. Following this pioneering research, the concept of drift has been used in explanations of accidents in aviation (Dekker, 2013), incidents in healthcare (Patterson and Wears, 2015), and events in the maritime sector (Oltedal and Engen, 2011), among others. In the various studies on drift, the concept refers to explanations of the relationship between the parts or levels of an organisational system (e.g. social and regulatory environment, structure and processes in organisations, cognitive practices of individuals) producing organisational deviances that can relate to misrepresentation, mistakes, surprise and disaster. The topic of drift is also related to the broader academic literature on fine-tuning in organisations, explaining how drift involves subtle changes and adaptations that are not necessarily negative developments. Feldman and Pentland (2003) wrote about how bureaucracy can foster flexibility, while Rerup and Levinthal (2006) and Farjoun (2010) summarised key aspects of the stability-flexibility duality. Such research has demonstrated the complex nature of institutional/management strategies associated with change in organisations, as well as the collaboration and information practices required for safety and reliability in high-risk contexts.

An important problem associated with major accidents and disasters is that some risks (or risk events), which are assessed as important by operational personnel and technical experts do not elicit strong institutional or public concern (Kasperson et al., 1988). Similar to the social and organisational factors emphasised in the explanations of drift, the social amplification of risk framework (SARF) (Kasperson et al., 1988; Pidgeon et al., 2003) has been applied within risk analysis to explain how hazards interact with psychological, social and cultural processes in ways that serve to amplify or attenuate risks. Much research has documented the first stage of risk amplification – namely, the direct behavioural and communicative responses to risk-related messages. Less work addresses the 'ripple effects' of risk amplification, although it is in this stage of the framework where the highest potential for great impacts exists.

Despite these similarities, little work has been done to combine the literature on accident theory and prevention within safety science to the social amplification of risk framework. Furthermore, most SARF-related research is directed toward the amplification of risks, whereas considerably less attention has been given to attenuation processes and impacts. In this article, we connect the literature across risk analysis and safety science by specifically combining the concept of risk attenuation with the concept of drift. We apply the social amplification framework to explain how social attenuation of risk can lead to drift in high-risk systems. Such systems are found across sectors in society; examples include transportation, nuclear power, energy and the chemical industry. These are typically large transnational systems characterised by tightening patterns of functional interdependencies, intensive knowledge requirements and expanding networks of cooperation and control (Le Coze, 2017; Goldin and Mariathasan, 2014; LaPorte, 1996). Our approach is informed by sociotechnical systems thinking (Le Coze et al., 2017), conceptualising safety as a nested hierarchy of levels. We see each level (e.g. individual, organisational, and social) as having discrete mechanisms, but also as being connected, and mechanisms can interact and develop in complex ways across levels.

Insights into the relationship between attenuation and drift can prove valuable by demonstrating the potentially high consequences of risk attenuation. When something goes disastrously wrong, it can be caused not by extraordinary failure and misfortune, but by the 'everyday' incremental changes and 'invisible' small steps that organisations take. Connecting risk attenuation with drift can provide additional insights into how deviance is normalised (Vaughan, 1997) or incubated (Turner and Pidgeon, 1997) in the production of disasters, as well as how shifts in values, such as those in the balance of societal concerns for safety relative to production, security, or innovation can justify shifting resources and activities in ways that impact risk. The latter represents a broadening of safety research, and our contribution sheds a necessary light on top management's contributions to major accidents, as the focus of safety research so far, including that on drift, has been and seems to continue to be on employees at the sharp end.

Next, we briefly present the social amplification framework, focussing on attenuation processes and impacts. Thereafter, we explain that some risk events have potentially high signal value for drift and connect the SARF framework and risk attenuation to three types of drift. We discuss the different types of drift based on previous research on major accidents and near misses. In the last sections, we discuss how drift can be identified as well as the implications for risk assessment and management.

8.1 The Social Amplification of the Risk Framework

The social amplification framework was introduced in 1988 (Kasperson et al., 1988) and represented a comprehensive attempt to integrate the technical analysis and social experience of risk. The framework explains the interaction between risk events and social processes (Renn et al., 1992). 'Social amplification of risk denotes the phenomenon by which information processes, institutional structures, social-group behaviour and individual responses shape the social experience of risk, thereby contributing to risk consequences' (Kasperson et al., 1988, p. 181). The processing of risk-related information can serve to heighten risk perception in different ways, but it can also lead to the weakening of risk perception (i.e. risk attenuation). Changes in social risk perceptions can shape behaviour and lead to both secondary and tertiary impacts that reach far beyond the initial impacts. Where amplification processes often start with an actual physical event or the recognition of an adverse effect, attenuation processes are more likely to start with, for example, expert concerns, new interpretations of technical data, failed equipment tests and expert meetings. These are all non-events in the sense of producing no materially adverse outcome (Weick and Sutcliffe, 2001). In organisations managing catastrophic risks, a typical risk event can be safety experts or investigators reporting that the organisation has experienced no accidents or serious incidents during the last year. These risk events are information communicated by, for example, safety professionals, managers or regulators. Certain characteristics of the risk event are transmitted and formed into a message (i.e. risk signal). The message then passes through social and individual 'stations of attenuation', in which information is interpreted according to the roles and rules of the social group, as well as the receiver's cognitive decoding and evaluation processes. Here the original message can be altered, as some risk signals are deleted or ignored while others are added. Following the previous example, the original message of no accidents or serious incidents may be altered into a message, which then communicates that the risk of major accidents is low. This leads to direct behavioural and communicative responses, such as the avoidance of safety procedures and rules or relaxed attitudes to safety.

The previously described pattern represents the first stage of the SARF. When the risk attenuation responses lead to secondary and tertiary effects – so-called 'ripples' – we enter into the second stage of the framework. The secondary impacts reach beyond the initial effects and those directly affected by the original event. Examples of such effects are increased production, reduced political pressure, a decline in risk regulation, increased trust in risk management and a focus on other risks. In turn, the secondary effects are perceived and interpreted by social groups and individuals in ways that produce third-order impacts. Tertiary ripple effects spread attenuation effects even further, and can affect other parties, technologies, sectors and different risk arenas. When risks are attenuated, each ripple of impact can also serve to hinder risk reduction.

8.2 Attenuation and Three Types of Drift

As argued in the introduction, risk attenuation can lead to drift by creating new patterns of behaviour in organisations. These can be subtle changes in practice at the sharp end of systems, or wider patterns of drift following changes in managerial behaviour. Such patterns may include adjustments in management priorities at the higher levels of an organisation, or even shifts in sectoral policy among similar, competitive organisations; even regulatory change. In addition, attenuation can result in drifting at even wider levels, such as when societies (e.g. authorities) potentially lose sight of the resources needed to remain within the boundary of safe practices (Pettersen and Schulman, 2014, 2016).

In relation to the amplification of risks, previous research has shown that some risk events that are highly dreaded and little understood hold a potential for high signal value (Slovic et al., 1984; Kasperson, 2012). These events can contribute to heightening risk perception and intensifying amplification. Such events can serve as a warning signal at the societal level or provide new information about similar risks. In addition, Slovic et al. (1984) found that the potential for second-order effects also indicate high-signal value. We argue that other risk events can contribute strongly to risk attenuation and potentially have a high signal value for drift. When it comes to risk attenuation, risk events with high-signal value are often well-known, understood and expected within existing knowledge or experience. The ripple effects resulting from attenuation take on different forms than those resulting from amplification. Whereas risk amplification may result in the loss of trust, stricter regulations or changes in market demand, the consequences of risk attenuation often represent the opposite tendencies: increases in production, relaxed regulations and higher trust in managers and risk regulators.

Sociologist William Freudenburg (1992) argued that 'improving safety' or 'protecting against fatal accidents' are goals that are hard to measure. Responsibility often becomes an issue of accounting measures: the number of inspections rather than the contribution of each inspection to safety, or the number of implemented corrective measures rather than concrete improvements in actual safety levels. Even in organisations with an above-average commitment to safety, the normal pattern following signal events contributing to attenuation would be for attentiveness and vigilance to weaken over time (Freudenburg, 1992). He explained that, only the occurrence of a disaster provides unquestionable proof that prevention activities are 'necessary' and that, in the absence of precluded events, safety measures can develop the impression of being 'non-productive'. Moreover, in the absence of truly frightening 'close calls', or in cases where close calls occur, but are interpreted to mean that current measures 'worked' and no further measures need be taken, the general tendency of organisations, both public and private, is to cut back on risk management expenditures until an accident provides proof that the cuts were too severe. Table 8.1 presents examples of risk events with potentially high-signal value for drift. The examples show how the original message of a risk event can be interpreted in ways that serve to decrease/ reduce the perception of risk, and, in turn, lead to drift.

Risk attenuation can happen at three different levels. At all three levels, this can have severe consequences for safety. The consequences of attenuation are less visible than those resulting from amplification and are therefore harder to identify. The consequences of attenuation might also include otherwise avoidable direct effects if risks were amplified (Kasperson et al., 2003).

TABLE 8.1

Risk Events with Potentially High-Signal Value for Drift

Original message of risk event	Interpreted message
Report that an organisation has had no accidents or incidents the last year	Risk of major accidents and disaster is low
News report that there were no interruptions in production in the last month	Risk managers are in control of the risks
Statement by regulator that accident statistics have a positive trend and that safety remains their first priority	Managers understand the risks and care about the people who may be harmed
Report that terrorists are planning to attack commercial airplanes in Europe	A new and possibly catastrophic risk has emerged requiring immediate attention and resources
Reappointment of regulators or corporate managers by elected politicians	The managers are in control of the risks and can be trusted
Report on operations show few enemy engagements	System design is over-controlling and unreasonable

TABLE 8.2

Three Types of Drift

Type of drift	Practical drift	Organisational drift	Societal drift
Level of change	Local	Systemic	Societal
Pattern of change	Shift in practice	Shift in strategies	Shift in values
Perception of change	Quick benefits	Benefits	Long-term benefits
Impact of change	Direct	Secondary ripple	Tertiary ripple

Based on Pettersen and Schulman (2014, 2016), we see the consequences of attenuation as three different types of drift: practical, organisational and societal (see Table 8.2). We discuss the different types of drift based on previous research on major accidents and near misses. These examples may include risk attenuation across both stages (i.e. direct responses, secondary and tertiary impacts) of the SARF framework, which are not easily separated in most real-life cases of major accidents. However, we will treat them as individual examples in order to illustrate each of the three types of drift and its characteristics relating to typical differences in level, pattern, perception and impact of change.

8.2.1 Type 1: Practical Drift

Practical drift (first column in Table 8.2) is the direct behavioural or communicative effect of risk attenuation. These are local and behaviourally anchored shifts in the logics of action, often at the sharp end of systems. Scott Snook (2000) researched the accidental shooting of two Black Hawk helicopters that were brought down in Iraq. In Snook's work, practical drift contributes to explaining why the two helicopters were shot down by friendly fire. Operation Provide Comfort (OPC) defended Kurds fleeing their homes in northern Iraq during the aftermath of the first Persian Gulf War (1991). The design and the rules of the operation were based on worst-case scenarios (e.g. with a lot of enemy engagement and designs for 'hot' war-type missions). However, most of the time there were few enemy engagements. Snook explained how these 'non-events' were interpreted as a message implying that the mission strategy and structure were over-controlling and unreasonable. This led to an attenuation of the risk of enemy engagement. When the rules did not match experiences, individuals adjusted their behaviour according to their perceptions (see last example in Table 8.1). In this case, the practical drift was a general loosening of planned behaviour. This resulted in a system with multiple sets of uncoordinated behaviours, each based on a different sub-unit's risk perception. In other words, locally pragmatic perceptions and responses were allowed to dictate action.

In Snook's (2000) analysis, it was the fact that people were expecting actions from others to comply with the original rules of engagement, which caused the coordination failure and the consequent shooting down of the helicopters, not the practical drift itself. Problems occurred when either party miscoded the situation or failed to act in accordance with planned behaviour and standard rules. These miscoding problems arose, according to Snook (2000), because individuals could not be simultaneously members of all groups. By definition they were not familiar with the change in perception and behaviour of others, leading them away from standardised procedures. Practical drift implies that it is a change unseen, unrecognised or unappraised elsewhere within the organisation – or beyond the organisation (Pettersen and Schulman, 2016), meaning that individuals will have misunderstandings about the current behaviour or 'state' of the organisation.

8.2.2 Type 2: Organisational Drift

Organisational drift (see Table 8.2) concerns an organisation's or a system of organisations' relationship to their larger environment (both inter-organisational relations and social context). Through organisational drift, safety goals are shifted as new values (e.g. productivity), which are then absorbed into leadership or the policy-determining structure of an organisation, leading to compromise on reliability at the system and/or local level. Vaughan's (1997) research on the catastrophic loss of the *Challenger* space shuttle is one such example. Although she did not use the concept of drift itself, she showed how drift could emerge through social processes contributing to the normalisation of deviance in NASA. In response to production pressure from management, deviances in work-related behaviour and equipment development were unfortunately tolerated. In addition, the perceptions of the shuttle program's acceptable risk level were reinforced and maintained by a 'culture of production'. In the case of NASA, the secondary ripple effects were the result of political pressure and policy adjustments attenuating risks at the management level. According to Farjoun (2005), organisational drift, understood here as the consequences of attenuation, is a well-documented pattern. His explanation of why the *Columbia* shuttle disaster happened 10 years after *Challenger*, emphasised social and institutional processes in similar ways to Vaughan. Farjoun suggested that, under prolonged periods of perceived safety, it is difficult to maintain a commitment to system safety as those safety goals will be compromised, particularly under conditions of scarce resources. Resource allocation to safety is reactive, and managerial concern for safety commonly increases after major failures. Farjoun claimed that this bias in commitment (e.g. resource allocation) may be very difficult to correct. The fluctuating nature of safety agendas makes learning and control difficult and risk attenuation more likely.

Also, organisational drift could be seen as a sectoral change among similar competitive organisations that establish policy adjustments as a common industry practice. A recent example according to Pettersen and Schulman (2014, 2016) shows that the mutual shift of organisations in the banking sector towards ever higher rates of return for shareholders, investors and depositors was what eventually led to riskier loans and investment policies. This led to a major drift away from previous risk management standards in banking practices within the whole banking sector.

8.2.3 Type 3: Societal Drift

Last, but not least, we have societal drift (last column in Table 8.2): a tertiary effect of risk attenuation relating to a shift in the balance of societal concerns for safety, relative to other values in society, such as production, security or innovation. For example, when a technology is viewed as reliable and the government receives information of many successes without failure, the impact of this can be an increased confidence coupled with political incentives to revise the likelihood of failure downwards in order to justify shifting resources to other activities (Heimann, 1997). These tertiary ripples, like changes in values, are less visible, and their causes can be interconnected as the amplification of one risk can lead to the attenuation of others. At the societal level, attenuation may influence what is culturally defined as a tolerable risk and shape public debates about hazards in relation to prevailing interests and beliefs. In addition, if degraded risk management is revealed following a serious unanticipated accident, the consequences of attenuation can result in distrust and a loss of credibility (Kasperson et al., 2003) that can prove very hard for authorities and safety regulators to regain.

An example of a signal event leading to societal drift was the terrorist attacks on 11 September 2001 (Pettersen and Bjørnskau, 2015). The attack was seen as a new and possibly catastrophic risk threatening civil aviation, requiring immediate attention and extraordinary resources. The attacks motivated regulatory restructuring as well as the redesign of key institutions in many Western societies. However, at the societal level, the amplification of terrorist attacks led to an emphasis on security values and displaced priorities given over to other risks. In civil aviation, security regulations were rapidly introduced, and the focus on terrorism overshadowed regulatory and managerial concerns for accident prevention. Among others, the potentially negative consequences of the newly introduced security measures were not sufficiently evaluated by responsible authorities (Pettersen and Bjørnskau, 2015).

8.3 Safety Attenuation

As seen, different stages of risk attenuation are triggers that may drive gradual departures from existing behaviour, and, over time, change the cultural patterns of risk perception. We define these patterns as *safety attenuation*, which is the migration of the cognitive and behavioural variables (which support safety) away from existing practices, leading to a weakening of safety margins at different levels.* When people are not aware of (or misunderstand) these departures, this can lead to 'errors' of assessment and management. Following this discussion, two important analytical questions should be asked when using the concept of drift in relation to risk attenuation (based on Pettersen and Schulman, 2014):

1. Drift with respect to what?

    Is the drift connected to a shift in psychology, protocols, practices or values? (e.g. goal displacement from safety to efficiency, a cognitive change from formal to schema-based decision-making).

2. Drift with respect to whom?

    Who is going to be making assessment and management errors because of it? Operators, directors, regulators, stakeholders or the public?

8.4 Implications for Risk Assessment and Management

All three types of drift resulting from safety attenuation relate to events that occur in a distant future (or distant places) or have delayed effects (Kasperson et al., 2003). This challenges risk analysis, which has usually

---

* The concept of safety attenuation is inspired by, and closely linked to, the notion of *reliability drift*. See Pettersen and Schulman (2016) for an explanation of the original concept.

focussed on 'visible' issues that can be meaningfully calculated (Pettersen, 2016). In addition, in the SARF (Kasperson et al., 1988), the boundaries of the traditional probabilistic risk analyses are highlighted and are considered to be too narrow and technical to address the socially complex and contextual nature of organisational factors such as attenuation and drift.

In high-risk organisations where major accidents and disasters rarely happen, major hazards are less clear and often hidden from view. Consequently, most risk events related to high-risk systems are not determined by their outcome of harm, but express non-events as well as organisational capacities for knowledge and control (see, as examples, the original messages in Table 8.1). A broader understanding of risk (Aven, 2014) that emphasises uncertainties and the knowledge supporting risk assessment results may serve to reduce the potential for attenuation and drift to occur. Such an understanding of risk involves shifting characteristics of risk from focussing on the probability of events and given consequences, to giving weight to uncertainty and knowledge in characterising risks (Aven and Renn, 2010). This alternative understanding of risk has implications for both the assessment and management of risk. Adopting this perspective of risk involves explicitly addressing uncertainties, as well as the relative strength/weakness of the knowledge upon which the assigned probabilities (and risk estimates) rest. This understanding of risk can prove especially important when the probability of a risk is judged to be low (Bjerga and Aven, 2015; Fjæran and Aven, 2010), which is often the case of the risk events leading to safety attenuation (see assessments as risk events in Table 8.1). For example, if risk here is understood as the combination of consequences and probability and the statistics of no previous accidents and incidents are presented as a low probability for future accidents, a risk manager could interpret the risk to be low (despite weak knowledge and the potentially high consequences of an accident). If, in this case, the risk manager bases her decision on the probability number and uses the risk assessment result as a specification for a decision of 'non-action' (e.g. allocation of fewer resources to safety), then this is an example of risk attenuation, which can lead to organisational drift. However, if the risk assessment was conducted based on a broad perspective of risk, the uncertainties and knowledge base underlying the assigned probability would have been explicitly addressed and communicated to the risk manager. This means that the risk manager can critically reflect on the risk assessment results and be aware of the potential for a low probability number to manifest in a decision that may act 'in favour' of safety attenuation and organisational drift. In turn, this may result in a different decision – or, at least, in a more informed decision on safety measures or considerations.

According to Aven and Zio (2011), the role of risk assessments and the way uncertainties are treated in them are of the utmost importance for the management and governance of risks. The degree to which uncertainty is recognised as a component of risk can have direct implications for the form and degree of precaution in regulatory intervention (Fjæran and Aven, 2010), and on the extent of tertiary ripple effects like societal drift. In this sense, safety attenuation can become a relevant concept for organisations and regulators to include in their assessment and management of major accident risk.

For organisations to respond to attenuation and identify drift, there is potential in seeing risk assessments and risk management from an increasingly integrated perspective. In the past, safety research has largely focussed on individual behaviour, human error and the responsibilities of the worker. Our analysis shows that drift is not just a phenomenon of local practice; an issue of the loosening of action at the sharp end in relation to an existing system of rules and procedures. As a consequence of attenuation, drift is often a sociotechnical phenomenon across levels, resembling an organisational condition of misrepresentation (Pettersen and Schulman, 2016) disposing an organisation or system to the emergence of vulnerabilities, which are also difficult for organisations to predict and design for. It also occurs both in a single organisation and across organisations (vertically and horizontally), depending on what is changing and who is making 'errors' because of it (Ibid). Overall, less research related to the organisational and systemic contributions to safety attenuation and drift has been done. Actually, the role of practical drivers may be over-explained in relation to the role of organisational or sectoral conditions. As previously discussed, conditions for attenuation involve aspects of social structure, group dynamics and the effects of power relations in or among organisations.

Taking the main points from the existence of attenuation and drift effects, we argue that there is a need to focus more attention on making the responsible organisations behave *responsibly* (Freudenburg, 2003). We need to change the focus from looking predominantly at the frontline workers and their relationships to technology, and instead direct more attention towards organisations and authorities. In today's global information society, this attention also includes looking at the role of other stakeholders, such as interest groups, media and NGOs, in setting the risk and safety agenda. In responsible organisations, top managers have an important role in keeping safety at the top of the strategic agenda, even though 'nothing happens'. Managers should pay attention to risk events with a potentially high signal value for drift (Table 8.1). In relation to risk attenuation, the fact that 'nothing happens' represents a warning signal that the organisation needs to pay attention to developing perceptions of risk, and the everyday behavioural changes that may follow. Additionally, the emergence of new and potentially catastrophic risks can displace priorities given over to existing risks. The latter demonstrates the synergies and tensions between different risk domains, such as safety and security (Pettersen and Bjørnskau, 2015). The potential for second-order effects of the risk attenuation, such as relaxing regulations or increasing productivity, may further increase the signal value of such events (non-events).

We argue that safety attenuation, understood as a sociotechnical and multilevel phenomenon, is a key concept for our times. Were it to be developed further, we could better address how processes of risk attenuation and its associated types of drift create consequences for safety outcomes. This concept, as well as the topic of this chapter, is believed to have relevance across critical sectors, such as transportation, nuclear power, the chemical industry and energy – all with tightening patterns of functional interdependencies, intensive knowledge requirements and expanding networks of cooperation and control (Le Coz,e 2017; Pettersen and Aase, 2008; Roe et al., 2002; Roe and Schulman, 2008; Goldin and Mariathasan 2015; LaPorte, 1996). Although the attenuation of risks is, in principle, unavoidable, drift can be managed by institutional permeability (Freudenburg, 2003), where information flows across levels in the organisation and internal processes are supported by regulatory oversight and inspection. This requires institutional support and managerial concern for long-term safety investments, as well as a recognition of the amount of human and organisational resources needed; it also necessarily involves a strategic shift away from focussing on the direct short term or anticipated benefits of increased production. In this sense, safety attenuation should be a relevant concept for organisations and regulators to include in their assessment and management of major accident risk.

Acknowledgements

We thank Paul Schulman for his key role in developing ideas and concepts about drift on which this research is based.

References

Aven, T. (2014). *Risk, surprises and black swans: Fundamental ideas and concepts in risk assessment and risk management.* Routledge, London.

Aven, T. and Renn, O. (2010). *Risk management and governance: Concepts, guidelines and applications.* Springer Science and Business Media, Berlin-Heidelberg, Germany.

Aven, T. and Zio, E. (2011). Some considerations on the treatment of uncertainties in risk assessment for practical decision making. *Reliability Engineering and System Safety,* Volume 96(1), 64–74.

Bjerga, T. and Aven, T. (2015). Adaptive risk management using new risk perspectives—an example from the oil and gas industry. *Reliability Engineering and System Safety,* Volume 134, 75–82.

Dekker, S. (2013). *Drift into failure.* Ashgate, London.

Farjoun, M. (2005). Organisational learning and action in the midst of safety drift: revisiting the space shuttle program's recent history. In Starbuck, W. and Farjoun, M. (Eds.), *Organisations at the Limit.* Blackwell Publishing, Malden, MA.

Farjoun, M. (2010). Beyond dualism: stability and change as a duality. *Academy of Management Review*, Volume 35(2), 202–225.

Feldman, M. and Pentland, B. (2003). Reconceptualising organisational routines as a source of flexibility and change. *Administrative Science Quarterly*, Volume 48(1), 94–118.

Fjæran, L. and Aven, T. (2010). On the link between risk perspectives and risk regulation—A comparison between two cases concerning base stations and wireless networks. *Reliability Engineering and System Safety*, Volume 95(6), 689–697.

Freudenburg, W. R. (1992). Nothing recedes like success? Risk analysis and the organisational amplifications of risk. *Issues in Health and Safety*, Volume 3(1), 1–35.

Freudenburg, W. R. (2003). Institutional failure and the organisational amplification of risks: The need for a closer look. *The Social Amplification of Risk*, Volume 102, 120.

Goldin, I. and Mariathasan, M. (2014). *The butterfly defect: How globalisation creates systemic risks, and what to do about it*. Princeton University Press, Princeton, NJ.

Heimann, L. C. F. (1997). *Acceptable risks: Politics, policy, and risky technologies*. University of Michigan Press, Ann Arbor, MI.

Kasperson, J. X., Kasperson, R. E., Pidgeon, N. and Slovic, P. (2003). The social amplification of risk: Assessing fifteen years of research and theory. *The Social Amplification of Risk*, Volume 1, 13–46.

Kasperson, R. E. (2012). The social amplification of risk and low-level radiation. *Bulletin of the Atomic Scientists*, Volume 68(3), 59–66.

Kasperson, R. E., Renn, O., Slovic, P., Brown, H. S., Emel, J., Goble, R. and Ratick, S. (1988). The social amplification of risk: A conceptual framework. *Risk Analysis*, Volume 8(2), 177–187.

LaPorte, T. R. (1996). High reliability organisations: Unlikely, demanding and at risk. *Journal of Contingencies and Crisis Management*, Volume 4(2), 60–71.

Le Coze, J. C. (2015). Reflecting on Jens Rasmussen's legacy. A strong program for a hard problem. *Safety Science*, Volume 71, 123–141.

Le Coze, J. C. (2017). Globalisation and high-risk systems. *Policy and Practice in Health and Safety*, Volume 15(1), 57–81.

Le Coze, J. C., Pettersen, K. A., Engen, O. A., Ylonen, M., Morsut, C., Skotnes, R., and Heikkila, J. (2017). *Sociotechnical systems theory and regulation of safety in high-risk industries*. White paper. Tampere: VTT Technical Research Centre 2017 (ISBN 978-951-38-8523-6).

Oltedal, H. A. and Engen, O. A. (2011). Safety management in shipping: Making sense of limited success. *Safety Science Monitor*, Volume 15(3), 1–19.

Patterson, M. D. and Wears, R. L. (2015). Resilience and precarious success. *Reliability Engineering and System Safety*, Volume 141, 45–53.

Perrow, C. (1984). *Normal accidents*. Princeton University Press, Princeton, NJ.

Pettersen, K. (2016). Understanding uncertainty: Thinking through in relation to high-risk technologies. *Routledge Handbook of Risk Studies*, Burgess, A., Alemanno, A., and Zinn, J. (Eds.). Routledge, Abingdon, New York, 39–48.

Pettersen, K. and Aase, K. (2008). Explaining safe work practices in aviation line maintenance. *Safety Science*, Volume 46(3), 510–519.

Pettersen, K. A. and Bjørnskau, T. (2015). Organisational contradictions between safety and security–Perceived challenges and ways of integrating critical infrastructure protection in civil aviation. *Safety Science*, Volume 71, 167–177.

Pettersen, K. A. and Schulman, P. R. (2014). *Realtrans theoretical framework.* Unpublished manuscript presented at the REALTRANS UC Berkeley workshop, June 2014, Berkeley, CA.

Pettersen, K. A. and Schulman, P. R. (2016). Drift, adaptation, resilience and reliability: Toward an empirical clarification. *Safety Science.*

Pidgeon, N., Kasperson, R. E. and Slovic, P. (Eds.) (2003). *The social amplification of risk.* Cambridge University Press, Cambrdige, UK.

Rasmussen, J. (1994). Risk management, adaptation, and design for safety. In Sahlin, N. E. and Brehmer, B. (Eds.), *Future risks and risk management.* Kluwer, Dordrecht, the Netherlands, 1–35.

Rasmussen, J. (1997). Risk management in a dynamic society: A modeling problem. *Safety Science,* Volume 27(2–3), 183–213.

Reason, J. (1997). *Managing the risks of organisational accidents.* Routledge, London.

Rerup, C. and Levinthal, D. (2006). Crossing an apparent chasm: Bridging mindful and less-mindful perspectives on organisational learning. *Organisation Science,* Volume 17(4), 502–513.

Renn, O., Burns, W. J., Kasperson, J. X., Kasperson, R. E. and Slovic, P. (1992). The social amplification of risk: Theoretical foundations and empirical applications. *Journal of Social Issues,* Volume 48(4), 137–160.

Roe, E. and Schulman P. (2008). *High reliability management—Operating on the edge.* Stanford University Press, Redwood City, CA.

Roe, E., Schulman, P., van Eeten, M. and de Bruijne, M. (2002). High reliability bandwidth management in large technical systems. *Journal of Public Administration Research and Theory,* Volume 15(2), 263–280.

Slovic, P., Lichtenstein, S. and Fischhoff, B. (1984). Modeling the societal impact of fatal accidents. *Management Science,* Volume 30(4), 464–474.

Snook, S. A. (2000). *Friendly fire.* Princeton University Press, Princeton, NJ.

Turner, B. A. and Pidgeon, N. F. (1997). *Man-made disasters.* Butterworth-Heinemann, Boston, MA.

Vaughan, D. (1997). *The Challenger launch decision: Risky technology, culture, and deviance at NASA.* University of Chicago Press, Chicago, IL.

Weick, K. and Sutcliffe, K. (2001). *Managing the unexpected: Assuring high performance in an age of complexity.* Jossey-Bass, San Francisco, CA.

9

Safety and the Professions: Natural or Strange Bedfellows?

Justin Waring and Simon Bishop

CONTENTS

9.1 Professions, Professionals, Professionalism.. 135
9.2 Professions, Risk and Safety 139
9.3 Managing Professional Safety 141
    9.3.1 Incident Reporting... 142
    9.3.2 Human Factors Analysis.. 143
    9.3.3 Safety Culture... 144
9.4 The Professionalisation of Safety?... 145
9.5 Conclusion 147
References.. 148

Introduction

The late sociologist Ulrich Beck (1992) not only reminded us that hazards and disasters are a universal feature of social life, but that the language of risk (and now safety) had evolved to become a prominent characteristic of contemporary society and culture. In an era of economic instability, environmental change and technological advancement, risks rarely escape public dialogue or political debate. Although mistakes and lapses in safety can happen in almost any setting or walk of life, it is often those safety incidents that happen in the most complex, specialised or advanced arenas of human activity that attract the most attention. Illustrative examples might include aviation and space travel, petrochemical and nuclear energy, the internet, major engineering projects and healthcare care (Perrow, 2011; Reason, 2016; Roberts et al., 2001; Vaughan, 1999). The organisations (and disasters) emanating from these settings have provided important insights about the sources of risk and reliability; understanding and managing risk in such settings has become a major focus for both academic work and developments in practice. What often characterises these settings is not only the complex nature of the given task or activity, but the highly specialised experts, knowledge workers or professionals who undertake and shape the organisation of these tasks (Vaughan, 1999). Take for example, Perrow's study of the Three Mile Island disaster, or Vaughan's analysis of NASA, both of which show the crucial role of highly trained engineers and technicians in the escalation of major system failure. In the health and social care sector, a series of high-profile scandals have highlighted the roles of doctors, nurses, social workers and therapists in harming, either directly or indirectly, the patients in their care (Waring, 2009). It has been estimated, for example, that as many as one in ten hospital patients experience some form of safety incident, making healthcare far more dangerous than more high-reliability sectors. As outlined in this chapter, these experts are highly heterogeneous, even with the same field of social activity, and operate within discrete socio-legal silos that exacerbate the potential for system-wide failure because of the problems of coordinating work across jurisdictional boundaries. It has often been claimed that such groups have particular expectations, demands, norms and beliefs around the organisation of their work. For example, Henry Mintzberg (1979) used the term 'professional bureaucracy' to describe a type of organisation that has a core of highly trained experts organised within discrete professional hierarchies and supported by administrative functions.

The social organisation of these experts and professionals is a significant reason why safety is both maintained and compromised. Yet, research and theory on organisational safety has almost talked past research and theory in the sociology of professions. The expertise and experience found within professional communities is crucial to recognising and resolving uncertainties and complications inherent in the performance of complex tasks (Abbott, 1988). For the most part, these complex tasks are located within organisational contexts. At the system level, professions also have dedicated regulatory procedures and strong cultures to promote safety and mitigate risk in the face of potential disasters. However, research also shows that professional status, hierarchies, boundaries and cultures can work to normalise risk or conceal potential dangers that might otherwise threaten their professional status (Vaughan, 1999). As such, analysis of organisational or system safety in these complex or specialised settings invariably requires attention to the social organisation of the experts and professionals that shape how these organisations function.

This chapter develops an introductory account of the interrelationship between organisational safety and the professions. In broad terms, professionals are a type of specialised occupation with extensive training and experience in a given area of expertise (Freidson, 1970). For sociologists, professions are significant because they have high degrees of institutionalised influence and power within the division of labour, often to the extent that they shape how specialist organisations and services work, based upon their control of expert knowledge. The first section of our chapter offers a more developed understanding of the role of professions within society.

The chapter then goes on to pose the question as to whether organisational safety and expert professions are natural or strange bedfellows. By this we simply mean: do professions always work to enhance or safeguard organisational safety, or might they actually work against organisational safety in the pursuit of maintaining their professional status. Drawing on research within health and social care, we first examine the ways professions have tended to shape the understanding of safety, and as consequence of their institutionalised power, influence how others think (or maybe do *not* think) about the safety of their own work. We also look at how professions have responded to the interventions of safety management, which through our own studies in the healthcare sector, can be seen as challenging many professions' ability to define and govern the safety of their work, and in turn, lead to a re-invention of professionalism. Finally, we ask whether the emergence of safety science and the specialist occupations of safety or risk management, might be regarded as a form of professionalisation revealing inter-professional tensions around the social definition and management of safety.

9.1 Professions, Professionals, Professionalism

The term 'profession' is widely used to denote some type of expert or specialist occupation; although the associated term 'professionalism' is used more colloquially to describe the desirable qualities of almost any occupational field. In the field of organisational studies, professions are often depicted as powerful institutional actors who can shape a given field of social activity (Scott, 2008). So, for example, the power of the medical profession is not only found in its ability to heal the sick, but more broadly to define the nature of illness and in turn to shape the organisation of services concerned with healing the sick.

To understand the special significance of the professions, and the implications for organisational safety, it is useful to draw upon theory and research found in the sociology of professions (MacDonald, 1995). As noted earlier, it is quite surprising that sociological research on the 'professions' and 'organisations' has often been developed along parallel tracks with only limited points of intersection and cross-fertilisation, despite the fact that most professions work in complex organisations. It is beyond the scope of this chapter to give proper credit to this extensive body of sociological research on the professions, but we will delineate five common sets of questions that are concerned with understanding: (i) why a certain type of occupation is distinct for others within the division of labour (i.e. what defines a profession), (ii) how have these occupations attained status and monopoly within the division of labour (i.e. professionalisation), (iii) what are the shared cultures and identities that unite members and demonstrate its virtues to society (i.e. professionalism), (iv) how do these occupations control the social organisation of work (i.e. professional power), and (v) how have institutional changes transformed their status and power (i.e. de-professionalisation).

In the early 20th century, sociological research on the liberal professions (i.e. doctors, lawyers, engineers) tended to be concerned with defining their essential characteristics or traits, such as extensive training, specialised expertise, code of conducts, licensure and high social standing (Car-Saunders and Wilson, 1964; Durkheim, 2013). Reflecting the work of Durkheim and Parsons, professions were seen as contributing to the functional progress of society through the application of expertise to tackle complex social problems. By the mid-20th century, professionals and experts were influential in almost all spheres of government, welfare, and industry, giving rise to what some have called the 'golden age' of professionalism and the ascendance of the 'professional society' (Perkins, 1990).

After this time, sociologists have developed more critical interpretations to explain not only what professions are, but how they come about, how they socialise their members, what impact they have on the division of labour and how do they maintain their boundaries (and power) within the division of labour. In contrast to earlier sociological studies that were concerned with determining what professions 'were', research from the mid-20th century became more concerned with what professionals 'did', how they 'interacted' with each other or their clients, and how these everyday practices both reflected and restated their institutionalised 'power' within society. Arguably, the most influential scholar in the field, Eliot Freidson, examined the likes of the medical profession to understand how these high-status experts shaped the social organisation of their own work, and the entire field of social activity (Freidson, 1970, 2001). Like others within the symbolic interactionist tradition, this line of enquiry brought to light the fine-grained, everyday experiences of professional work, highlighting in particular how the boundaries and privileges of professional status were cultivated and maintained through everyday interactions (MacDonald, 1995). One of Freidson's central arguments was that, by virtue of their exclusive access to (and application of) expert knowledge, professional occupations are typically accorded high degrees of 'autonomy' in the social organisation of work. At one level, this autonomy relates to a form of technical discretion, so that only the profession can define the character and standards of the work. At another level, this autonomy extends to the social organisation of the work, to the extent that professions determine who can and cannot undertake certain tasks. Significantly, Freidson describes how this autonomy is state-sanctioned or protected by law in the form of professional licensure.

> So long as a profession is free of the technical evaluation and control of other occupations in the division of labour, its lack of ultimate freedom from the state, and even its lack of control over the socio-economic terms

of work do not significantly change its essential character as a profession. (Freidson, 1970, p. 25)

In parallel with their autonomy, Freidson distinguished professions by their capacity to 'self-regulate', which stems from the assumption that only members of a profession can determine and assess the standards of conduct, because no other worker, i.e. manager or client, has the relevant knowledge to assess the intrinsic quality of the work. Self-regulation is manifest, for example, in setting the educational and moral standards of professional entry, and monitoring adherence to professional standards in everyday practice.

> The profession is the sole source of competence to recognise deviant performance and to regulate itself in general. Its autonomy is justified and tested by self-regulation. (Freidson, 1970, p. 137)

For Freidson, professionalisation (the process of acquiring state-sanctioned autonomy) is interpreted as a strategy of professional control and monopoly in the division of labour. In short, professions draw upon their expertise to both define the nature of social or technical problems, and, in doing so, determine the way services are organised to address these problems. As such, professions are highly powerful social actors.

Other scholars have developed Freidson's (1970) ideas describing, for example, professionalisation as a strategy of 'occupational closure'. Larson (1979) uses the concept of a 'professional project' to describe how occupational groups have used their claims to expert knowledge to maximise and control social and economic rewards. Research across a number of professional domains has further examined how everyday interactions help to defend and challenge professional boundaries (or who can and cannot perform certain tasks), and how these interactions confer certain privileges to those who are seen as being legitimate enough to undertake said tasks. Abbott (1988) further argues that this exclusivity can be seen as the basis of their jurisdictional boundary within the system of work. Abbott's work is significant because it suggests that professions are located within a dynamic ecosystem of work, where different occupations interact, and at times compete to provide specialist services, using their expertise (and other strategic resources) to create, extend or maintain their jurisdiction. This perspective brings to light the inter-professional dimensions of professional practice. This includes identifying how complex social problems often require multiple solutions – and therefore multiple specialists – but also how professions are formed within and defined by jurisdictional negotiation, competition and hierarchy. This critical and interpretive tradition shows how professional groups can organise themselves and their work to serve their own political interests. This does not preclude wider social benefits stemming from the process of professionalisation; the ethics.

In his later work, Freidson (2001) describes 'professionalism' as being a 'third logic' or way of organising the provision of specialist services that differs from the logic of 'bureaucracy' and the 'market'. For Freidson, professionalism has a number of inherent advantages, especially as an alternative to the dehumanising effect of bureaucracy, and the overriding pursuit of profit with markets. By the late 20th century, however, it became generally accepted that the zenith of professional power was coming to an end, with more managerial and market influences shaping the organisation of expert work. A number of prominent ideas were invoked to explain this change, including the rise of the consumer society, the information technology revolution, the growth of corporatism and financial capitalism, shifting political ideologies and the spread of managerialism (Freidson, 2001). These far reaching institutional changes have transformed the prevailing systems of professional work. In particular, the managerialisation of professional work is often associated with the growth of corporate interests and business practices in a number of 'professional service' areas, such as law, accounting and medicine (Muzio et al., 2013). This has seen large, even global, professional service firms come to replace professional partnerships, and introduce more explicit and commercial modes of organising work. Parallel changes can also be seen to affect public service professionals, such as teaching, social work, nursing and medicine, where policymakers have embraced business and management practices to make public service organisations more efficient, customer-focussed and commercially aware (Ferlie et al., 1996). For many commentators, bureaucratisation and managerialisation has challenged professional autonomy and power, with managers seeking to determine, supervise and assess both the 'content' and the 'context' of professional work, that is, what they do and how they do it. For example, performance management systems are now commonplace across professional services with managers setting and routinely monitoring the performance objectives for professional work, such as the number of legal cases won, the number of pupils with good grades, the number of patients treated, and so on.

Of interest here, is the way advances in 'safety science' play into the dynamic relationship between managers and professionals. Many of the tools and techniques developed to increase safety involve measurement and management of technical activities. This often involves treating professional groups as relatively homogenous 'technicians' and failing to recognise that such management strategies are inherently political in terms of their scope to determine the basis of professional conduct, challenge professional jurisdictions and subjugating professionals to bureaucratic oversight. At the same time, certain occupational groups have participated in the development of safety science, including involvement in evidence creation, engaging with safety discourse and promoting safety initiatives and used these new forms of 'diagnosis and treatment' to cultivate their own 'professional' status and influence. That is, the techniques or risk management become a frontier for investigating the tensions between occupational groups, as manifest in the day-to-day organisational practice. To better understand the changing relationships between safety science and the professions, it is important to look closer at the way questions of risk and safety have typically been analysed in relation to the professions.

9.2 Professions, Risk and Safety

In his seminal essay on 'mistakes at work', Everett Hughes (1951) observed that all human endeavour has a chance (risk) of mistake, and that in certain fields of activity this is more pronounced and profound. Because of this, he suggested that special relationships often form around certain social activities, which help to reduce and absorb the risk of failure, that is, the relationship between specialist and lay; where the latter delegates the risk of mistake to the former. This type of relationship is central to professional work, whereby professionals use their expert knowledge to define the possibility for risk, and offer the means for its reduction. This is usually one justification for the exclusivity of professional jurisdiction. However, Hughes points out that different social groups will have different expectations of acceptable and unacceptable risk. When an activity is delegated to an expert, mistakes are likely to be seen more severely than if the risk had been run by the lay. The specialist, however, may argue that it is only they who have the skills to perform the work; and only they who have the knowledge to define mistakes. Hughes suggests that this leads to specialists developing symbolic rituals to facilitate the identification of mistakes, and also to contain social exposure; while the specialised groups themselves develop collective rationales and defence mechanisms to control their own psychological, moral and financial risks.

The control of risk is therefore a central part of professional self-regulation, based on the assumption that only members of a given profession possess the knowledge and experience to define and assess the standards of professional work. A number of scholars have laid out the historical foundations of the 'regulatory bargain' between the state and the professions, and the formal institutions and functions of a profession's regulatory body (Ogus, 1995; Dingwall and Lewis, 1983; Salter, 2001; Waring et al., 2010). For example, most professions are governed by a formal agency that determines the standards of professional education, the technical and moral codes for professional conduct and the procedures for assessing compliance, which collectively inform the basis of qualification (entry) and disqualification (exit). In the wake of prominent scandals and inquiries into professional practice, especially in healthcare, social work and (recently) financial services, there is a growing belief that these professional bodies are not always 'fit for purpose' (Waring et al., 2010). In the UK, for example, the inquiry into the doctor Harold Shipman, who murdered over 200 of his patients, found that the regulatory body placed the interests of the profession ahead of those of the public (Smith, 2004). It often seems that these regulators work to conceal evidence of wrongdoing in order to maintain the reputation and jurisdiction of the profession. Over the last 20 years, there have been concerted efforts to make the formal mechanisms of professional regulation more transparent and inclusive of non-professional voices (Waring et al., 2010).

Following Hughes, sociologists have also looked at the more everyday practices of professional regulation in order to understand how issues of professional risk and safety are governed through a range of informal, culturally significant activities. These studies often build on the work of scholars such as Howard Becker, whose study of professional socialisation for doctors found that much of the learning relates, not only to technical knowledge and its application, but what it means to 'be' a professional in terms of shared values, norms and identities (Becker et al., 2002). For example, Arlukes (1977) study of the surgical 'death round' (a kind of audit of outcomes) found that these activities worked to normalise mistakes and protect the reputation of surgeons. This included allowing surgeons to emphasis the complexity of particular cases and the need for learning, rather than call into question the proficiency or standards of professional practice. Similarly, Bosk's (2003) study of surgical trainees shows that mistakes were constructed along three lines, where judgement and technical mistakes were remembered, but forgiven, as a 'regrettable but inevitable part of the baptism under fire that is house officer training' (Bosk, 1986, p. 466). However, normative mistakes around professional demeanour or inter-personal conduct were assessed more harshly as revealing qualities un-becoming of a professional. Some 20 years later, Rosenthal (1995) looked further at the localised practices for governing professional error and risk, finding a range of in-house activities that reinforced the notion of professional collegiality and protected the profession from external scrutiny.

Such studies repeatedly show that a prominent feature of professional regulation is an associated or underlying set of cultural beliefs and assumptions that mistakes and risks are an inevitable feature of professional work. Further, these should be contained within the profession, including the identification and management of potential wrongdoing so as not to undermine the collective reputation and standing of the profession. This means that many areas of professional practice, with important implications for both individual and organisational safety, have not received the type of inquiry or analysis that might be expected in other industries.

9.3 Managing Professional Safety

The last 30 years have seen profound changes in the social organisation of expert work, with changes in the formal regulation of professions as well as more explicit attempts to manage the risk and safety of professional conduct. A significant driver of change, especially for government policy, has been the emergence of major scandals in a number of professional service fields, from banking and finance, to health and social care. At the level of formal professional regulation, policymakers have shown greater willingness to question the principles of 'self-regulation' requiring both more transparent systems of regulation and imposing new supra-regulatory bodies to regulate the regulators (Waring et al., 2010). An example in the UK is the Professional Standards Agency to promote and oversee the standards of professional regulation. In addition, many professional bodies have introduced more continuous and developmental forms of appraisal or 'revalidation' to move away from an entry/exit model of regulation, towards one that regularly re-licences professional members. Reflecting upon these changes, it has been suggested that many professions are moving away from a mode of 'state-endorsed collegial self-regulation' to a form of 'state-enforced bureaucratic regulation' (Waring et al., 2010).

Looking closer at the level of organisational safety and risk, an array of management interventions has been introduced into professional domains to promote and assure the standards of work, in line with changing expectations of business, government and consumers. One way of framing these interventions is through Michael Power's (1997) notion of the 'audit explosion' and the 'audit society'. The audit society, Power argues, is characterised by the wide-scale adoption of accounting and auditing practices across almost all areas of social and political life. Explicit performance expectations or targets are used to direct and motivate behaviour and parallel systems of assessment are introduced to routinely monitor compliance, with these mechanisms playing a central role in the managerialisation of expert work. Significantly, the 'rituals of verification' associated with this audit culture may not only promote performance improvements aligned with corporate or policy ambitions but can also have unintended consequences and unanticipated risks. For example, the scandal surrounding Mid-Staffordshire NHS Hospital in the UK showed that substandard levels of care and even patient death could be attributed to a culture of *doing the system's business,* or prioritising targets over compassionate professional care (Francis, 2013).

It is amongst these broader changes to professional regulation that attempts to improve the management of risk and safety have become established features of most professional service organisations. In broad terms, these have been concerned with identifying, stratifying and managing the risks that impact upon, or are a consequence of, professional practice.

For many safety-critical services the influence of safety science, ergonomics and human factors on the systems of risk and safety management is apparent. As outlined in other chapters within this collection, this consists of a re-conceptualisation of safety that includes a greater appreciation of the relationship between the frontline professional practice, and the latent factors located within the organisation of professional work (Reason, 2016). It has often involved the introduction of formal risk reporting and learning systems for identifying, documenting and analysing the threats to safety. And, also, an emphasis on promoting a 'safety culture' that encourages professionals to prioritise safety; be mindful to risk, and respond positively to safety enhancing interventions (e.g. see Waterson, 2014). However, the implementation of risk and safety management systems within professional organisations can be complicated by the established cultures and customs of professional self-regulation, professional resistance to management, and the inherently complex uncertainties of professional work. To elaborate on the challenges of managing the safety of professional work, we draw upon our own research within the UK healthcare system to look at three key strands of safety management and the implications this has for professional groups: incident reporting, human factors analysis and culture management.

9.3.1 Incident Reporting

A prominent feature of the UK patient safety agenda, as with other safety critical industries, has been the introduction of incident reporting systems to routinely collect data on actual or potential safety events. As with other industries, the underlying assumption is that professionals should document and communicate their experiences of safety incidents in order to inform learning around the common and system level factors that might have contributed to a specific, and similar incidents. In the NHS, these incident reporting systems have taken a number of forms (e.g. paper-based and electronic) and have been modified to reflect the requirements of clinical specialities, with the information being collected at both individual hospital and national levels. As we will discuss, a key requirement for implementing such systems has been to tackle a supposed 'blame culture', whereby professionals are inclined to blame one another, or expect to be blamed for safety events, which invariably discourages professionals from being open and reporting their experiences (Department of Health, 2000).

Research on the introduction of these reporting systems suggests that many technical and practical factors can complicate reporting, such as the design of forms, the language used or the way they are introduced to staff (Waring, 2005). However, our research also reveals a number of more systemic barriers to the successful implementation of these systems associated in particular with the character and culture of professional work (Waring, 2005, 2009). First, healthcare professionals – especially doctors – often appear to accept, and even *expect* certain levels of risk in their work, brought about in part by the inherent uncertainty of physical illness as well as the inherent risks of treatment, such as surgery, or the side effects of medication. As such, it is often seen as unnecessary to report or signal those risks that are to be expected, and those which need to be accommodated or tolerated within professional practice. Second, many professions have well-established customs and procedures for identifying and learning from safety issues, from the informal 'corridor conversations' between professionals, to departmental audit meetings and more formal 'fitness to practice' investigations (Waring and Bishop, 2010). In contrast to formal incident reporting, these rely on (and support) professional relations of trust and hierarchy. Third, many doctors in our research show an antipathy or even hostility towards administrative tasks, considering incident reporting to be at best a bureaucratic burden, and at worst the work of managerial informers (Waring, 2005). Resistance to incident reporting seemed to stem from doctors' concerns about managerial use of incident reporting to routinely survey and monitor professional standards (specifically professional mistakes). As such, there was a concern that incident reporting would afford managers new opportunities to challenge and govern professional work. The findings of this work suggest that a variety of professional cultural factors complicate the introduction of incident reporting, over and above the concern with a blame culture.

9.3.2 Human Factors Analysis

Alongside incident reporting, many professional services have seen the introduction of more developed procedures for investigating and analysing safety events, often informed by the principles of human factors. In the UK's healthcare system, this is associated with the introduction of 'root cause analysis' investigations that embrace the principles of 'systems thinking' to uncover and foster learning around the latent factors that negatively influence the safety of professional work (Nicolini et al., 2011). This has involved the adoption of various managerial techniques, from 'risk matrices' to stratify incidents to 'timeline' and 'fish-bone' diagrams to help elaborate the interlinked factors leading up to an incident. Our research on the use of these approaches reveals the strong influence of organisational and managerial expectations about the focus and nature of learning. That is, investigations, analysis and outcomes often seem to be driven by the formal reporting requirements of statutory bodies, and, in particular, concerned with restoring a sense of organisational legitimacy in the wake of a given incident, over and above supporting positive growth, learning and change (Nicolini et al., 2011; Waring and Currie, 2009). Alongside identifying causes, incident investigations are also concerned with 'tidying up' incidents, establishing organisational defence to complaints and/or litigation and in contrast to the aspirations of policies, allocated responsibility to individuals or groups. It has been suggested that incident analysis in healthcare has not (yet) developed the types of learning and improvement expected by policies or seen in other sectors (Department of Health, 2000). The reasons for this might, in part, be due to the bureaucratic and administrative pressures placed on healthcare managers, but our research also reveals that it is often difficult for managers to investigate professional practice, with hostility towards the investigation process, and an assumption that they are used as a form of 'witch-hunt' to blame individual professionals. Again, tensions between professionals and managers complicate efforts at organisational learning. And yet, we also find that clinicians are often eager to talk about, and learn from, the challenges and mishaps experienced in their work, but typically through following more informal and collegial rituals that tend to happen at the backstage of the workplace (Waring and Bishop, 2010). The challenge presented by these systems is that they can inhibit more system-wide learning and attention to systemic factors and are often associated with protecting professional reputations (Rosenthal, 1995; Waring and Bishop, 2010).

9.3.3 Safety Culture

As noted above, a significant feature of the UK's patient-safety agenda has been the goal of replacing a 'blame culture' with a 'safety' or 'learning' culture (Department of Health, 2000; Francis, 2013). This aspiration reflects a large body of research on high-reliability organisations that show how particular cultural attributes are associated with positive safety outcomes, including mindfulness to risk and learning from safety breaches (Roberts et al., 2001). In the healthcare sector there has been much interest in transforming professional cultures, especially fostering cultures that are more compassionate, open and safety conscious (Davies et al., 2000). Attempts to change professional cultures for patient safety include, for example, educational interventions, directive checklist and guidelines and incentives (or fines) to promote and reward (or penalise) (un-)desirable behaviours (McDonald et al., 2006). What becomes apparent from these examples is that they make certain assumptions about professional cultures, that is, they are formed through learning and socialisation, that they can be re-directed through rules and standards and that they are open to extrinsic manipulation. Although all of these may have some role in shaping professional culture at the surface, this may not lead to the changes intended. Culture change initiatives are often met with high degrees of cynicism and organisational game-playing, which, in fact, support deeply embedded aspects of professional culture and identity formed through many years of socialisation and specialisation. Research suggests the use of crude behavioural incentives, that is, rewards and punishments, to shape professional culture can have negative consequences, with both managers and professionals motivated to chase targets rather than attend to the core goal of safety (McDonald et al., 2006). Equally, attempts to standardise professional practice through guidelines and checklists are seen to over-simplify the complex realities and uncertainties of professional work, and even exacerbate the potential for risks to go undetected through re-directing professional attention (McDonald et al., 2006). Strategies used to promote a safety culture for professionals therefore serve to further alienate professionals from the goals of safety or improvement, because the goals and values do not easily align with those of professionals, and instead reinforce opposition to organisational management.

9.4 The Professionalisation of Safety?

Looking across the attempts to introduce reporting and learning systems within the healthcare sector, our research highlights a number of interconnected issues related to the socio-cultural and political dimensions of professional work. As suggested above, professional groups have distinct cultures and ways of organising that reflect and reinforce political interests. A major issue for the management of professional safety and risk by non-professionals is that it exposes areas of professional practice to external scrutiny. Not only does this create the possibility for managers or others to re-organise professional work in the name of safety, but, more fundamentally, it allows non-professionals to call into question the foundations of professional legitimacy or jurisdiction. In other words, if professional jurisdiction is premised on the ability of a profession to not only define and resolve a given social problem but also to do this with regard to the safety of that activity, then evidence of professional misconduct (or poor safety practice) challenges the ability of the professional to maintain this jurisdiction. One interpretation of the challenge outlined above might be that new reporting and learning systems open up new jurisdictions for managers, rather than professionals, to determine, review and report on the standards of professional practice, albeit in a wider organisational context. This brings to light a type of inter-occupational, or inter-professional conflict over the management of risk and safety within professional services.

Our discussion therefore considers the possibility that the field of safety and risk management might itself begin in the processes of professionalisation, and that one possible explanation for the challenges of managing professional safety is that it involves a form of inter-professional conflict within the wider system or ecology of expert work (Abbott, 1988). To elaborate on this idea, we highlight a number of developments in the field of safety management that might be interpreted as contributing towards a professional project of the 'safety' profession themselves. First is the development and formation of a distinct body of expert knowledge around the field of risk and safety. Drawing on the field of safety science and human factors, for instance, there is now a coherent and distinct corpus of knowledge comprising both abstract, theoretical and conceptual precepts around the classification and understanding of safety, as well as more applied and methodological knowledge around the management of safety issues, reflected in the work of James Reason, Richard Cook, Jens Rasmussen and other leaders in the safety field cited extensively in this volume. In other words, there has been a re-definition, diagnosis and treatment of safety as a social issue, with a distinct field of technical knowledge presented as the primary means of addressing this issue. Second, and in parallel with the above, this knowledge is increasingly communicated and transmitted through formal and accredited education and training programmes, underpinning distinct qualifications and occupational designations in the area of safety and risk management. This provides a basis for social recognition or even 'license' to establish the standards of professional entry and competence, as well as to control who has access to this expertise and promotes the institutionalisation of patient safety as a profession in its own right. Third, is the formation of bodies or professional associations to maintain and promote the standards of education and practice for those working in the field of safety and risk management.

Moving beyond a descriptive interpretation, the professionalisation of safety management not only involves the designation of a new occupational specialism, but the carving out of a legitimate domain of occupational closure. In other words, organisations not only employ safety specialists to promote their safety, but they often give these specialists exclusivity or significant influence over how safety is managed, and, in doing so, limit the scope for other occupations to inform technical decisions about safety. For example, safety specialists are typically responsible for overseeing and administering organisational incident reporting and risks management procedures. Drawing on Abbott (1988), safety sciences' jurisdictional claims over 'safety' as a distinct aspect of an organisation's functioning; requiring a particular form of technical or managerial expertise, invariably creates the types of inter-professional conflict or boundary disputes seen in other professional systems. This can be seen especially where it intersects with the performance and standards of other professionals and occupations. Drawing on our own studies, we suggest that the occurrence of increasing jurisdictional battles between safety and risk management and traditional clinical professional (particularly medical) groups are occurring at multiple levels. At the level of the healthcare workplace, we find safety as a central theme in the negotiation of change at work (Bishop and Waring, 2016), with safety managers citing requirements for increasing managerial oversight and transparency in the name of quality improvement and safety, while clinical professionals focus arguments over resources and staffing. At the level of regulation and legal mechanisms, debates increasingly focus on the presence of, and compliance with, systems for managing risk and safety. More than an additional facet of managerialism, these battles can be seen as creating space for 'safety science' as an emerging professional domain.

Significantly, this focusses inter-professional conflict around arguably one of the most sensitive issues for any professional group, namely: the safety and standards of their conduct, upon which their own claims to professional legitimacy and jurisdictions are typically based. As such, the emergence of safety science as a distinct occupational jurisdiction undermines the long-held custom of professional self-regulation by casting light on areas of professional practice long concealed from lay groups, and in the process safety specialists call into question the intrinsic safety of professional practice, casting doubt on its jurisdictional legitimacy. This might therefore enable the development of a new professional jurisdiction of safety specialists at the expense of previously self-regulating professions and experts. Returning to the work of Hughes (1951), if a profession is seen by lay groups as unable to absorb and reduce the risk of mistake, then their distinct legitimacy to address these issues within the division of labour will be undermined. It is unsurprising therefore that the emergence of safety science as a specialist occupational field, and its development within established professional service domains, can be the site of inter-professional and organisational conflict where questions of culture and power complicate the promotion of safety.

9.5 Conclusion

Through introducing and summarising key debates within the sociology of professions, this chapter aims to open up new thinking about the relationships between professions and organisational safety. As described, professionals typically work in organisational contexts, and represent an important source of expertise, competence and skill through which safe working practices are organised and delivered. We also suggest that, by virtue of the many types of ambiguities and complex tasks professionals address, issues of risk and safety are an inherent feature of professional work. Taking a sociological perspective offers additional insight about the relationship between knowledge, risk and power; where experts are often accorded high degrees of status or power because of their ability to manage risk and uncertainty based upon their control of expert knowledge. It also suggests that the emergence of new specialists in the field of risk and risk management could challenge the jurisdictions of established experts and professionals, as new ways of analysing and managing risk are introduced into the organisation of work. Drawing upon our own research, we show how healthcare professions, who have traditionally defined the understandings and regulation of safety in their work, now face the advent of new clinical risk and patient safety specialists; in turn, this has questioned the status and power of professionals. These new safety specialists appear to be challenging the jurisdictions of established healthcare professionals through questioning the very essence of professional knowledge and practice: a profession's ability to determine and manage uncertainty and risk.

References

Abbott, A. (1988). *The system of professions: An essay on the division of labor.* Chicago, IL: University of Chicago Press.

Arluke, A. (1977). Social control rituals in medicine: The case of death rounds. *Health care and health knowledge.* London: Croom Helm, 10725.

Beck, U. (1992). *Risk Society: Towards a new modernity.* London: Sage.

Becker, H. S. (Ed.). (2002). *Boys in white: Student culture in medical school.* Transaction Publishers.

Bishop, S. and Waring, J. (2016). Becoming hybrid: The negotiated order on the front line of public–private partnerships. *Human Relations,* Volume 69(10), 1937–1958.

Bosk, C. L. (1986). Professional responsibility and medical error. *Applications of Social Science to Clinical Medicine and Health Policy,* 460–477.

Bosk, C. L. (2003). *Forgive and remember: Managing medical failure.* Chicago, IL: University of Chicago Press.

Carr-Saunders, A. M. and Wilson, P. A. (1964). *The professions.* Frank Cass.

Davies, H. T., Nutley, S. M. and Mannion, R. (2000). Organisational culture and quality of health care. *Quality and Safety in Health Care,* Volume 9(2), 111–119.

Department of Health (2000). *An organisation with a memory.* London: TSO.

Dingwall, R. and Lewis, P. S. C. (1983). *The sociology of the professions: Lawyers, doctors and others.* New York: Macmillan; St Martin's Press.

Durkheim, E. (2013). *Professional ethics and civic morals.* London, UK: Routledge.

Ferlie, E. (1996). *The new public management in action.* Oxford: Oxford University Press.

Francis, R. (2013). *The report of the Mid Staffordshire NHS Foundation Trust public inquiry.* London: TSO.

Freidson, E. (1970). *Profession of medicine: A study of the sociology of applied knowledge.* Chicago, IL: University of Chicago Press.

Freidson, E. (2001). *Professionalism, the third logic: On the practice of knowledge.* Chicago, IL: University of Chicago Press.

Hughes, E. C. (1951). Mistakes at work. *Canadian Journal of Economics and Political Science/Revue canadienne de économiques et science politique,* Volume 17(3), 320–327.

Larson, M. S. (1979). *The rise of professionalism: A sociological analysis,* (Volume 233). University of California Press.

Macdonald, K. M. (1995). *The sociology of the professions.* Sage.

McDonald, R., Waring, J. and Harrison, S. (2006). Rules, safety and the narrativisation of identity: A hospital operating theatre case study. *Sociology of Health and Illness,* Volume 28(2), 178–202.

Mintzberg, H. (1979). The professional bureaucracy. *Organisation and Governance in Higher Education,* 50–70.

Muzio, D., Brock, D. M. and Suddaby, R. (2013). Professions and institutional change: Towards an institutionalist sociology of the professions. *Journal of Management Studies*, Volume 50(5), 699–721.

Nicolini, D., Waring, J. and Mengis, J. (2011). Policy and practice in the use of root cause analysis to investigate clinical adverse events: Mind the gap. *Social Science and Medicine*, Volume 73(2), 217–225.

Ogus, A. (1995). Rethinking self-regulation. *Oxford Journal of Legal Studies*, Volume 15(1), 97–108.

Perkin, H. (1990). *The rise of professional society: England since 1880*. Psychology Press.

Perrow, C. (2011). *Normal accidents: Living with high-risk technologies*. Princeton University Press.

Power, M. (1997). *The audit society: Rituals of verification*. Oxford: OUP.

Reason, J. (2016). *Managing the risks of organisational accidents*. Routledge.

Roberts, K. H., Bea, R. and Bartles, D. L. (2001). Must accidents happen? Lessons from high-reliability organisations. *Academy of Management Executive*, Volume 15(3), 70–78.

Rosenthal, M. M. (1995). *The incompetent doctor: Behind closed doors*. Open University Press.

Salter, B. (2001). Who rules? The new politics of medical regulation. *Social Science and Medicine*, Volume 52(6), 871–883.

Scott, W. R. (2008). Lords of the dance: Professionals as institutional agents. *Organisation Studies*, Volume 29(2), 219–238.

Smith, J. (2004). *The Shipman Inquiry—Fifth report: Safeguarding patients: Lessons from the past—proposals for the future*. London: The Stationery Office.

Vaughan, D. (1999). The dark side of organisations: Mistake, misconduct, and disaster. *Annual Review of Sociology*, Volume 25(1), 271–305.

Waring, J. J. (2005). Beyond blame: Cultural barriers to medical incident reporting. *Social Science and Medicine*, Volume 60(9), 1927–1935.

Waring, J. J. (2009). Constructing and re-constructing narratives of patient safety. *Social Science and Medicine*, Volume 69(12), 1722–1731.

Waring, J. J., and Bishop, S. (2010). "Water cooler" learning: Knowledge sharing at the clinical "backstage" and its contribution to patient safety. *Journal of Health Organisation and Management*, Volume 24(4), 325–342.

Waring, J. and Currie, G. (2009). Managing expert knowledge: Organisational challenges and managerial futures for the UK medical profession. *Organisation Studies*, Volume 30(7), 755–778.

Waring, J., Dixon-Woods, M. and Yeung, K. (2010). Modernising medical regulation: Where are we now? *Journal of Health Organisation and Management*, Volume 24(6), 540–555.

Waterson, P. (Ed.). (2014). *Patient safety culture: Theory, methods and application*. Ashgate.

10

Visualising Safety

Jean-Christophe Le Coze

CONTENTS

10.1 A Very Short 'Safety Tour' of the 1980s–1990s 152
    10.1.1 High-Risk Systems 152
    10.1.2 Sociotechnical Systems 154
    10.1.3 Safety Models 155
    10.1.4 Cognitive and Social Properties of Enduring Visualisations 157
        10.1.4.1 Cognitive Properties 157
        10.1.4.2 Social Properties 159
10.2 A Safety Tour in the 2010s 161
    10.2.1 High-Risk Systems: Updating Perrow's Matrix 162
    10.2.2 Sociotechnical View: Challenging Rasmussen's
        Hierarchical View 163
    10.2.3 Safety Models: Characterising Rasmussen's Boundaries 164
10.3 Discussion 167
10.4 Conclusion 168
References 169

Introduction

In safety research, one finds a great number of drawings supporting the conceptualisation of many topics investigated in this domain. The aim of this chapter is to argue that visualisations have played and continue to play a central role.* Literacy, media, visual and science studies, or interaction, artefact and information design have introduced and discussed the importance of drawings, graphics or pictures, formulated in expressions such as *visual literacy* or *visualisation* in the past 30 years. It is an immense field, with its bestsellers (e.g. *Information Is Beautiful*, McCandless, 2009) growing in their

---

* In this text, I will use the words 'drawing', 'visualisation', 'picture', 'graphics', 'visual image', 'artefact' or 'representation', which all have to be understood as being distinguished from writings, texts or mental images (Mitchell, 2015). Of course, writings and texts *are* visual, but their logic is nevertheless different from drawings (more about this later).

importance within the context of big data (Lupton, 2015). In safety research, some works have explored the importance of visualising. For instance, as in cognitive engineering, stemming from the need to design computer interfaces (e.g. Bennett and Flach, 2011); in ethnographic studies, based on sociomaterial sensitivities (e.g. Haavik, 2011), but also in the field of graphic design, when retrospectively commenting on engineering decisions that led to disasters (e.g. Tufte, 1997, 2006). In this chapter, I want to discuss a complementary topic targeting the conceptual developments of safety research. Indeed, topics such as causality, accident, sociotechnical systems or cognition have all been investigated while relying very much on drawings.

The first part consists of tracing a brief history of the field of safety on the basis of a selection of a restricted number of drawings. This illustrates just how much can be conceptualised and synthesised graphically, illustrating the role that visualisations have played, and will continue to play, in this domain. The second part follows the same pattern of introducing pictures, but this time, while presenting new propositions of graphical models that address and translate empirical and conceptual contemporary trends in safety research such as constructivism, complexity, strategy or globalisation. While borrowing insights in a literature as diverse as literacy, media and science studies; or interaction, artefact and information design, this article supports the idea that visualisation will continue to play an important role.

10.1 A Very Short 'Safety Tour' of the 1980s–1990s

One way to convince of the importance of visualisation is to propose a very brief tour of safety research. Such a simplified tour can easily be organised on the basis of a selection of drawings that have structured the field. Without exaggerating, one could say that, because of the enduring influence of some of these drawings, safety as a research domain is as much a product of these powerful visualisations as it is of its texts (Le Coze, 2016). I select four images, one on the identification and comparison of high-risk systems, another on the sociotechnical dimension, and two safety models.

10.1.1 High-Risk Systems

One can start with Perrow's classification of high-risk systems in 1984. Perrow's book had several merits (Perrow, 1999). One was to single out some organisations for their specific potential externalities: technological disasters. The rich text of Perrow was accompanied with a matrix that summarised, but also further conceptualised, his argument. Of course, some would argue that it is difficult to prove or to assert anything without empirical support; how influential this figure has been (likewise for any of the following drawings

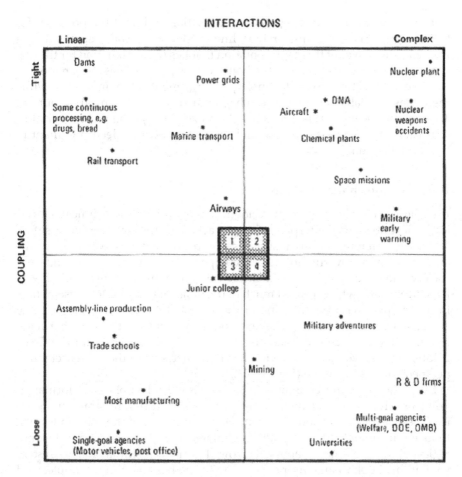

INTERACTIONS

FIGURE 10.1
Perrow's matrix of high-risk systems. (With permission from Princeton University Press.)

I have chosen in this chapter), but it is undeniable that it translated much of what Perrow has achieved with his text beyond words (Figure 10.1).

The top right quadrant is dedicated to systemic accident proneness (normal accident) with nuclear plants and weapons, space missions or aircraft because of their intersection of complexity and tight coupling. Not only did this picture contribute for the first time to an identification of high-risk systems (e.g. dams, marine transport, rail transport, mining, chemical plants, nuclear plants and airways), it also classified them according to their potential for unexpected surprises: normal accidents. It brought something more than the text could. It grouped together the high-risk systems, which had first been analysed in the book and organised them in relation to each other, perfectly illustrating the spatial properties that only drawings can afford to authors.

It did so while introducing other organisations without the potential for technological risks (e.g. universities, line production). And, perhaps it is in the heuristic power of this representation that lies the explanation for the reason why the understanding of the book *Normal Accident* has been restricted for a very long time to its technological argument (Hopkins, 2001), while there was available, chapter after chapter, a full sociotechnical account of accidents underlying this argument (Le Coze, 2015a). The technological or structural argument of coupling and interactions was indeed dominant in this visualisation.

10.1.2 Sociotechnical Systems

Although Perrow's analysis of accident was entirely sociotechnical, empirically describing the interplay of many actors in the production of safety for many different, high-risk systems (e.g. professional associations, insurance, legal and government actors, company executives, unions and civil society), he never graphically illustrated this dimension of his work. It was Jens Rasmussen who captured much of the imagination of safety researchers in this respect with his drawing of the sociotechnical view. Incrementally designed over the 1980s–1990s as far as one can infer from the available material; inspired by a control theoretic and cybernetics background (Le Coze, 2015b), over the years this drawing has remained one of the most acclaimed perspectives on this issue (Figure 10.2).

The principle of this column is that safety is a product of how information flows up and down between levels representing different strata of societies. These levels correspond to scientific disciplines, and the column is open to its environment, such as political climate, changing market and technological change. As noted above, Perrow described a similar macro-system view in his book in an empirical way (with the help of many examples and case analyses), but it was a narrative analysis that was not rendered as a drawing. And, that so much can be grasped about sociotechnical systems through Rasmussen's view illustrates just how well thinking is a result of creating, supporting or developing ideas graphically. Meaning is available without much text and it is very quickly perceived. Rasmussen was an artful designer of striking visual heuristics in this field, something to be linked with his engineering training (Kant, 2017).

It is an aspect of this researcher that also applied to his contribution to our understanding of human error, cognition and accident analysis. His approach to error was a naturalistic one, emphasising the trade-offs one operates when confronted with complex tasks. Indeed, despite much effort dedicated (most of the time) to the design of work situations by engineers and managers, in real circumstances, operators have had to adapt in order to meet the constraints of both the context (which are never fully anticipated by designers) and the objectives of their activities.

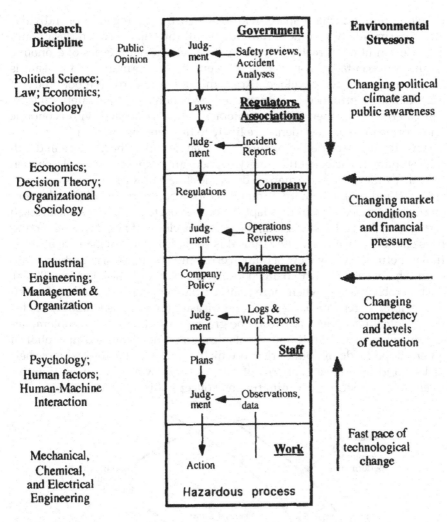

FIGURE 10.2
Rasmussen's sociotechnical view. (With permission from Elsevier, Rasmussen, 1997.)

10.1.3 Safety Models

It is on this basis that Rasmussen created a model based on the principle that individuals exhibit a degree of freedom when performing their daily routines, but also when meeting unexpected and unplanned problems. When a very high number of actors behave in adaptive ways at different times and in different locations, the possibility of migration beyond the boundaries of safe performance is likely to happen under specific and distributed circumstances. This idea is materialised as a series of curved lines defining an envelope representing a space, within which the aggregation of individuals' experiments (when these individuals fill the gaps between design and real-life conditions) creates a dynamic between the boundaries of economic failure, unacceptable workload and acceptable performance. To be safe is to remain within the envelope. Excess migration towards the boundary of acceptable performance is counteracted by actions such as 'safety culture' campaigns. When they fail to counteract trends associated with economic and workload issues, accidents are likely to happen (Figure 10.3).

What Rasmussen tried to picture was what he called the 'defence in depth fallacy', namely the belief that safety barriers are permanently in place when in reality they are in a permanent dynamic state. For Rasmussen, safety barriers depend on the activity of many actors who are likely, due to the constraints described above, to adapt beyond expected plans. As Rasmussen wrote, 'if a shortcut is successful, the person is clever, if not, it is an error and he is blamed' (Rasmussen, 1987). It is this very idea that provided the intellectual impetus for a new view of human error and reliability of cognition – concepts which were, in turn, changed into a notion of resilience (Le Coze, 2017a).

The problem arises when more than one barrier evolves according to such a principle, thereby creating a path for disaster (as visualised above). As for the sociotechnical view, much can be grasped intuitively when contemplating this elegant, dynamic and simple drawing. The visualised metaphor of an envelope made of boundaries in which a number of actors explore their tasks – and then create patterns of interactions – works very well to intuitively convey a strong, meaningful and conceptualised view of safety.

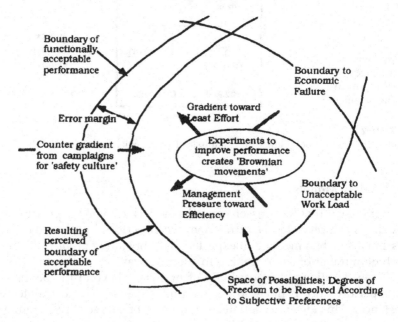

FIGURE 10.3
Rasmussen's safety model. (With permission from Elsevier, Rasmussen, 1997.)

Behind this contribution was in fact an analytical and graphical alternative to Reason's highly popular view of accidents through what later became known as the Swiss cheese model (Reason, 1990a). In the period of reflection following the technical disasters of the 1980s (Bhopal, Chernobyl, Piper Alpha, *Challenger*, etc.) it became apparent that the defence-in-depth concept was an important one; Reason had ruminated over the subject of catastrophe beyond human error in the 1980s with the help of sociological contributions from Turner or Perrow. The model consists of a series of basic elements: decision-makers, line management, preconditions, productive activities and finally defences. They represent a series of 'planes', where 'several factors are required to create a "trajectory of opportunity" through these multiple defences' (Reason, 1990b, 33). This initial contribution has been followed by many variations of the model over the years, but the basic graphical ideas have remained the same (Figure 10.4). As with Rasmussen, the drawing efficiently brings together simple shapes of rectangles, holes and an arrow, relying on a perspective that materialises a sense of depth in relation to the social and engineering layers and prevention practices.

10.1.4 Cognitive and Social Properties of Enduring Visualisations

As shown, many core problems of safety research (comparing high-risk systems, grasping sociotechnical systems or conceptualising safety) have been addressed with the help of drawings. In this respect, I want to emphasise two sets of properties which lie behind their appeal and success. One concerning their cognitive value; the other, their social one (Le Coze, 2016). Although it is best to think about these properties as 'sociocognitive' rather than separately, this distinction is maintained here.

10.1.4.1 Cognitive Properties

From a cognitive point of view, drawings are artistic, aesthetic artefacts. They are constructed, and often based on a combination of metaphors to offer both generic and normative perspectives. First, from a cognitive artistic (or aesthetic) perspective, and, contrary to what many would probably be inclined

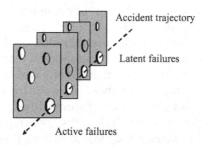

FIGURE 10.4
Defence-in-depth model. (Adapted from Reason, 1997.)

to think, visualisations do not 'simplify' reality so much as actively provide their own conceptual lenses. As Lynch, a science sociologist, wrote about observation practices in science which rely on visualisations, 'the notion of simplification is too simple, and it glosses over features of transformational practices which, when examined in their own right, do not seem to be matters of only filtering or selection' (Lynch, 1990, p. 160).

Visual practices are constructive. They imply creative and imaginative observers and designers who bring their own distinctive angle to construct phenomena through a picture. This is vividly captured by Tufte, a theorist of graphic design: 'Those who discover an explanation are often those who construct its representation' (Tufte, 1997, p. 9). It seems obvious here with the visualisations introduced above that they do not translate reality *'as it is'* (Glasersfeld, 1995), but actively construct useful or suggestive ways of thinking about complex processes with the help of graphics.

Second, the selected drawings above associate on paper (in quite a restricted space) an incredibly vast amount of these complex processes, whose description practically requires much empirical work to be translated into text (within a traditional ethnographic mode). But, by substituting writing with drawing, authors can avoid the serial, sequential or linear properties of texts, and instead produce abstractions, which are easily and almost immediately understandable as wholes.

Kress, a literacy academic, has made this point extremely clearly, contrasting the logic of writing with the logic of drawing: 'The organisation of writing (...) is governed by the logic of time, and by the logic of sequence of its elements in time, in temporally governed arrangements. The organisation of the image, by contrast, is governed by the logic of space and by the logic of simultaneity of its visual/depicted elements in spatially organised arrangements' (2003, p. 2). The effect of this logic is actually well known, as the following popular expressions convey: 'a picture is worth a thousand words', or, 'a good sketch is better than a long discourse' (an expression attributed to Napoleon).

In this respect, one is entitled to wonder what came first... a text rendered into a drawing, or a drawing which was then explained by a text? Moreover, this question is not at all trivial. It is a rather deep epistemological one. A very similar conundrum divides mathematicians, physicists, historians and philosophers on the relationship between mathematics and pictures. 'Einstein's relativity theory itself could be better understood in geometric-pictorial terms', said Galison, a historian of science (2002, p. 304). What is the relationship between mathematical formulas and drawings? Which comes first? Are mathematical formulas more scientific and rational than drawings?

Note also, and this is my third point, that many would be tempted to think that the representations introduced above are quite unsophisticated, as they are only in black and white, composed of simple lines and shapes, and combined with a small amount of text. Today, designers equipped with computers can produce much more elaborate visualisations than these, particularly within the context of big data (Meirelles, 2015). However, by imaginatively and creatively indicating classification principles (Figure 10.1), structuring specific kinds of relationships between several dimensions (Figure 10.2), characterising safety patterns or movements (Figure 10.3) or event trajectory (Figure 10.4), they exhibit this property of overcoming the restrictions of what Tufte defines as *'flatland'* through metaphors (e.g. individual behaviour within an envelope made of boundaries, Figure 10.3; an arrow through holes in barriers, Figure 10.4, see Le Coze, 2013).

'The world is complex, dynamic, multidimensional; the paper is static, flat. How are we to represent the rich visual world of experience on mere flatland?' (Tufte, 1990, p. 9). How, first, to identify them to comparatory high-risk systems? How to represent sociotechnical realities? How to think about safety as a complex dynamic? All of these important questions are formulated with the help of graphical artefacts. A fourth aspect that contributes to their relevance for safety research, as well as their enduring success, is that they combine both generic and normative properties. The classification of Perrow, the sociotechnical view, or the safety model of Rasmussen and Reason can be applied to (m)any safety critical systems and offer assessment principles.

10.1.4.2 *Social Properties*

From a social perspective, these representations are (empowering) inscriptions, boundary and (value-laden) performative objects, notions which are derived from Science and Technology Studies (STS). First, they are inscriptions because they allow scientists to frame what it is to be thinking about a specific phenomenon; here, high-risk and sociotechnical systems, but also safety. By circulating among peers, these inscriptions help stabilise a research network agreeing to recognise in these drawings some of the properties of the objects under study. In this respect, I believe that these visual artefacts have played a considerable role in establishing safety as an independent field of investigation – even while there are so many other existing disciplines (e.g. engineering, psychology, sociology, etc.) contributing to it. Swuste's history of safety is also strongly structured by an approach of visualisations (2016).

Therefore, these inscriptions can also be seen as empowering their users because of the way in which they allow them to make things controllable; as scientists are describing phenomena in a way that can then be used concretely by a range of different actors. Latour refers to the past to illustrate this. When maps of the world were established by Europeans, they allowed them to master ocean routes (Latour, 1987). Drawings in this respect can be seen as reductions of the world – in the good sense of the term – that capture essential features of the world and thus empower their users. Moreover, when the Swiss cheese visually advocates that 'human errors' leading to disasters are only the manifestations or symptoms of latent conditions created by strategic actors of companies, they divert and limit blame away from these front-line actors.

In this respect, they incorporate politics. They belong to these value-laden artefacts that promote a certain view of/for the social world (Winner, 1980). In Perrow's case, nuclear power plants are situated in the accident-prone top right corner of his 2 × 2 matrix, which implies that societies would be better off abandoning these technologies. The same can be said of Rasmussen's view of sociotechnical systems, in which he puts the government at the top, above private companies, translating his political view of the relationship between the public and private sectors. And the Swiss cheese is also a socio-political statement about the unfair and unrealistic simplification that 'human error' represents when it comes to understanding accident and finding a culprit.

Second, the success of these representations is an indication of the effect that boundary objects have, in particular their interpretive flexibility (Star, 2010; Trompette and Vinck, 2010). This property allows people with different backgrounds, experiences and expertise to collaborate. Many of these drawings have been central to the bridge between practitioners and researchers; one naturally thinks, for instance, of the lasting success of Reason's model. Not everyone will make sense of the drawings in the same way, but they have the ability to channel discussions and debates and serve as a basis to elaborate and imagine solutions to problems.

It is in this context that one understands the critics of the use of the Swiss cheese model as a tool for accident investigation in aviation (e.g. Shorrock et al., 2004). These drawings are indeed also performative; namely, they orient actions. The argument of Shorrock et al., is indeed that by paying attention to remote slices and their holes, as implied by the Swiss cheese model (namely by moving away from more proximal problems such as pilot cognition), investigators could miss areas for improvements. This is an example of criticisms that illustrate the concrete relationship that is established between the world, visualisations and actions. By challenging the practical implications of the Swiss cheese visualisation, Shorrock et al. refer directly to its performative dimension.

Enduring visualisations in safety research are therefore constructive, artistic (or aesthetic), generic, normative, metaphorical, but they also exhibit those properties defined as (empowering) inscription, boundary; they are (value-laden) performative objects. Table 10.1 summarises these points. Unpacking the multifaceted properties of such artefacts as developed in this chapter reveals some of the reasons behind their sustained influence over the years, as well as their pervasiveness in scientific publications in the field of safety.

Of course, to come back to the earlier warning of this chapter, a history of safety through few representations is an immensely simplified account of 30 to 40 years of scientific production. There are some other examples of graphic renderings of high-risks systems with different principles; there are many other cases of sociotechnical views, and there are also other safety/accident models. Some of these will be mentioned again to attenuate the selective aspect of this chapter, but this very brief retrospective serves one purpose, namely: to insist on the centrality of drawing complex phenomena.

TABLE 10.1

Properties (Cognitive and Social) of Enduring Visualisations in Safety Research

- **Constructive**: These drawings are the products of creative and imaginative observers and designers who innovate to depict objects not really in an objective way, but in a constructivist one (reality does not 'correspond' to the pictures) implying a historically situated observer/designer
- **Metaphorical**: These pictures rely in this respect on metaphors available at the time to provide enlightening perspectives on complex processes which contribute to how appealing these pictures are
- **Artistic (or aesthetic)**: These visualisations are produced through choices of diverse graphic principles and picture composition (shape, colour, size, perspective, proportion, movement, etc.) which contribute to their success
- **Simultaneous and spatial**: Contrary to writing which is linear, visualisation provides an immediate grasp of wholes on the basis of design, metaphorical and artistic choices in a specific spatial mode
- **Generic**: These drawings can be applied across high-risk systems which offer a very wide potential use providing the general character of model that science very often expects
- **Normative**: These graphical representations offer principles to assess safety by explicitly indicating dimensions to be considered
- **(Empowering) inscription**: These visual artefacts constitute the materialised support to frame and sustain a conceptualisation among peers creating networks of users and discriminating scientific objects (i.e. safety), empowering their users
- **Boundary object**: These graphic objects offer opportunities for actors of different backgrounds, experiences and perspectives (and in particular researchers–practitioners) to communicate, coordinate and cooperate on the interpretation of complex phenomena thanks to the interpretive flexibility of visualisations
- **Performative**: When used to interpret issues, these visualisations translate into concrete actions and practices, and contribute to enact a world
- **Value-laden artefacts**: Visualisation incorporates politics; they are not value-laden artefacts but rather translate and convey a socio-political perspective of the object under study

There are also topics not covered in this chapter that have also been clearly supported by images, for instance, human error and cognition, or the problem of causality in accidents. Because of space restrictions, it is not possible to fully explore the other, already existing representations of similar issues or topics, which are not illustrated here, and I now turn to alternative propositions.

10.2 A Safety Tour in the 2010s

In this second part, three new visualisations are proposed in order to question or challenge, thirty years later, our approach to the topics of high-risk and sociotechnical systems and safety models in the current context. They are not to be seen as a substitute for outdated ones, but rather as alternative views reflecting contemporary changes, trends or issues. These visualisations incorporate both conceptual and empirical aspects. The argument is that certain key notions such as networks, complexity, constructivism, strategy, globalisation or systemic and existential risks must be part of the drawings; supporting current developments and challenges in the field of safety. These three propositions are now presented in turn.

10.2.1 High-Risk Systems: Updating Perrow's Matrix

The first proposition offers a different perspective on the identification and comparison of high-risk systems and disasters. It considers new risk categories that have emerged in the past 30 years (namely systemic and existential risks), and puts them together in a structured space based on two dimensions other than interactivity (linear/complex), and coupling (loose/ linear) (Le Coze, 2018). In fact, although Perrow subsequently produced one of the first books on what became the category of systemic risks, considering together natural, terrorist but also cyber risks along with technological ones (Perrow, 2007), he did not visually update his earlier matrix accordingly.

The other options for identifying and comparing some high-risk systems suffer from similar limitations (Schulman, 1993; Amalberti, 2005; Hollnagel, 2009), namely that they do not expand the range of issues in the background of high-risk systems over the past 20 years. Take, for instance, the systemic risks associated with globalisation and its consequences (OECD, 2003). Increased flows of people, goods, money and information has shaped a new world of risks; of systemic risks (Beck, 1996; Goldin and Mariathasan, 2014). This notion has become a standard way of addressing the risks associated with globalisation in the aftermath of a series of catastrophes in the early 21st century due to terrorism or natural events (such as tsunami and hurricanes), but also due to pandemics or finance.

Existential risks (Bostrom and Cicovik, 2008) are connected to concepts such as the anthropocene and transhumanism. Because of the extent to which the planet is transformed by human activities as framed by the concept of anthropocene, the chance of societal collapse is higher than ever due to potential ecological disasters. Because of the new technological prospects of synthetic biology, nanotechnology and artificial intelligence, the possibilities of extreme adverse consequences for humanity as a whole, are, according to many authors, realistically plausible. But a comet, an asteroid or a supervolcano eruption would do as well. These are existential risks.

These categories of risks established in the last 10 to 20 years (in parallel with the more familiar notion of technological disasters) create an intensity continuum, ranging from major to extreme risks. And, because of the globalised nature of many of these risks, the notion of a scale of governance becomes an important category when considering the issue of preventing such risks. As a consequence, an alternative proposition to Perrow's matrix of risk is illustrated below (Figure 10.5).

Note that interactivity (linear or complex) or coupling (tight or loose) are no longer the core dimensions. First, major accidents happen whether or not

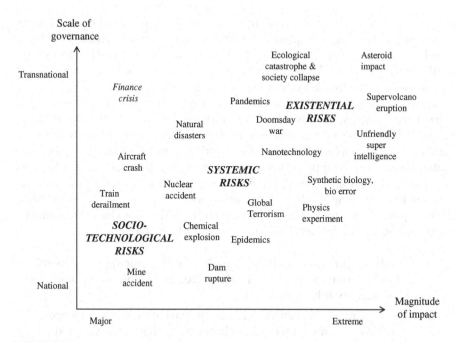

FIGURE 10.5
Categories of risks in relation to intensity and scale of governance. (With permission from Elsevier, Le Coze, 2018.)

systems are tightly coupled or complex. Second, they have to be considered as a result of globalised processes in relation to systemic and existential risks. There are indeed complex interactions between these different categories (which are not explored here for space reasons, see Le Coze for a discussion [2018]).

10.2.2 Sociotechnical View: Challenging Rasmussen's Hierarchical View

The same can be said about Rasmussen's sociotechnical system view. Although a very popular one, the view has not been updated despite much challenging conceptual and empirical trends in relation to its core features. In this respect, it is interesting to note that his sociotechnical view shares principles with many other views (e.g. Pauchant and Mitroff, 1992; Moray, 1994; Evan and Manion, 2003). First, one always finds a decomposition of dimensions distinguishing technology, work, staff or team/group, organisation and management and government. Rasmussen is therefore not the only one to imply this kind of nested and hierarchical approach to the topic, an orientation generally derived from a system view. But to conceptualise and draw this nested nature of sociotechnical systems, different kinds of geometric figures have been used, such as circles, squares, rectangles or triangles. They exploit the properties of these shapes to represent the idea of layers or levels that are more or less hierarchically structured, nested or embedded. Superposed rectangles, squares inside squares, triangles divided by lines, circles in circles: several possibilities have been experimented with graphically (see Figure 10.6). Each of these combinations of shapes has been used to represent sociotechnical systems by one or more authors.

One proposition is to move away visually from these kinds of hierarchical structures to stress certain aspects of our current understanding of sociotechnical systems that rely on contemporary notions such as constructivism, network and complexity. These three notions are important as they indicate two alternative approaches to safety: first, a horizontal, distributed and circular (rather than vertical, hierarchical and linear) one. Second, an anti-dualist (challenging objectivity and anthropocentrism) one. Based on these notions, the proposition is to proceed with the following visual transformations, using Hollnagel's cyclical model as a building block:

1. 'Socialising and materialising' the view by introducing the principle that individuals, artefacts and nature are interacting (anti-dualism – introducing non-humans).
2. Representing the observers/scientists as part of the system/networks (anti-dualism – introducing the observers in their observations).
3. Dropping the top-down (vertical) representation in favour of a horizontal 'polycentric' or 'acentric' view and showing the complexity of the self-organised – negative–positive feedback loops – dynamic of phenomena in a context of globalisation (thinking horizontally and in circularity).

The figure can be read in many different ways, whether one starts with the scientists, or with the process operators, technology or civil society (Figure 10.7).

10.2.3 Safety Models: Characterising Rasmussen's Boundaries

A similar line of reasoning is followed with Rasmussen's migration model (Figure 10.3) or Reason's defence-in-depth model (Figure 10.4). While for the comparison of high-risk systems (Figure 10.1) one had to introduce other

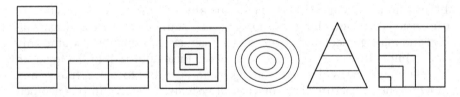

FIGURE 10.6
Several combinations of geometric figures to represent sociotechnical systems.

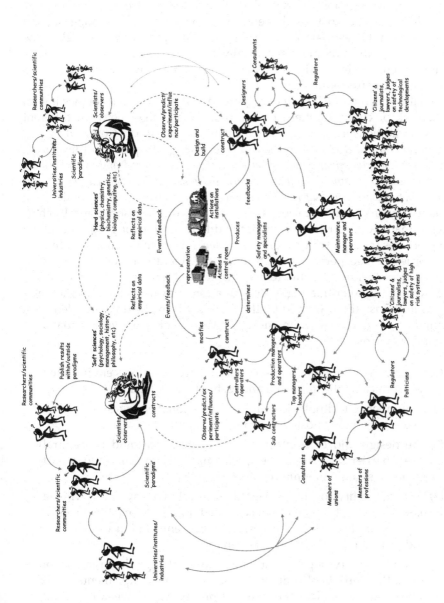

FIGURE 10.7

Socio-natural-technical system (SNTS). (With permission from Elsevier, Le Coze, 2018.)

categories of risks (Figure 10.5), and for the sociotechnical view (Figure 10.2) one had to consider new important core notions (Figure 10.6), for the safety model, one has only to introduce new analytical insights. By capitalising on decades of investigations of technological disasters (Le Coze, 2010; Hopkins, 2012), it is possible to visually refine what was powerfully suggested in Rasmussen and Reason's drawings, but unspecified from certain angles. The following visualisation more specifically characterises the place of strategy by top executives in relation to other actors populating safety critical systems and their environment.

It retains much of the appeal of a migration of practices, but couples it with a feedback or circular pattern on the basis of a merger between different research traditions in safety management and sociology (Le Coze, 2013, 2016). Key notions are introduced in the drawing with the help of short texts such as, *'strategy adaptation in organisation environment', 'technical and organisational change', 'status of technical and human barriers', 'signals and whistle-blowers', 'influence of safety department', 'safety internal and external oversights'.* These notions are dynamically linked in several loops of interactions that help sociologically characterise (namely, through the activities of different actors or groups of actors) the processes through which high-risk systems manage to remain within the boundaries of safe performance, through the interactions of a diversity of actors across hierarchies, differentiations, specialisations and organisations (Figure 10.8) while considering strategy as a core dimension (Le Coze, 2019b).

This idea, visually conveyed by the drawing, is explicitly associated with a number of sentences commenting systemic, constructive, circular and dynamic patterns:

(1) Strategy adaptations (by leaders) in the organisation's environment (economical, political, social and technological) lead to

(2) A number of technological and organisational changes at different levels, which may positively or negatively affect

(3) The design and/or implementation of (technical and procedural) safety barriers by those at the operational level (in teams and departments), a situation monitored and controlled by

(4) First, an ability to process signals (possibly conveyed by 'whistle-blowers') about specific safety-related problems or the negative impacts of changes to the design or implementation of (technical and procedural) safety barriers, relying on

(5) Second, a safety department which can challenge the organisation about the impacts of changes to the design and implementation of safety barriers and/or the status of processing of (weak or strong) signals. This department is backed up by

(6) Third, safety (external or internal) reviews which can play a role of 'organisational redundancy' for the internal safety department (or service) on these very same issues.

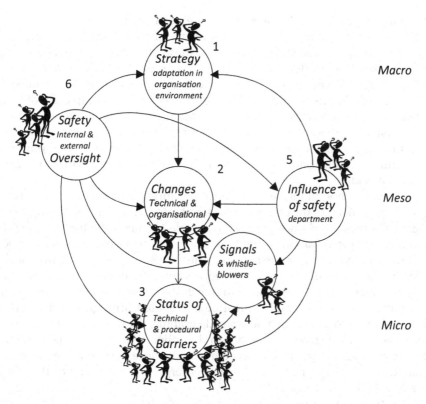

FIGURE 10.8
A systemic and dynamic sensitising model of safety (SDSMS). (With permission from Elsevier, Le Coze, 2013.)

From the perspective of the underlying feedback or circular pattern of this visualisation, one should distinguish (1) and (2) as the actions that translate into specific consequences represented in (3) and (4), which, in turn, require reactions (5) and (6) to remain within the boundaries of safe performance. This simple loop of actions-consequences-reactions is, in fact, highly complex in reality, for the reason indicated above.

10.3 Discussion

These three visualisations seek to contribute to a renewal of how safety is approached through their cognitive potentialities defined in this chapter as constructivist, metaphorical, artistic (or aesthetic), normative and generic, but also through their social potentialities described as (empowering) inscriptions, boundary and performative, value-laden artefacts (Table 10.1).

By trying to graphically incorporate several ideas that structure mindsets in safety research such as globalisation, systemic and existential risks, strategy, constructivism, complexity or networks, these alternative pictures propose to accompany the production of knowledge in this field as done in the 1980s.

Replacing missing aspects in previous drawings such as visualising reflexive individuals within feedback loops to stress the complexity of their interactions within sociomaterial networks spread across time and space, these visualisations insist on features not represented as such thus far. But of course, these new drawings cannot be substitutes for texts that offer the possibility of refining and producing analytical developments that drawings cannot provide.

Moreover, although the new visualisations introduced in this chapter have already been used in the context of personal safety interventions in high-risk organisations, they do not pretend to be able to match the heuristic power and quality of the drawings that they suggest to replace; they remain modest propositions. Meeting the requirements of visual quality that Tufte identifies in the following quote is indeed a strong challenge for anyone, and only few pictures have been able to match these expectations. 'A sure sign of a puzzle is that the graphic must be interpreted through a verbal rather than a visual process (...) in a non-puzzle graphic, the translation of visual to verbal is quickly learned, automatic and implicit – so that the visual image flows right through the verbal decoder initially necessary to understand the graphic. As Paul Valery wrote, "seeing is forgetting the name of the thing one sees"' (Tufte, 1983, p. 153).

What these new propositions do most of all therefore is to make a point about the need to keep on drawing and visualising the complex problems associated with the prevention of disasters, while keeping track of conceptual and empirical changes in safety research. Not only does this prove to be intellectually stimulating and challenging, it helps to interact with a diversity of actors including actors in safety-critical organisations.

10.4 Conclusion

The contention of this chapter is that visualisations are central in safety research. As shown in literacy and science studies or artefact and information design, visualisations generate and enact. They generate specific arguments by conveying meaning through their combined metaphorical and aesthetic sides. They enact in the sense that they provide support for action, and bind different social actors together in more or less powerful networks. Complex issues such as identifying and comparing high-risk systems, conceptualising sociotechnical systems and capturing safety dynamics have all been supported by drawings.

Their existence has strongly participated in the maturing of safety as an independent field of intellectual developments and practices, establishing bridges between researchers and practitioners. In this chapter, the selection of some of these important drawings produced during the 1980s was the opportunity to contrast them with alternative ones that reflect contemporary trends such as globalisation, and more explicitly incorporate important concepts such as complexity, constructivism or strategy. While clearly maintaining a link with past visualisations by explicitly modifying some of their characteristics, these new propositions would also like to sensitise readers, researchers and practitioners to new options of framing safety.

References

Amalberti, R., Auroy, Y., Berwick, D. and Barach, P. (2005). Five system barriers to achieving ultrasafe health care. *Annals of Internal Medicine*, Volume 142(9): 756–764.

Beck, U. (1996). World risk society as cosmopolitan society? Ecological questions in a framework of manufactured uncertainties. *Theory, Culture and Society*, Volume 13(4): 1–32.

Bennett, K. and Flach, J. M. (2011). *Display and interface design. Subtle science, exact art.* Boca Raton, FL: CRC Press/Taylor & Francis Group.

Bostrom, N. and Ćirković, M. M. (eds.) (2008). *Global catastrophic risks.* Oxford: Oxford University Press.

Evan, M. W. and Manion, M. (2002). *Minding the machines: Preventing technological disasters.* Prentice Hall, Upper Saddle River, NJ.

Galison, P. (2002). Images scatter into data. Data gather into images, in Latour, B., Weibel, P (eds) *Iconoclash: Beyond the Image Wars in Science, Religion and Art,* Karlsruhe: MIT Press and ZKM, pp. 300–323.

Goldin, I. and Mariathasan, M. (2014). *The butterfly defect: How globalisation creates systemic risks, and what to do about it.* Princeton, NJ: Princeton University Press.

Haavik, T. (2011). On components and relations in sociotechnical systems. *Journal of Contingencies and Crisis Management*, Volume 19(2): 99–109.

Hollnagel, E. (2009). *The ETTO principle: Why things that go right sometimes go wrong.* Farnham, UK: Ashgate.

Hopkins, A. (2001). Was Three Mile Island a normal accident? *Journal of Contingencies and Crisis Management*, Volume 9(2): 65–72.

Hopkins, A. (2012). *Disastrous decisions: The human and organisational causes of the Gulf of Mexico blowout.* Sydney, Australia: CCH.

Kant, V. (2017). Supporting the human life-raft in confronting the juggernaut of technology: Jens Rasmussen, 1961–1986. *Applied Ergonomics*, Volume 59: 570–580.

Kress, G. (2003). *Literacy in the new media age.* London: Routledge.

Latour, B. (1987). *Science in action.* Cambridge, MA: Harvard University Press.

Le Coze, J.C. (2010). Accident in a French dynamite factory: An example of an organisational investigation. *Safety Science*, Volume 48: 80–90.

Le Coze, J.C. (2013). New models for new times. An anti-dualist move. *Safety Science*, Volume 59: 200–218.

Le Coze, J.C. (2015a). 1984–2014. Normal accident. Was Charles Perrow right for the wrong reasons? *Journal of Contingencies and Crisis Management*, Volume 23(4): 275–286.

Le Coze, J.C. (2015b). Reflecting on Jens Rasmussen's legacy, a strong program for a hard problem. *Safety Science*, Volume 71: 123–141.

Le Coze, J.C. (2016). Trente ans d'accidents. Le nouveau visage des risques sociotechnologiques. Toulouse, France: Octarès.

Le Coze, J.C. (2017). Globalization and high risk systems. *Policy and Practice in Health and Safety*, Volume 15(1): 57–81.

Le Coze, J.C. (2018). An essay: Societal safety and the global 1,2,3. *Safety Science*, Volume 110: 23–30.

Le Coze J.C, (2019a). Vive la diversité! High reliability organisations (HRO) silience engineering (RE). *Safety Science*, Volume 117: 469–478.

Le Coze, J.C. (2019b). Safety as strategy. Mistake, failure and fiasco in high-risk systems. *Safety Science*, Volume 116: 259–274.

Lupton, D. (2015). *Digital sociology: An introduction*. Abingdon: Routledge.

Lynch, M. (1990). in Lynch, M., Woolgar, S. (eds.) *Representation in scientific practice*. Cambridge, MA: MIT Press.

Mc Candless, D. (2009). *Information is beautiful* (new edition). London: Collin.

Meirelles, I. (2015). *Design of information. An introduction to the histories, theories, and best practices behind effective information visualisations*. Osceola, WI: Rockport Publishers.

Mitchell, W.J.T. (2015). Image Science. Iconology, Visual Culture and Media Aesthetics Chicago, IL: University of Chicago Press.

Moray, N. (1994). Error reduction as a systems problem. In M. Bogner (Ed.), *Human error in medicine* (pp. 67–91). Hillsdale, NJ: Lawrence Erlbaum Associates.

Pauchant, T. and Mitroff, I. (1992). *Transforming the crisis-prone organization: Preventing individual, organizational, and environmental tragedies*. San Francisco, CA: Jossey-Bass.

Perrow, C. (1999). *Normal accident. Living in high risk technology*. Princeton, NJ: Princeton University Press.

Perrow, C. (2007). *The next catastrophe. Reducing our vulnerabilities to natural, industrial and terrorist disasters*. Princeton, NJ: Princeton University Press.

Rasmussen, J. (1987). Approaches to the control of the effects of human error on chemical plant safety, CCPS/AIChE. *International Symposium on Preventing Major Chemical Accidents*.

Rasmussen, J. (1997). Risk management in a dynamic society: A modeling problem. *Safety Science*, Volume 27: 183–213.

Reason, J. (1990a). The contribution of latent human failures to the breakdown of complex systems. *Philosophical transactions of the Royal Society of London B.*, Volume 327: 475–484.

Reason, J. (1990b). *Human error*. Cambridge: Cambridge University Press.

Reason, J. (1997). Managing the risk of organisational accidents. Aldershot, Hampshire, England: Ashgate.

Schulman, P. (1993). The analysis of high reliability organizations: A comparative framework. In K. Roberts (Ed.), *New challenges in understanding organisations* (pp. 33–54). New York: Macmillan.

Shorrock, S., Young, M., Faulkner, J. (2004). Who moved my (Swiss) cheese? The (R) evolution of human factors in transport safety investigation. ISASI 2004. *Human Factors in Investigations*, Volume 28(3): 10–13.

Star, S.L. (2010). Ceci n'est pas un objet frontière. Réflexions sur l'origine d'un concept. *Revue d'anthroppologie des connaissances*, Volume 4(1): 18–35.

Swuste, P. (2016). Is big data risk assessment a novelty? *Safety and Reliability*, Volume 36(3): 134–152.

Trompette, P. and Vinck, D. (2009). Retour sur la notion d'objet frontière. *Revue d'anthropologie des connaissances*, Volume 3(1).

Tufte, E.R. (1983). *The visual display of quantitative information*. Cheshire, CT: Graphics Press.

Tufte, E.R. (1990). *Envisioning information*. Cheshire, CT: Graphics Press.

Tufte, E.R. (1997). *Visual explanations: Images and quantities, evidence and narrative*. Cheshire, CT: Graphics Press.

Tufte, E.R. (2006). *Beautiful evidence*. Cheshire, CT: Graphics Press.

Winner, L. (1980). *Do artifacts have politics? Daedalus*, Volume 109(1): 121–136.

Von Glasersfeld, E. (1995). *Radical constructivism. A way of knowing and learning*. London: Taylor & Francis.

11

The Discursive Effects of Safety Science

Johan Bergström

CONTENTS

11.1 Case One: Resilience Engineering.. 174
    11.1.1 Discursive Effects of Resilience Engineering 176
11.2 Case Two: Just Culture 178
11.3 A Critique of the Engineering Discourse.. 181
    11.3.1 The Need for a Descriptive Rather Than Normative Agenda 181
    11.3.2 Theoretic Language Suited (and Not Suited) for Discourse.... 182
11.4 Concluding Remarks.. 183
References.. 183

Introduction

Safety Science is striking an interesting balance between, on the one hand, the pragmatic desire to improve safety management practices and performance, and, on the other, to critically analyse the nature of high-risk organisational life. In this chapter I will argue that the desire to introduce academic notions into organisational practices and regulatory frameworks will have discursive effects; discursive effects that will perhaps appear as unintended consequences to the academic community. I will also argue that rather than adding to a moral language of how accidents happen because of the failure to follow a particular theory, there is a need for safety science to step up its role as a critical school of thought dedicated to shedding light on dimensions of organisational discourse, ethics and power; dimensions often influenced by safety scientists themselves.

First, my Foucauldian-inspired view on discourse is the set of rules determining what language is possible and impossible to use for a certain discussion (in this case, relating to organisational safety and risk) in a limited time and space. According to Foucault, a discursive object of knowledge is not the result of a scientific discovery; not a piece of an a priori existing puzzle of reality waiting to be put together. Instead, Foucault argues, objects of knowledge are established as a result of historical processes, which (in a limited time and space) shape the conditions of possibility for certain objects of knowledge (e.g. resilience or safety culture) to determine what language is possible and what language is impossible to think, speak and write in a given (safety) discourse (Bergström, 2017; Foucault, 1981, 2002; Henriqson, Schuler, van Winsen and Dekker, 2014; van Winsen, Henriqson, Schuler and Dekker, 2014).

There are several examples of the discursive interplay between safety science and high-risk organisational practice, i.e. of how academic concepts have been introduced as discursive objects of knowledge to use to think, talk about and do organisational safety management. These include, but are not limited to, the theory of High-Reliability Organisations (Rochlin, La Porte and Roberts, 1987; Weick and Sutcliffe, 2007), which has evolved into a management concept marketed by academic institutions as well as organisational consultants; the notion of safety culture, which is today sold (again, by academics as well as consultancy firms) as the normative organisational homogeneity promising the full safety commitment by all organisational levels (Antonsen, 2009; Henriqson et al., 2014; Silbey, 2009); more recently Just Culture, by its discursive journey from the academic literature (Dekker, 2007; Reason, 1997), via the management consultancy business[*] and into the regulation of occurrence reporting in European aviation through the new EASA regulation 376/2014,[†] and Resilience Engineering, which for several years has formed a community dedicated to changing organisational safety practices. My argument is, again, that such discursive journeys will come with discursive effects, and in this chapter I will analyse in greater detail two of the cases introduced above: Resilience Engineering and Just Culture, to make the point that such discursive effects might take the form of a moral language that disempowers precisely those who the theories seem to try to empower at first glance.

11.1 Case One: Resilience Engineering

Over the last 15 years, resilience has made its way from being a concept merely used in several academic disciplines (including material physics and engineering, psychology, health, and ecology) (Alexander, 2013; de Bruijne, Boin and van Eeten, 2010) to an established object of knowledge in a wide safety and security discourse. Such discourse ranges from the micro to the macro; from human to societal resilience (Bergström and Dekker, 2014), and in between we find the rather recently established academic interest into the resilience of high-risk organisations (Braithwaite, Wears and Hollnagel, 2017;

---

[*] Perhaps most successfully by David Marx at Autcome Engenuity: https://www.outcome-eng.com

[†] https://www.easa.europa.eu/document-library/regulations/regulation-eu-no-3762014

Hollnagel, 2011; Hollnagel, Braithwaite and Wears, 2013; Hollnagel, Nemeth and Dekker, 2008; Hollnagel, Nemeth and Dekker, 2016; Hollnagel, Woods and Leveson, 2006; Nemeth and Herrera, 2015; Nemeth and Hollnagel, 2014; Wears, Hollnagel and Braithwaite, 2015), which will be the focus of this position chapter.

Scholars of Rasmussen's sociotechnical safety science, rooted in a tradition of *Cognitive Engineering* (Le Coze, 2015a, 2015b), see high-risk organisations as complex systems, which are constantly adapting to balance multiple goal conflicts. Such complex systems are inherently vulnerable to 'drift toward failure as defences erode in the face of production pressure' (D D Woods, 2003). It was with these words that David Woods, in his 2003 *Testimony on the future of NASA to the Senate Committee on Commerce, Science and Transportation*, introduced Resilience Engineering as a school of thought dedicated to 'help organisations maintain high safety despite production pressure' (p. 2). In his testimony, Woods connected the problem(s) of complexity as they played out in the case of *Columbia*, to potential solutions offered by focussing on system resilience.

In October 2004; one year after his testimony to the U.S. Senate, David Woods gathered, together with around 20 prominent safety scholars, for a first symposium dedicated to the topic of Resilience Engineering in Söderköping, Sweden. The symposium resulted in the first volume in an ongoing series of edited books on the topic of Resilience Engineering. The perhaps most informative read for any student interested in the core ideas embedded in the formation of Resilience Engineering as a school of thought is Woods and Hollnagel's prologue (Woods and Hollnagel, 2006) to the first edited Ashgate volume (Hollnagel et al., 2006). In no more than six pages, Woods and Hollnagel make the argument for the need for Resilience Engineering:

> Resilience engineering is a paradigm for safety management that focusses on how to help people cope with complexity under pressure to achieve success. It strongly contrasts with what is typical today – a paradigm of tabulating error as if it were a thing, followed by interventions to reduce this count. A resilient organisation treats safety as a core value, not a commodity that can be counted. Indeed, safety shows itself only by the events that do not happen! Rather than view past success as a reason to ramp down investments, such organisations continue to invest in anticipating the changing potential for failure because they appreciate that their knowledge of the gaps is imperfect and that their environment constantly changes. One measure of resilience is therefore the ability to create foresight – to anticipate the changing shape of risk, before failure and harm occurs. (p. 6)

In this short paragraph, Woods and Hollnagel manage to express a dedication, a frustration, a new view, a path and even a hint on how to measure resilience. Resilience Engineers, they argue, are dedicated to *help* people and organisations cope with complexity (the classic problem for Rasmussian scholars). Just as Woods argued in his testimony to the U.S. Senate, complexity is a source of risk. However, instead of drawing the conclusion that such complexity then needs to be avoided, or its associated risks accepted, 'Resilience Engineering ... represents the optimist stance and its agenda is to develop ways to control or manage a system's adaptive capacities based on empirical evidence' (Woods and Branlat, 2011). In short, the promise made by Resilience Engineering scholars is that resilience can be just that: engineered into organisational structures and processes.

Since its first symposium Resilience Engineering has grown, becoming established as a field of thought in safety science. In June 2017, the seventh symposium was held in Liege, Belgium. When writing this chapter, eight edited books (typically including refined versions of a selected number of symposium proceedings) have been published in the Resilience Engineering series and in 2015 the academic journal, *Reliability Engineering and System Safety* published the first special issue on the topic of Resilience Engineering (Nemeth and Herrera, 2015), which further marks its status as a now established research discipline in safety science. While the field itself seems surprisingly uninterested in referencing how it distinguishes itself or coincides with seemingly similar schools of thought (Hopkins, 2013), other scholars are now starting to make that connection (Haavik, Antonsen, Rosness and Hale, 2016; Le Coze, 2016; Pettersen and Schulman, 2016).

11.1.1 Discursive Effects of Resilience Engineering

The most apparent discursive effect of Resilience Engineering is the recent (2015) introduction of resilience into the EASA regulatory framework for a so-called Crew Resource Management (CRM) training of airline pilots. Introduced after a number of spectacular aviation disasters in the 1970s and 1980s, CRM training seeks to improve cockpit collaboration and make the best use of all possible resources for problem solving (Helmreich, Merritt and Wilhelm, 1999). The programme, promising improved information management, communication, leadership, situational awareness and decision making has, since its introduction to aviation, been spread to numerous other high-risk industries in a way which has even been labelled as a sort of 'aviation imperialism' (Dekker, 2008).

Early on the Resilience Engineering community made references to teamwork training and improvements, a process of which I have myself taken part (Bergström, Henriqson and Dahlström, 2011; Dekker and Lundström, 2006; Furniss, Back, Blandford, Hildebrandt and Broberg, 2011; Gomes, Borges, Huber and Carvalho, 2014; Lundberg and Rankin, 2013; Saurin, Wachs, Righi and Henriqson, 2013), so it should perhaps be seen as a great success that some ten years after introducing the notion of resilience to a safety science community, it was enacted into the regulatory framework for aviation team-training.

Figure 11.1 provides an extract from pages 13 and 14 of the EASA document *Annex II to ED Decision 2015-022-R**, introducing Resilience Engineering into the CRM regulatory framework:

Before going to the more analytical discussion on the potentially discursive effects of this regulatory change, some initial observations can be made.

GM5 ORO.FC.115 Crew resource management (CRM) training
RESILIENCE DEVELOPMENT

    (a) The main aspects of resilience development can be described as the ability to:

        (1) learn ('knowing what has happened');
        (2) monitor ('knowing what to look for');
        (3) anticipate ('finding out and knowing what to expect'); and
        (4) respond ('knowing what to do and being capable of doing it').

    (b) Operational safety is a continuous process of evaluation of and adjustment to existing and future conditions. In this context, and following the description in (a), resilience development involves an ongoing and adaptable process including situation assessment, self-review, decision and action. Training in resilience development enables crew members to draw the right conclusions from both positive and negative experiences. Based on those experiences, crew members are better prepared to maintain or create safety margins by adapting to dynamic complex situations.

    (c) The training topics in (f)(3) of AMC1 ORO.FC.115 are to be understood as follows:

        (1) Mental flexibility

            (i) The phrase 'understand that mental flexibility is necessary to recognise critical changes' means that crew members are prepared to respond to situations for which there is no set procedure.
            (ii) The phrase 'reflect on their judgement and adjust it to the unique situation' means that crew members learn to review their judgement based on the unique characteristics of the given circumstances.
            (iii) The phrase 'avoid fixed prejudices and over-reliance on standard solutions' means that crew members learn to update solutions and standard response sets, which have been formed on prior knowledge.
            (iv) The phrase 'remain open to changing assumptions and perceptions' means that crew members constantly monitor the situation, and are prepared to adjust their understanding of the evolving conditions.

        (2) Performance adaptation

            (i) The phrase 'mitigate frozen behaviours, overreactions and inappropriate hesitation' means that crew members correct improper actions with a balanced response.
            (ii) The phrase 'adjust actions to current conditions' means that crew members' responses are in accordance with the actual situation.

FIGURE 11.1

Extract from pages 13 and 14 of EASA's Annex II to ED Decision 2015-022-R.

---

* The entire decision with its annexes is available from: https://www.easa.europa.eu/doc ument-library/agency-decisions/ed-decision-2015022r

First, the regulatory formulations make clear that operational resilience is located at the level of pilot mental processes (mental flexibility) and behaviour (performance adaptation). Second, that this operational resilience can be improved through training. Third, there is an interesting use of normative language which is open to interpretation. Examples include, in the same sentence, the use of words such as 'overreactions', 'inappropriate hesitation', 'improper action' and 'balanced response'. This is a behaviourist approach to resilience with little support in academic traditions, except for the recognition that adaptive strategies (work as it is actually done) often take place beyond written procedures (work as it is imagined) (Patterson, Cook, Woods and Render, 2006).

Crew Resource Management training has the unfortunate history of seeing its training categories turned into moral categories of accountability demanding. In numerous studies Dekker and his colleagues have shown how categories traditionally used in CRM training including situational awareness, complacency, decision-making and even just CRM itself as a broad category, have been used to describe the failures (loss of situational awareness, poor decision-making, breakdown of CRM) of pilots within the causal constructions of accidents (Dekker and Hollnagel, 2004; Dekker, Nyce, van Winsen and Henriqson, 2010; Dekker and Woods, 2002; van Winsen et al., 2014). I fear that the notion of resilience might yet become an additional category that holds pilots morally accountable for their actions.

As said above, the language by which resilience is introduced to the EASA regulations is highly normative. It adds new categories of how pilots might exceed the moral boundaries of their profession, including an *over-reliance* on standard procedures, *overreactions*, *inappropriate* hesitation, *improper* action and *unbalanced* responses. Discursively; rather than empowering pilots to influence their discretionary space, resilience then provides means to further responsibilise pilots for the behaviour of the system in which they are embedded. The language responsibilises pilots not only for the performance of the system, but for their beliefs in resilience theory. Accidents are avoided by acting in accordance with resilience theory (by properly adapting to changing conditions, not over-relying on standard procedures and responding in a balanced manner and not being inappropriately hesitant) and accidents then become the ultimate punishment for not believing in the theory. Several safety theories have been advocated for in this way, including resilience in the case of the Fukushima nuclear disaster (Hollnagel and Fujita, 2013).

11.2 Case Two: Just Culture

The notion of Just Culture was introduced by Reason (1997), as one of the four components of a given Safety Culture. Reason argued that, by agreeing on a set of principles for drawing the line between acceptable and unacceptable actions, a just culture can be 'engineered' into the management of high-risk organisations. Following Reason's argument, management practices to do just that (including definitions and decision algorithms) started to get developed (e.g. Marx, 2001). A couple of years later, a critique was formulated by Dekker (2007) arguing that Just Culture should not be about a backward-looking drawing of the line between acceptable and unacceptable actions in order to determine when to retributively punish behaviour, but rather, it should be about a forward-looking restoration of relations between all involved actors. In his 2007 book, Dekker asks not where the line between acceptable and unacceptable behaviour should be drawn, but who should draw it. Dekker later developed his view to incorporate Braithwaite's ideas of justice as a restorative, rather than a retributive process (Dekker and Breakey, 2016; Woodlock, 2017).

Interestingly; while Reason and Dekker are in clear disagreement on 'what' a Just Culture is and 'how' they might engineer one, both authors have contributed to the legitimisation of Just Culture as an academic object of knowledge for safety practitioners and regulators to interpret and use. And use it they have – especially within healthcare* and aviation. As with the case of resilience above, the notion of Just Culture has made its way into the regulatory framework governing European aviation. In the EU Regulation 376/2014 on occurrence reporting in aviation, Just Culture is introduced as: 'a culture in which front-line operators or other persons are not punished for actions, omissions or decisions taken by them that are commensurate with their experience and training, but in which gross negligence, wilful violations and destructive acts are not tolerated' (Article 1, paragraph 12).

Since law makers are now incorporating the concept of Just Culture into their legal framework for occurrence reporting in European aviation, Dekker's question of 'who' gets to draw the line seems to have been settled – at least in the European aviation domain. The discursive effects of the academic term Just Culture is the above legal definition, which, in future court cases, will be referred to by judges in their drawing of the line between acceptable and unacceptable behaviour. If Just Culture was safety scientists' and practitioners' way to try and keep the judiciary out by establishing their own procedures for exercising organisational justice and accountability, things are now different. Just Culture is no longer for organisations to engineer. Instead, the law has claimed the authority to make the call. Perhaps it did so because it was threatened by an apparent desire of the academic community's to establish justice procedures for themselves (based on the

---

* U.S. healthcare organisations are the typical target costumers for Just Culture tools and programmes, such as https://www.outcome-eng.com/just-culture-training/. For the American Nurses Association's position statement on the notion of Just Culture, please visit: http://nursingworld.org/psjustculture

observation that the judicial system supposedly has a lack of understanding when it came to such high-risk work) (Wachter and Pronovost, 2009). Interestingly, while Just Culture is often introduced as a way to empower sharp-end staff to report even their own errors without fear of retribution, the discursive effect of safety science's contribution has rather been an empowering of the judicial system's authority to make the call. And, as Woodlock (a sociologist of law) concludes: sharp-end staff are well aware of who gets to draw the line. In his interview-studies of how Swedish aircraft maintenance engineers perceive the European regulation on occurrence reporting, he concludes that:

> Beyond social group cohesion, what emerges from … (the results) … is that EASA governance is experienced as a coercive force and is described as an asymmetrical relationship of power. The safety focussed regulations are perceived as instruments of force – as a disciplinary power – primarily serving lawmakers. Complete subordination to the rules is experienced … as an expectation of EASA. This clearly suggests an experience of over-persuasion as the 'strategy' for securing AME compliance. This is not perceived … as consensual or promoting dialogic engagement for nurturing trust. Rather, it is understood as an explicit exercise of power, which in turn generates anxiety. (Woodlock, 2017, p. 71)

There is one more implication of this. Not only does regulating Just Culture policy hold the risk of empowering the actors of the judiciary system (law makers, lawyers, judges) while disempowering sharp-end staff; what it also seems to do is disempower the academic community to gain much influence on the matter. While the academic community has provided a term (Just Culture), conflicted or not in its academic meanings, and given it academic legitimacy; the law has now claimed its legal meaning to govern its legal use.

The legal appropriation of the two terms discussed in this chapter (Resilience Engineering and Just Culture) are discursive consequences of academics' efforts to see their concepts influence safety management practices. There is no doubt that introducing the notion of resilience into the EASA training regulation and Just Culture into the regulation on occurrence reporting, has taken tremendous efforts and has all been done with the best of intentions to inform the business with contemporary safety research concepts. However, introducing resilience into the training curriculum will have discursive effects beyond the theoretical ideas of resilience that were espoused back in 2005; effects that hold the risk of further responsibilising sharp-end operators for the safety and risks of airline operations. Furthermore, the legal definition of Just Culture does not empower organisations to establish procedures for justice and accountability for the judicial system to keep out of, but rather for the legal system to claim the power to draw the line.

11.3 A Critique of the Engineering Discourse

11.3.1 The Need for a Descriptive Rather Than Normative Agenda

Academic institutions – and their scientists – have a unique role in that they can take the position of the critical observer. Rather than coming up with safety management answers, programmes or solutions, safety scientists can ask challenging questions and point out the discursive effects of current organisational practices (including the configuration of power relations, drifting system behaviour and how organisations solve ethical conflicts). Indeed, such questions can have policy implications, but *engineering* is not their primary aim. When the focus of resilience was introduced to Ecosystem Science, it was founded as the curious study of how ecosystems both absorb and adapt to stress (Holling, 1973), which set the stage for future studies of biodiversity, phase-shifts, systemic memory and micro-level revolts. In its later socio-ecological development this branch of resilience research seems tempted to go normative (Gunderson, 2010), but in safety science, I argue that we still have to stay descriptively curious rather than normatively dedicated to *engineer* resilience or justice.

In order to stay true to its roots in complexity theory, resilience research needs to avoid the reductionist trap of locating system resilience at the level of adaptive, sharp-end staff. The complexity theory's focus is how the different system levels interact and ask questions such as how/why/when do brittle systems require resilient behaviour and how/why/when does resilient behaviour preserve brittle system structures? These questions are rooted in the interactions and relations *between* organisational levels, rather than the behaviour *at* organisational levels (Bergström and Dekker, 2014). In a descriptive and critical research agenda, sharp-end staff should not be asked to rely on their mental flexibility and adaptive talents in order to stay resilient, or to simply trust their organisation to treat them in a non-punitive way following the submission of an occurrence report. Instead, questions should be raised whether (and how) the system provides the adaptive resources and capacities (typically discussed in the Rasmussian school as a 'discretionary space' or 'margin of manoeuvre' (Cook and Rasmussen, 2005; Dekker, 2012; Woods and Branlat, 2011)) needed for the sharp-end staff to adapt to the situations and challenges that they encounter in their daily lives. Resilience and justice would then not primarily be a matter of training targeted at mental processes and motivation (e.g. to report), but of the logics of organisational work design and structures of organisational power.

While the underlying political/moral agenda of the normative application of resilience carries discursive effects (which I think should be avoided), I still find the descriptive focus, after the 2004 Söderköping Symposium summarised as a 'confused consensus' (Dekker, 2006), highly interesting and very relevant for future research. It includes the modelling of drift, studying the gap between work as it is imagined and work as it is actually done, and how organisations respond to the failures they encounter. Similarly, a curious perspective to organisational justice can be a critical research approach to the capillary and hierarchical relations of power governing organisational life in high-risk organisations; including the willingness to report concern, learning processes outside of the formalised bureaucracy and rituals of ensuring justice. An obvious future research focus here is how the change of the rules of the game through the EU Regulation 376/2014 changes (if at all) how the game is actually played.

11.3.2 Theoretic Language Suited (and Not Suited) for Discourse

Just as discourse governs what language is possible to use in speech and writing, it also silences other language. Above, I have already introduced how academic concepts have occasionally made their way into regulatory discourse; perhaps through the promise of being able to 'engineer' resilience (D. D. Woods, 2003) and Just Culture (Reason, 1997) into organisations. There is path dependency to the notion of engineering in both constructs. The Rasmussian school leading up to Resilience Engineering is founded in the school of cognitive engineering; in the 1980s it was dedicated to the design of human–machine interfaces (Le Coze, 2015b). On the other hand, Reason developed a theory of safety management largely based on some of the classic engineering principles of industrial design (Larouzee and Guarnieri, 2015). With both Resilience Engineering and Just Culture carrying a normative language of how organisations 'ought' to be engineered, perhaps the enactment of their language into regulatory regimes should be seen as the ultimate success story of discursive effects? My warning here is that such discursive effects might change power relations in ways that safety scientists neither expect nor desire.

Not only does discourse enable language; it also silences language. So, what academic body of knowledge on safety is not enacted into regulatory discourse? Indeed, it seems like less optimistic and more sociological (rather than engineering) studies of the mechanisms leading up to disasters are the ones that do not make much of an impact beyond the academic community. Vaughan's study of the *Challenger* disaster (Vaughan, 1996) is one such example. Her study shows how a 'culture of production' in combination with a 'structural secrecy' led to the 'normalisation of deviance' up to the point which NASA lost track of the differentiation between conducting experiments, and high-risk operations. The 'normalisation of deviance' is similar to Turner's theory of the 'incubation period' (Turner and Pidgeon, 1997), or Snook's notion of 'practical drift' (Snook, 2000). All these theories of the organisational dynamics precipitating disaster – even though they are clearly rooted in case studies of how large-scale organisations lose control of, and track over, their safety management processes – seem to be noticeably absent in the language used in the discourses of safety management and regulation. The same is true for Perrow's work on how catastrophic potential is inherent in structures of technological complexity and organisational/societal power, rather than in any manager's or operator's motivation (Perrow, 1999). While Perrow once, in what seems like a strike of optimism, concluded that 'Normal Accident Theory and High-Reliability Theory took the theory of accidents out of the hands of economists and engineers and put it into the hands of organisational theorists' (Perrow, 1994, p. 220), I would argue that the discursive impact on safety management and regulation has nonetheless remained in the hands of the engineers.

11.4 Concluding Remarks

In this chapter, I have outlined how the desire to *engineer* resilience and justice might have discursive effects beyond what the dedicated safety science scholars ever intended. This normative desire to contribute to the management of high-risk organisations runs the risk of configuring sharp-end staff in a relationship with their institutions and legal frameworks in which they are responsibilised for the resilience and safety of their wider organisation by the reliance on their adaptive capacities and willingness to always share safety-related information despite the legal systems authority to exercise accountability. Further, the legal appropriation seems to disempower the safety science scholars in that the legal definitions triumph further academic debate and influence.

A more promising future agenda of high-risk organisational resilience research is a descriptive one of *how* organisations distribute degrees of freedom; of *what* interactions (communication, configuration) and relations (power, degrees of expertise, gender) across organisational levels contribute to the dynamics of negotiating organisational safety. I would especially like to see the ethical question of whether the safety management practices used are *fair* to the people configured at various organisational levels. Getting there seems hard in a regulatory regime which seems to give discursive power to a language with its roots firmly in engineering traditions.

References

Alexander, D. E. (2013). Resilience and disaster risk reduction: An etymological journey. *Natural Hazards and Earth System Science*, Volume 13(11), 2707–2716.

Antonsen, S. (2009). Safety culture and the issue of power. *Safety Science*, Volume 47(2), 183–191.

Bergström, J. (2018). An archaeology of soceital resilience. *Safety Science*, Volume 110(C), 31–38.

Bergström, J. and Dekker, S. (2014). Bridging the macro and the micro by considering the meso: Reflections on the fractal nature of resilience. *Ecology and Society*, Volume 19(4).

Bergström, J., Henriqson, E. and Dahlström, N. (2011). *From crew resource management to operational resilience*. Paper presented at the Proceedings of the 4th Symposium on Resilience Engineering, Sophia Antipolis, France, 8–10 June 2011, Paris.

Braithwaite, J., Wears, R. L. and Hollnagel, E. (2017). *Resilient health care*, Volume 3: *Reconciling work-as-imagined and work-as-done*, (Volume 3). Boca Raton, FL: CRC Press.

Cook, R. and Rasmussen, J. (2005). "Going solid": A model of system dynamics and consequences for patient safety. *Quality Safety Health Care*, Volume 14(2), 130–134.

de Bruijne, M., Boin, A. and van Eeten, M. (2010). Resilience: Exploring the concept and its meanings. In L. Comfort, A. Boin and C. Demchak (Eds.), *Designing resilience: Preparing for extreme events*. Pittsburgh, PA: University of Pittsburgh Press, 13–32.

Dekker, S. (2006). Resilience engineering: Chronicling the emergence of confused concensus. In E. Hollnagel, D. Woods, and N. Leveson (Eds.), *Resilience engineering, concepts and precepts* (pp. 77–92). Aldershot: Ashgate.

Dekker, S. (2007). *Just culture, balancing safety and accountability*. Aldershot: Ashgate.

Dekker, S. (2012). *Just culture: Balancing safety and accountability* (2nd ed.). Alsershot: Ashgate.

Dekker, S. and Hollnagel, E. (2004). Human factors and folk models. *Cognition, Technology and Work*, Volume 6(2), 79–86.

Dekker, S. W. A. (2008). Crew resource management gold rush: Resisting aviation imperialism. *ANZ Journal of Surgery*, Volume 78(8), 638–639.

Dekker, S. W. A. and Breakey, H. (2016). 'Just culture': Improving safety by achieving substantive, procedural and restorative justice. *Safety Science*, Volume 85, 187–193.

Dekker, S. W. A. and Lundström, J. (2006). From threat and error management (TEM) to resilience. *Human Factors and Aerospace Safety*, Volume 6(3), 261.

Dekker, S. W. A., Nyce, J. M., van Winsen, R. and Henriqson, E. (2010). Epistemological self-confidence in human factors research. *Journal of Cognitive Engineering and Decision Making*, Volume 4(1), 27–38.

Dekker, S. W. A. and Woods, D. D. (2002). Maba-maba or abracadabra? Progress on human—automation co-ordination. *Cognition, Technology and Work*, Volume 4(4), 240–244.

Foucault, M. (1981). The order of discourse. In Y. Robert (Ed.), *Untying the text: A poststructuralist reader* (pp. 48–79). London: Routledge, Kegan and Paul.

Foucault, M. (2002). *The archaeology of knowledge*. London: Routledge.

Furniss, D., Back, J., Blandford, A., Hildebrandt, M. and Broberg, H. (2011). A resilience markers framework for small teams. *Reliability Engineering and System Safety*, Volume 96(1), 2–10.

Gomes, J. O., Borges, M. R. S., Huber, G. J. and Carvalho, P. V. R. (2014). Analysis of the resilience of team performance during a nuclear emergency response exercise. *Applied Ergonomics*, Volume 45(3), 780–788.

Gunderson, L. (2010). Ecological and human community resilience in response to natural disasters. *Ecology and Society*, Volume 15(2).

Haavik, T. K., Antonsen, S., Rosness, R. and Hale, A. (2016). HRO and RE: A pragmatic perspective. *Safety Science*. doi:10.1016/j.ssci.2016.08.010.

Helmreich, R. L., Merritt, A. C. and Wilhelm, J. A. (1999). The evolution of crew resource management training in commercial aviation. *International Journal of Aviation Psychology*, Volume 9(1), 19–32.

Henriqson, É., Schuler, B., van Winsen, R. and Dekker, S. W. (2014). The constitution and effects of safety culture as an object in the discourse of accident prevention: A Foucauldian approach. *Safety Science*, Volume 70, 465–476.

Holling, C. S. (1973). Resilience and stability of ecological systems. *Annual Review of Ecology and Systematics*, Volume 4, 1–23.

Hollnagel, E. (2011). *Resilience engineering in practice: A guidebook*. Farnham, Surrey, England; Burlington, VT: Ashgate.

Hollnagel, E., Braithwaite, J. and Wears, R. L. (2013). *Resilient health care*. Aldershot: Ashgate.

Hollnagel, E. and Fujita, Y. (2013). The Fukushima disaster—Systemic failures as the lack of resilience. *Nuclear Engineering and Technology*, Volume 45(1), 13–20.

Hollnagel, E., Nemeth, C. P. and Dekker, S. (2008). *Resilience engineering perspectives, remaining sensitive to the possibility of failure*, (Volume 1). Aldershot: Ashgate.

Hollnagel, E., Nemeth, C. P. and Dekker, S. (2016). *Resilience engineering perspectives: Preparation and restoration*, (Volume 2). Farnham: Ashgate.

Hollnagel, E., Woods, D. and Leveson, N. (2006). *Resilience engineering, concepts and precepts*. Aldershot: Ashgate.

Hopkins, A. (2014). Issues in safety science. *Safety Science*, Volume 67, 6–14.

Larouzee, J. and Guarnieri, F. (2015). *From theory to practice: Itinerary of reasons swiss cheese model*. Paper presented at the ESREL 2015.

Le Coze, J.-C. (2015a). Reflecting on Jens Rasmussen's legacy (2) behind and beyond, a 'constructivist turn'. *Applied Ergonomics*. doi:10.1016/j.apergo.2015.07.013.

Le Coze, J.-C. (2015b). Reflecting on Jens Rasmussen's legacy. A strong program for a hard problem. *Safety Science*, Volume 71, 123–141.

Le Coze, J. C. (2016). Vive la diversité! High reliability organisation (HRO) and resilience engineering (RE). *Safety Science*. doi:10.1016/j.ssci.2016.04.006.

Lundberg, J. and Rankin, A. (2013). Resilience and vulnerability of small flexible crisis response teams: Implications for training and preparation. *Cognition, Technology and Work*. doi:10.1007/s10111-013-0253-z.

Marx, D. (2001). *Patient safety and the just culture: A primer for health care executives*. NY: Columbia University.

Nemeth, C. P. and Herrera, I. (2015). Building change: Resilience engineering after ten years. *Reliability Engineering and System Safety*, Volume 141, 1–4.

Nemeth, C. P. and Hollnagel, E. (2014). *Resilience engineering in practice: Becoming Resilient*, (Volume 2). Farnham: Ashgate.

Patterson, E. S., Cook, R. I., Woods, D. D. and Render, M. L. (2006). Gaps and resilience. In S. Bogner (Ed.), *Human error in medicine* (2nd ed.). Hillsdale, NJ: Lawrence Erlbaum Associates, 255–310.

Perrow, C. (1994). The limits of safety: The enhancement of a theory of accidents. *Journal of Contingencies and Crisis Management*, Volume 2(4), 212–220.

Perrow, C. (1999). *Normal accidents: Living with high-risk technologies*. Princeton, NJ: Princeton University Press.

Pettersen, K. A. and Schulman, P. R. (2016). Drift, adaptation, resilience and reliability: Toward an empirical clarification. *Safety Science*. doi:10.1016/j.ssci.2016.03.004.

Reason, J. (1997). *Managing the risks of organisational accidents*. Farnham: Ashgate.

Rochlin, G. I., La Porte, T. R. and Roberts, K. H. (1987). The self-designing high-reliability organisation: Aircraft carrier flight operations at sea. *Naval War College Review*, Volume 40(4), 76–90.

Saurin, T. A., Wachs, P., Righi, A. W. and Henriqson, E. (2013). The design of scenario-based training from the resilience engineering perspective: A study with grid electricians. *Accident Analysis and Prevention*. doi:10.1016/j.aap.2013.05.022.

Silbey, S. S. (2009). Taming prometheus: Talk about safety and culture. *Annual Review of Sociology*, Volume 35(1), 341–369.

Snook, S. A. (2000). *Friendly fire, The accidental shootdown of U.S. Black Hawks over northern Iraq*. Princeton: Princeton University Press.

Turner, B. A. and Pidgeon, N. F. (1997). *Man-made disasters*. Boston: Butterworth-Heinemann.

van Winsen, R., Henriqson, E., Schuler, B. and Dekker, S. W. (2014). Situation awareness: Some conditions of possibility. *Theoretical Issues in Ergonomics Science*, 1–16. doi:10.1080/1463922X.2014.880529.

Vaughan, D. (1996). *The challenger launch decision*. Chicago: The University of Chicago Press.

Wachter, R. M. and Pronovost, P. J. (2009). Balancing "no blame" with accountability in patient safety. *New England Journal of Medicine*, Volume 361(14), 1401–1406.

Wears, R. L., Hollnagel, E. and Braithwaite, J. (2015). *Resilient healthcare: The resilience of everyday clinical work* (2nd ed.). Farnham: Ashgate.

Weick, K. E. and Sutcliffe, K. M. (2007). *Managing the unexpected, resilient performance in an age of uncertainty* (2nd ed.). San Francisco: Jossey-Bass.

Woodlock, J. (2017). *Hammering in the Head of 'the Other' Aviation Profession: Swedish Aircraft Maintenance Engineers Experiences of EU Civil Aviation Regulation*. (MSc), Lund University, Lund. Retrieved from http://lup.lub.lu.se/student-papers/record/8905118.

Woods, D. and Hollnagel, E. (2006). Prologue: Resilience engineering concepts. In E. Hollnagel, D. Woods and N. Leveson (Eds.), *Resilience engineering, concepts and precepts* (pp. 1–6). Aldershot: Ashgate.

Woods, D. D. (2003). Creating foresight: How resilience engineering can transform NASAs approach to risky decision making. Testimony on the future of NASA to Senate Committee on Commerce. *Science and Transportation, John McCain, Chair*, Washington, DC. 9 October 2003, 10–04.

Woods, D. D. and Branlat, M. (2011). Basic petterns in how adaptive systems fail. In E. Hollnagel, J. Pariès, D. D. Woods, and J. Wrethall (Eds.), *Resilience engineering in practice—A guidebook* (pp. 127–143). Farnham, Surrey: Ashgate.

12

Investigating Accidents: The Case for Disaster Case Studies in Safety Science

Jan Hayes

CONTENTS

12.1 Academic Investigations of Accidents 188
    12.1.1 Accidents as a Source of Lessons to Be Learned 188
    12.1.2 The Link between Cases and Action 190
12.2 The San Bruno Pipeline Rupture 191
    12.2.1 The Accident Scenario 191
    12.2.2 Organisational Analysis in Action 192
    12.2.3 What Went Wrong 193
        12.2.3.1 Linking Actions to Consequences 193
        12.2.3.2 Focus on Worker Safety 194
        12.2.3.3 Safety as Compliance 194
        12.2.3.4 The Role of Engineering Professionals 194
    12.2.4 Dealing with Uncertainty 195
    12.2.5 Why Were These Accidents Not Anticipated? 196
12.3 Challenges for the Future 197
    12.3.1 Is Accident Analysis Falling Out of Favour? 197
    12.3.2 Time and Space for Analysis 198
    12.3.3 Addressing Top Management 198
12.4 Conclusion 199
Acknowledgements 200
Note 200
Bibliography 200

Introduction

Learning from accidents is a popular practice for organisations that operate complex sociotechnical systems and a popular topic of safety research. Since the 1990s, high-reliability researchers have increased the focus on learning from small faults and failures (Weick, Sutcliffe and Obstfeld, 1999).

These small departures from perfect, or at least acceptable, performance can be seen as latent failures (Reason, 1997) that contribute to disaster incubation (Turner, 1978).

In parallel with the study of small, local failures, many organisations are keen to learn from major failures further afield. Early attention focussed mainly on sector-specific accidents; however, the advent of the theory of organisational accidents also brought to the fore the possibility of learning across sectors (Reason, 1997). Organising and patterns of work are now understood to be largely common across design, construction, operation, maintenance and management of complex systems, despite major variations in technology, thus making failures in one sector equally relevant in others.

Disasters due to major failures of sociotechnical systems are mercifully rare. In the wake of an aircraft crash, chemical plant fire, pipeline rupture or oil well blowout, major investigations are undertaken and documented in publications released into the public domain by standing investigating agencies (e.g. National Transportation Safety Board, 2011) and especially appointed commissions (e.g. Baker, 2007). These reports (and the associated transcripts of interviews and other documentation) provide a rich source of material for analysis by academics interested in a deep understanding of why accidents happen and so how they might be prevented (e.g. Hopkins, 2008, 2012; Quinlan, 2014; Snook, 2000; Vaughan, 1996).

Accident analysis is a well-established branch of safety science and so is worthy of critical reflection. In this chapter, I start by reviewing the history of academic studies of accidents and their causes, before exploring the link between accident narratives and learning. The chapter then turns to a summary of one particular accident narrative, that of the San Bruno pipeline rupture, before looking at some of the challenges to producing effective accident narratives in the future.

12.1 Academic Investigations of Accidents

12.1.1 Accidents as a Source of Lessons to Be Learned

In the late 1970s, British sociologist Barry Turner turned his attention to the role of the social in accident causation. He set out to compare reports of investigations into large-scale accidents and industrial disasters, coming to the conclusion that '[i]t is better to think of the problem of understanding disasters as a "sociotechnical" problem, with social, organisational and technical processes interacting to produce the phenomena to be studied' (1978, p. 3). At the time, not many people heard this advice, but Turner's 1978 book is now seen as the beginning of research into what Reason would two decades later call, 'organisational accidents' (1997).

Turner's aim was to describe how and why accidents occur rather than to set out a prescription for accident prevention. Nevertheless, there are implications for learning in order to prevent accidents inherent in his work. In contrast, only a few years later, Perrow (1984) came to the conclusion that, for some types of systems, accidents are inevitable or 'normal'. His theory is essentially a dead end when it comes to accident prevention, as no improvements in engineering or organising can sufficiently control the potential for disaster in systems that are both highly complex and tightly coupled.

Both Turner and Perrow studied multiple accidents and sought to determine general themes from them. Other researchers worked with a similar goal but were grounded in different academic traditions. Jens Rasmussen's sociotechnical systems model of risk management extending from individual tasks to government policy became the basis for AcciMaps (Rasmussen, 1997), a systematic way of studying accident causation. James Reason's work in industrial psychology on the causes of human error developed into his model of organisational error based on the metaphor of Swiss cheese (Reason, 1997).

In the 1990s, other researchers also started to immerse themselves in information regarding a single accident; they produced detailed, plain language accounts, which put the reader in the shoes of those involved. With many of these researchers being sociologists, their work often draws on traditions of ethnography more common in anthropology. Two of the most influential are Diane Vaughan and Andrew Hopkins.

Vaughan's study of the *Challenger* disaster gave rise to the term 'the normalisation of deviance'. This phrase captures the way in which a series of seemingly minor, routine changes in how work is done can lead to catastrophe. Many are familiar with this key concept, but I suspect far fewer have actually read the 575 pages of Vaughan's incredibly detailed and fascinating account of life at NASA before the *Challenger* disaster (1996). Methodologically, Vaughan saw her task as 'reconstructing the history of decision making' with her account 'portray[ing] an incremental descent into poor judgement' (1996: xiii). Further, she describes the work as a 'sociology of mistake' showing 'how mistake, mishap and disaster are socially organised and systematically produced by social structures' (1996, p. xiv).

Methodologically, Hopkins' series of books describing disasters such as Longford (2000), Texas City (2008) and Deepwater Horizon (2012) use Rasmussen's AcciMap concept to provide structure. He draws on high-reliability theory in addition to his background as a structural Marxist in order to spot ways in which senior management – often inadvertently – influences the way in which work is organised, thereby leading to frontline errors. His recommendations address factors such as senior management financial incentive schemes (Hopkins and Maslen, 2015) and organisation design (Hopkins, 2008, 2012).

In producing these influential studies, Hopkins (2016, p. 93) draws on the large volume of material produced by major accident inquiries, calling them 'a priceless source of information about organisational cultures and the way they impact on safety'. A high-quality accident analysis that robustly addresses the organisational circumstances leading to disaster is much more than simply a rereading of an investigation report from a new perspective. It requires the analyst to be immersed in the organisation and to produce an account that is the result of 'outward inquiry and inward reflection' (Charmaz and Mitchell, 1996, p. 286). Vaughan calls her study of the *Challenger* disaster a 'historical ethnography' (Vaughan, 1996, p. xiii). Hopkins describes the method he uses in his series of accident analyses as 'desktop ethnography' (Hopkins, 2016, p. 94). Ethnography is most closely associated with anthropology and the studies in which researchers live with remote tribes in order to develop an understanding of their customs. Today, many social scientists use other ways to 'observe' particular groups, including studying written texts (such as company procedures and communications), and transcripts of interactions (such as testimony given as evidence in hearings) and yet these studies are no less immersive and ultimately no less ethnographic in nature than earlier fieldwork in more remote and exotic locations (Silverman, 2001).

Even though an accident researcher may never interact directly with the organisation(s) involved in an accident, writing a successful account requires that they develop a good sense of what it was like to work there before the accident happened. The intent is to see through the eyes of those present and produce a detailed account that gives readers an understanding of the layers of reality before the disruption caused by the accident itself. As in all ethnographic writing, the authorial voice is critical to the reader's experience. Hopkins rarely writes in first person and yet through his matter-of-fact recounting of events in deliberately non-technical and candid language, he speaks directly to those who work in hazardous industry. He challenges them to make changes in their organisations to ensure that the accidents of the past are not repeated. Hopkins calls this approach to writing 'public sociology' (Hopkins, 2016, p. 97).

Such analyses can extend well beyond those which were directly involved in the specifics that directly precipitated disaster. In asking 'why' an accident occurred, there is no limit to how far the analysis can go, provided relevant data is available. This is increasingly important as modern work is characterised by changed employment practices including subcontracting and outsourcing, privatisation of publicly owned organisations and other similar phenomena. In this way, we can draw a direct link between the economic and business environment in which work has been carried out, the structure within which work was done and the work practices that have developed at all levels.

12.1.2 The Link between Cases and Action

Best practical use of the deep and vibrant studies of accidents produced by academia requires an understanding of the link between cases and action taken in the field. Action-oriented safety decision-making is partly driven by rules, but to focus only on rules is to ignore the extent to which safety depends on the professional judgement of those involved in all functions and at all levels (Hayes, 2013, 2015; Hayes and Maslen, 2015; Klein, 2003). When it comes to learning how to do one's job, there is a qualitative leap in learning processes when moving from the rule-driven analysis of a novice to the fluid, continuous action of the expert (Dreyfus and Dreyfus, 1986). Knowledge of particular cases is critical as experts 'operate on the basis of intimate knowledge of several thousand concrete cases in their areas of expertise' (Flyvbjerg, 2006, p. 222). For the purposes of disaster prevention, it is therefore desirable that relevant professionals can draw on rich case studies to inform their everyday actions, in particular about what the consequences of their choices could be. This raises an important question regarding the primary intended audience for studies of particular disasters.

Pidgeon and O'Leary (2000) characterise the ability to link actions to potential consequences as having a good 'safety imagination'. Of course, in the case of major disasters, the link is somewhat indirect. In a well-functioning system, no single decision will lead to immediate disaster, but a valuable lesson from disaster case studies for practicing professionals in hazardous industries is that disasters arise when people are going about their work in apparently ordinary ways. As Vaughan says of her analysis of the *Challenger* disaster, 'it exposes the incrementalism of most life in organisations and the way that incrementalism can contribute to extraordinary events that happen' (Vaughan, 1996, p. xiv).

The link between the interpretation of a given situation and action can be described as a process of sensemaking (Weick, 1995). Confronted with a choice to be made, people will ask themselves, 'What's the story here?' and then 'Now what should I do?' The second question 'has the force of bringing meaning into existence' (Weick, Sutcliffe and Obstfeld, 2005, p. 410). When drawing on disaster case studies as part of that process, lessons from past disasters can directly influence current professional practice.

12.2 The San Bruno Pipeline Rupture

To illustrate some of these issues, it is instructive to examine a specific disaster case study to consider how the analysis was done, what it found, and how that knowledge was used. Our analysis of the San Bruno pipeline failure provides such an example (Hayes and Hopkins, 2014).

12.2.1 The Accident Scenario

Just after 6 PM on 9 September 2010, a large fire erupted in the residential area of San Bruno in northern California. Initial news reports speculated that the fire was the result of a plane crash (since the area is only five miles away from San Francisco Airport), but it soon became clear that the source of fuel was natural gas escaping from a ruptured high-pressure gas transmission line which runs under the suburb. The pipeline is owned and operated by the Pacific Gas and Electric Company (PG&E). Supply of gas to the fire was isolated after 90 minutes, although the firefighting effort continued for two days. As a result of the fire, eight deaths and numerous injuries occurred, all among local residents. Thirty-eight houses were destroyed and 70 were damaged.

The failure point was soon identified as a longitudinal seam weld (i.e. a lengthwise weld turning a sheet of steel into a cylindrical pipeline) that was poorly made when the line was fabricated and installed in 1956. Apparently, no effective integrity testing was done before the line was put into service, and it had not been tested or physically inspected in the intervening decades.

Immediately prior to the pipeline rupture, field maintenance work was going on at the upstream Milpitas Terminal. The work interfered with the operation of the terminal inlet pressure control system which led to an increase in gas pressure in the downstream system and so failure of the faulty weld. Despite the operating upset, the pressure in the line remained below the designated maximum allowable operating pressure (MAOP). The regulatory requirements for establishing the MAOP by testing were 'grand-fathered' for old pipelines. Such pipelines were deemed to be exempt from more recent requirements for pressure testing. Instead, compliance could be achieved by fixing the MAOP at the maximum operational pressure to which the system had been exposed in the previous five years. PG&E had used this method to maintain the MAOP thus avoiding the need for any expensive and technically difficult integrity testing.

In summary, the pipeline was faulty when installed and had been in service for over 50 years without ever being subjected to any integrity testing or inspection. It is instructive for everyone with responsibility for similar systems to consider how this happened.

12.2.2 Organisational Analysis in Action

The base data for the organisational analysis was drawn from the National Transportation Safety Board (NTSB) investigation. Given the severity of the consequences, the NTSB produced a detailed report on the accident, but the source data for our analysis goes much further. Primary sources included three investigation reports – the National Transportation Safety Board report (National Transportation Safety Board, 2011), the Report of the Independent Review Panel (California Public Utilities Commission, 2011) and the California Public Utilities Commission Investigation Report (California Public Utilities Commission, 2012).

This material was augmented by nearly 600 pages of transcripts of formal NTSB interviews conducted with 14 key people in the days immediately following the incident; transcripts of formal NTSB hearings held in March 2011 totalling over 500 pages and 400 exhibits on the NTSB website comprising internal memos, procedures and other documents from PG&E. Assimilating such a large volume of material took two years from late 2012 to late 2014. This was not easy. Nevertheless, reviewing it all was an exercise in ethnography in that it felt as if I was there in the organisation with these people in the days and weeks before the disaster took place.[1]

In structuring my account of what happened and why, I write primarily for a professional engineering audience, aiming to find and highlight the collective professional experience in the details of the behaviour of a few specific individuals. It is rare for those involved in an accident to be acting with malice, but rather from the best of intentions. In my rendering of organisational life, I aim to connect artefacts such as individual views, pithy comments, routine engineering reports, technocratic procedures, and bland, jargon-filled corporate documents in a way that reveals unstated assumptions and organisational confusion, lack of logic, ambivalence and inconsistency.

In this case, the analysis was completed without direct interaction from any individuals or organisations involved in the events leading up to the accident; however, the results have been socialised with many audiences. Of particular note is that before the book was published, the results of the analysis were presented nine times to pipeline sector groups in Australia and the United States; in academic safety forums as well as to groups from other hazardous industries. Questions and discussion from these audiences refined my thinking. As is my primary aim for research of this kind, many practicing engineers and managers have seen echoes of their own organisations in the failures that cumulated at PG&E with such catastrophic consequences.

12.2.3 What Went Wrong

Few people come to work deliberately intending to do a poor job, especially those working in a hazardous industry environment such as a high-pressure gas pipeline company. Nevertheless, there are many ways in which work done at PG&E did not achieve the desired outcomes and, in fact, contributed to catastrophe. Four areas are discussed further below.

12.2.3.1 Linking Actions to Consequences

No organisation that operates a system with the potential for disaster would accept a design where a single failure or error could lead to catastrophe. Organisations always intend that engineering and procedural risk controls combine to form 'layers of protection'. This approach is ubiquitous because we all know that, despite our best efforts, risk controls are not 100% reliable. The Swiss cheese model is the iconic representation of the idea that accidents happen when weaknesses in systems accumulate so that all controls fail – sometimes with catastrophic consequences.

PG&E's operations included many latent errors (Reason, 1997). The ultimate sleeper was of course the faulty weld, but its presence remained hidden and it ultimately triggered disaster, as a result of a suite of other failures in organisational systems and practices that accumulated over the years. To prevent this situation, all staff members with safety-critical roles must understand the potential consequences of their actions. They must have sufficient 'safety imagination' to link their day-to-day work with the potential for a serious accident (Pidgeon and O'Leary, 2000).

12.2.3.2 Focus on Worker Safety

As with many other companies, PG&E public statements in annual reports and the like placed a strong emphasis on safety but failed to distinguish between personal safety and the potential for major accidents. This is a common organisational error. Major accident prevention (or process safety, as it is sometimes called) is about keeping the process under control – or, more colloquially, keeping the hydrocarbons in the pipes, or otherwise contained. Failure to do so can lead to catastrophic consequences for workers and the public. Personal safety on the other hand concerns hazards that, while not directly related to the process fluid itself, can injure or even kill workers, for example, falling from a height whilst doing maintenance. These are the so-called 'slips, trips and falls'. PG&E focussed on worker safety, but it did not treat public safety in the same way.

12.2.3.3 Safety as Compliance

In the absence of a real understanding of the potential for disaster, it is clear from the many statements about safety made by PG&E personnel in NTSB hearings that PG&E's strategy for managing public safety was based simply on compliance with codes and standards.

Compliance with safety rules (in the form of regulations or internal company procedures) is clearly a key strategy in ensuring safe outcomes. The problem comes when compliance with rules is seen as being an end in itself. Other research has highlighted the trap of framing safety only in terms of compliance (Hale and Borys, 2013a,b) which can lead to a short-term view of requirements and a 'tick in the box' mentality, rather than the long-term perspective necessary to ensure long-term public safety. Perhaps most importantly in this case, removing the moral dimension from discussion of safety puts compliance issues on par with other organisational goals, such as cost and schedule.

12.2.3.4 The Role of Engineering Professionals

Another significant downside of a compliance-based approach to safety is that it writes out of the picture the roles of engineering judgement and professional expertise that are also critical to achieving great safety performance. Other research (Hayes, 2013) has shown the extent to which major hazard industries rely on the sense of professionalism of key staff and their identification as members of a professional group. This sits alongside their view of themselves as company employees. One function of such a view of working life is to foster an awareness of the public trust held by professionals in hazardous industries and the sense of responsibility for one's actions that accompanies that. Statements made by PG&E technical staff in interviews and hearings are driven by their response to questioning so they are somewhat constrained, and yet there is little sense from them that they feel a sense of professional responsibility for the results of their actions. Responses are, almost without exception, framed by reference to company systems, rather than reference to any broader understanding of responsibilities or context.

Members of a profession are often characterised by qualities such as a sense of public trust, an ability to act independently of their employer, the tendency to value technical expertise and adhere to a code of professional ethics (Middlehurst and Kennie, 1997). It is these attitudes that are hugely beneficial to organisations; as such, individuals provide an additional valuable perspective on what makes the system safe, rather than mindlessly acting in line with management priorities at all times, irrespective of the specific circumstances.

12.2.4 Dealing with Uncertainty

The integrity management algorithms applied to the PG&E network constitute a form of planning. Planning is a straightforward instrumental activity (a means to an end) in cases where uncertainty is low. Integrity management is a very different case. For a complex network of facilities such as PG&E's system, many things are not certain. Data for old pipelines may be missing. Not all lines have been tested or inspected, so their condition is uncertain. More than that, sometimes data itself might be unreliable in ways that are not obvious. Effective planning to achieve an important goal under these conditions is difficult. Under conditions of high uncertainty, Clarke has highlighted the extent to which planning can become a symbolic, rather than instrumental, activity and it appears that this was the case at PG&E. When planning takes on a primarily symbolic role, the purpose of the plan becomes: 'asserting to others that the uncontrolled can be controlled' (1999, p. 16). Clarke (1999) also points to the manner in which symbolic plans represent a fantasy – in the sense of a promise that will never be fulfilled – which are often couched in a special vocabulary which then shapes discussion.

This explains both how and why the integrity management system at PG&E had taken on a symbolic, rather than functional, role. PG&E's integrity management system was flawed in many ways, and was not effectively performing the function for which it was originally intended. 'Knowing' about the system was grounded in elaborate algorithms and graphs purported to show that risk was on the decline, and yet this analysis was only tenuously linked to the actual level of danger. This is not to suggest that, in these circumstances, companies or individuals are deliberately fabricating this type of plan in order to deceive themselves or others, but rather that, in the face of significant uncertainty, earnest attempts to plan can lose touch with reality. In the case of San Bruno, this included problems with records of physical data where dummy values had been used in the place of missing information for old pipelines. Over time this data came to be seen as an accurate record of reality. Those responsible for integrity management could not foresee problems because the model itself had taken on a life of its own and become, for all intents and purposes, the reality of what was being managed, rather than the physical system that it was originally intended to represent.

12.2.5 Why Were These Accidents Not Anticipated?

The San Bruno pipeline rupture was not anticipated by senior management at PG&E who were focussed on maximising profit instead of running a safe gas company. The evidence shows that a focus on maximising return by cutting costs over a period of years had led to the degeneration of PG&E's physical facilities. This extended to a lack of maintenance work and reduced levels of inspection. Resource constraints in the engineering area led to a lack of effective engineering support and analysis. As a result, the organisation was unable to retain corporate memory in the face of slowly developing problems.

Preventing long-term accidents such as this requires a senior management team whose members are vigilant in their focus, ensuring that safety-critical maintenance and inspection work remains up to date, despite the fact that, if they get this wrong, the chances are high that it will be those who come after them who will pay the price. One way to maintain senior management focus on long-term public safety is to link it to remuneration. Bonus payments are typically a major component of remuneration packages at senior company levels. Despite much organisational rhetoric that safety is number one priority, in fact, bonus payments are typically linked to financial measures such as share price, return-on-investment or profit. It is therefore hardly surprising that senior executives favour financial performance over long-term safety when it comes to management decision-making.

In summary, the analysis attempts to provide material to assist operations and maintenance personnel, inspection engineers, risk managers and senior managers to critically consider their professional practice. Members of all these professions in the pipeline sector and in other hazardous industries are invited to consider whether their own organisations might be vulnerable to the same kinds of failures seen at San Bruno, which had such devastating consequences. Feedback received to date is that, for some people at least, the work achieves that – in fact, one pipeline engineer told me that the book had changed his life. In my view, the time and effort expended in the analysis are thus justified.

12.3 Challenges for the Future

12.3.1 Is Accident Analysis Falling Out of Favour?

Seeking lessons from accidents has been popular for many decades, although Hollnagel's work on Safety I and Safety II has highlighted the danger of only looking for safety management improvements in purely reductionist style accident analysis (Hollnagel, 2014). In this typology, the purpose of efforts in safety management under Safety I thinking is to reduce the number of adverse events to an acceptable level. Strategies are to find, fix and also prevent abnormal system states. Safety I is built on what Hollnagel (2014, p. 63) calls 'the law of reverse causality' where every effect has a cause that can be identified and eliminated. He calls for a shift away from studying failures because:

- Safety is defined as the absence of something that is in the negative, whereas it is typically not possible to count how many fewer things go wrong, meaning that it is effectively impossible to judge whether a given system is safe or not.
- Studying safety in this way requires that we understand cause and effect relationships. In complex systems this is often not possible in advance.
- Accident analysis in this mode assumes that systems are decomposable whereas they are often not. It also assumes bimodality (the system either works or it doesn't).

In contrast, Hollnagel has postulated the idea of Safety II, a condition where as much as possible goes right, that is, the ability to succeed is as high as possible. In this mode, it is understood that things that go right and things that go wrong both happen in the same way, that is, as a result of performance adjustments. Safety efforts should therefore be focussed on understanding and reinforcing positive performance adjustments.

In fact, as should be obvious from the earlier discussion, accident investigations do not have to be as limited as Hollnagel holds them out to be. His characterisation of accident analyses seeking 'monolithic explanations' (2014, p. 140) clearly does not apply to the examples quoted earlier that go to great lengths to address complexity in causation and link the conditions that produced each accident to the everyday. It should be noted that Hollnagel himself does not propose that safety science or safety practitioners *abandon* accident and incident investigation, but rather change their focus to ensure that such analyses do not take a Safety I approach. When it comes to organisational accident analyses this is very much a call from behind.

Hollnagel's call to learn from successes is welcome; yet he makes it by attacking a narrow version of the accident analysis tradition that focusses on simple explanations and linear analysis. Poor analysis of this type is far from the universal, and so accident analysis does not deserve such sweeping criticism. There are fashions in safety research as with everything else. It would be a pity if, as a result of a push to learn from success, the popularity of in-depth disaster case studies wanes.

12.3.2 Time and Space for Analysis

The digital age has made it possible for researchers to access enormous volumes of data on any major disaster that has a major inquiry. Reviewing this information to extract a coherent and useful narrative on what it was like to be in a given organisation prior to disaster is the key job of the accident researcher. It may seem that having such material to draw on will promote more academic accident case studies addressing the social causes of disaster. However, the truth is that such analyses are time consuming to produce and may not be completed for several years after the accident occurs. There are two key reasons for this. The first is that there is sometimes a delay in making information available due to legal proceedings. This is unfortunate, but it is also a fact of life when litigation with a view to blame and punishment takes priority over investigations seeking to produce lessons aimed at preventing repetition. This problem has been well canvassed elsewhere (Cedergren and Petersen, 2011; Dekker, 2007).

Another important issue is the academic time and effort involved in such studies. It is evident that most disaster studies of this kind are the work of individuals, rather than groups of researchers. Apart from Vaughan and Hopkins' work described earlier, other accidents that have been studied in this genre include the accidental shoot down of U.S. Black Hawk helicopters (Snook, 2000), the Uberlingen air traffic control disaster (Brooker, 2008) and the Montara blowout (Hayes, 2012). Most of these detailed, ethnographic studies of single accidents seem to have been done by lone analysts. Long-term immersion in literally thousands of pages of evidence is a luxury that perhaps many academics simply cannot afford. Speculation based on my own experience suggests that such work is popular with industry, but is not necessarily the best investment of time in terms of maximising academic outputs.

We can only hope that current trends in academic priorities allow time for such long-term immersive studies as the deep understanding and practical value of the results certainly justify the overall investment of time and effort.

12.3.3 Addressing Top Management

As a learning tool, the purpose of the stories contained in a case study of any given accident is for individual professionals to see themselves in the stories of mayhem and disaster. This means acknowledging that disasters arise from the most banal of circumstances – everyday work. Those who are involved in disaster are members of the same community of professional practice as the rest of their profession. The strongest lessons come from professionals understanding this, and then going on to develop a chronic sense of unease – or, more positively, fostering their safety imagination.

The biggest challenge for accident analysis is to extend that experience to the highest levels of management. Despite decades of safety culture research that emphasises the role of safety leadership, little normal operations research addresses the practices of senior management. Addressing this level of management directly is left largely to studies of disaster. Until this can be effectively addressed in normal operations research, disaster analysts have the incredibly important role of speaking truth to the most powerful actors in most organisations and pointing out that what they do has a direct impact on safety outcomes – in other words, a direct impact on the lives of their workers.

Addressing senior management does not require a different story, but rather an extension of the same detailed analysis to address the actions of the senior managers involved. A good case study is not designed to be 'all things to all people' but rather, 'different things to different people' (Flyvbjerg, 2006, p. 238). While not all investigations extend into senior management and sometimes source material is hard to come by, drawing a direct link between senior management choices and the lives of workers and the public is one of the most valuable contributions of academic disaster analysts.

12.4 Conclusion

Case studies of disasters are always popular. From personal experience of using case study material with a wide range of audiences, I can say that the more experienced the audience, the more engaged they are with this type of material. This doesn't mean they are necessarily experienced in the technology that failed, but rather experienced in organisational and professional life. This is consistent with Dreyfus and Dreyfus's work, which emphasises the role of stories and individual cases in expert learning. Experienced professionals see themselves in the stories of disaster and understand that accidents arise out of seemingly normal work practices. Small interactions and short moments can have large consequences (Weick, Sutcliffe and Obstfeld, 2005, p. 410).

One criticism often levelled at stories of single accidents is that the extent to which specific lessons can be generalised is unclear, but this misses the point of the analysis. Accident analyses are not seeking generalisations or a single truth, but rather 'the case story itself is the result' (Flyvbjerg, 2006, p. 238). As Stake says, 'we do not study a case primarily to understand other cases. Our first obligation is to understand this one case' (1995, p. 4). In assisting those who work in hazardous industry to understand one case, we can impact professional practice and therefore many other important decisions. 'Encapsulating a crisis into a single, simple story, however shocking or frightening this story might be, reduces the scale of the events and subsequently enables individuals and groups to incorporate the events into their mental world' (Boudes and Laroche, 2009, p. 392). In this way, a store of cases supports professionals in hazardous industry to use the past to inform the future.

Academic studies of accidents with a focus on human and organisational causes require significant effort to produce, but professionals in many fields find them invaluable for learning. The field faces some challenges from fashions in research that are currently pushing research away from accident investigation, and, as a result, pushing away the resources required to do a good job. We can only hope that more academics are attracted to this field, so that more research is produced that supports professionals working at all levels, from the control room to the board room, to be able to relate hindsight to foresight and thus make disasters less likely in the future.

Acknowledgements

This work was funded by the Energy Pipelines Cooperative Research Centre, supported through the Australian Government's Cooperative Research Centres Program. The cash and in-kind support from the Australian Pipeline Industry Association Research and Standards Committee is gratefully acknowledged.

Note

1. The analysis of the San Bruno disaster summarised in this chapter is drawn from a volume that I co-authored with Andrew Hopkins but we did not do the detailed accident analysis together. The book addresses two pipeline failures and Andrew Hopkins's work largely comprises the chapters addressing the Enbridge pipeline failure at Marshall, Michigan. I undertook the detailed analysis of the San Bruno disaster.

Bibliography

Baker, J. (2007), *The Report of the BP U.S. Refineries Independent Safety Review Panel*.

Boudes, T. and Laroche, H. (2009), 'Taking off the heat: Narrative sensemaking in post-crisis inquiry reports', *Organisation Studies*, Volume 30(4), pp. 377–396.

Brooker, P. (2008), 'The Uberlingen accident: Macro-level safety lessons', *Safety Science*, Volume 46, pp. 1483–1508.

California Public Utilities Commission, (2011), Report of the Independent Review Panel, San Bruno Explosion, Prepared for the CPUC Revised Copy 24 June 2011, California Public Utilities Commission.

California Public Utilities Commission, (2012), Consumer Protection and Safety Division Incident Investigation Report, 9 September 2010 PG&E Pipeline Rupture in San Bruno, California, released 12 January 2012, California Public Utilities Commission.

Cedergren, A. and Petersen, K. (2011), 'Prerequisites for learning from accident investigations—A cross-country comparison of national accident investigation boards', *Safety Science*, Volume 49, pp. 1238–1245.

Charmaz, K. and Mitchell, R.G. (1996), 'The myth of silent authorship: Self, substance, and style in ethnographic writing', *Symbolic Interaction*, Volume 19, pp. 285–302.

Clarke, L. (1999), *Mission Improbable: Using Fantasy Documents to Tame Disaster*, Chicago, University of Chicago Press.

Dekker, S. (2007), *Just Culture: Balancing Safety and Accountability*, Aldershot, Ashgate.

Dreyfus, H.L., and Dreyfus, S.E. (1986), *Mind Over Machine*, New York, The Free Press.

Flyvbjerg, B. (2006), 'Five misunderstandings about case-study research', *Qualitative Inquiry*, Volume 12(2), pp. 219–245.

Hale, A. and Borys, D. (2013a), 'Working to rule, or working safely? Part 1: A state of the art review', *Safety Science*, Volume 55, pp. 207–221.

Hale, A. and Borys, D. (2013b), 'Working to rule, or working safely? Part 2: The management of safety rules and procedures', *Safety Science*, Volume 55, pp. 222–231.

Hayes, J. (2012), 'Operator competence and capacity—lessons from the Montara blowout', *Safety Science*, Volume 50, pp. 563–574.

Hayes, J. (2013), *Operational Decision-Making in High-Hazard Organisations: Drawing a Line in the Sand*, Farnham, Ashgate.

Hayes, J. (2015), 'Investigating design office dynamics that support safe design', *Safety Science*, Volume 78, pp. 25–34.

Hayes, J. and Hopkins, A. (2014), *Nightmare Pipeline Failures: Fantasy Planning, Black Swans and Integrity Management*, Sydney, CCH.

Hayes, J. and Maslen, S. (2015), 'Knowing stories that matter: Learning for effective safety decision-making', *Journal of Risk Research*, Volume 18(6), pp. 714–726.

Hollnagel, E. (2014), *Safety I and Safety II: The Past and Future of Safety Management*, Farnham, Surrey, Ashgate.

Hopkins, A. (2000), *Lessons from Longford: The Esso Gas Plant Explosion*, Sydney, CCH.

Hopkins, A. (2008), *Failure to Learn: The BP Texas City Refinery Disaster*, Sydney, CCH.

Hopkins, A. (2012), *Disastrous Decisions: The Human and Organisational Causes of the Gulf of Mexico Blowout*, Sydney, CCH.

Hopkins, A. (2016), *Quiet Outrage: The Way of a Sociologist*, Sydney, Wolters Kluwer.

Hopkins, A. and Maslen, S. (2015), *Risky Rewards: How Company Bonuses Affect Safety*, Aldershot, Ashgate.

Klein, G. (2003), *The Power of Intuition: How to Use Your Gut Feelings to Make Better Decisions at Work*, New York, Currency/Doubleday.

Middlehurst, R. and Kennie, T. (1997), 'Leading professionals: Towards new concepts of professionalism', In J. Broadbent, M. Dietrich, and J. Roberts, eds. *The End of the Professions? The Restructuring of Professional Work*, London, Routledge, pp. 49–66.

National Transportation Safety Board, (2011), Pipeline Accident Report: Pacific Gas and Electric Company Natural Gas Transmission Pipeline Rupture and Fire, San Bruno, California, 9 September 2010 (NTSB/PAR-11/01 PB2011-916501), accident report, National Transportation Safety Board.

Perrow, C. (1984), *Normal Accidents: Living with High-Risk Technologies*, New York, Basic Books.

Pidgeon, N. and O'Leary, M. (2000), 'Man-made disasters: Why technology and organisations (sometimes) fail', *Safety Science*, Volume 34, pp. 15–30.

Quinlan, M. (2014), *Ten Pathways to Death and Disaster: Learning from Fatal Incidents in Mines and Other High Hazard Workplaces*, Sydney, Federation Press.

Rasmussen, J. (1997), 'Risk management in a dynamic society: A modelling problem', *Safety Science*, Volume 27(2–3), pp. 183–213.

Reason, J. (1997), *Managing the Risks of Organisational Accidents*, Aldershot, Ashgate.

Silverman, D. (2001), *Interpreting Qualitative Data: Methods for Analysing Text, Talk and Interaction*, London, Sage.

Snook, S.A. (2000), *Friendly Fire: The Accidental Shootdown of U.S. Black Hawks over Northern Iraq*, Princeton, Princeton University Press.

Stake, R.E. (1995), *The Art of Case Study Research*, Thousand Oaks, Sage.

Turner, B.A. (1978), *Man-Made Disasters*, London, Wykeham Publications (London) Ltd.

Vaughan, D. (1996), *The Challenger Launch Decision: Risky Technology, Culture and Deviance at NASA*, Chicago, University of Chicago Press.

Weick, K.E. (1995), *Sensemaking in Organisations*, Thousand Oaks, Sage Publications.

Weick, K.E., Sutcliffe, K.M., and Obstfeld, D. (1999), 'Organising for high reliability: Processes of collective mindfulness', In R.I. Sutton and B.M. Staw, eds. *Research in Organisational Behaviour*, Volume 21, Stamford, JAI Press Inc., pp. 81–123.

Weick, K.E., Sutcliffe, K.M., and Obstfeld, D. (2005), 'Organising and the process of sensemaking', *Organisation Science*, Volume 16(4), pp. 409–421.

13

Towards Actionable Safety Science

T. Reiman and K. Viitanen

CONTENTS

13.1 Conceptualizing the Gap between Safety Scientists and
Practitioners..204
    13.1.1 The Gap as a Knowledge Issue 204
    13.1.2 The Gap as an Issue of Differing Cultures.................................206
13.2 Integrative Model...208
13.3 Case Examples...210
    13.3.1 AcciMap 211
    13.3.2 Safety Culture...213
13.4 Discussion..219
References...220

Introduction

Safety science aims to produce knowledge that decision-makers and safety practitioners can utilise. A safety scientist should contribute to safety practice. Likewise, safety practitioners should offer safety scientists research problems, challenges and phenomena to interpret and conceptualise into something manageable. Thus, a safety scientist and a safety practitioner should have a natural co-habitat, each contributing to the other one's work. However, practitioners seem to view safety science as disconnected from everyday problems, too theoretical and fragmentary to be of any practical use. Similarly, safety scientists often wonder why safety practitioners are not interested in scientific findings, or why are they stuck with models or practices that science has proved invalid. The reality of the safety scientist and safety practitioner co-habitat does not appear to meet the ideal.

    This chapter's motivation is the authors' recent ventures from safety science into safety practice, the challenges associated with this context shift and the challenges the authors have experienced as researchers interacting with safety practitioners. There was a need to make sense of these two mental worlds. The central thesis of the chapter is that bridging the gap between science and practice is not only a question of knowledge and its creation but also, and more importantly, it is a question of assumptions. Assumptions that form in daily work, that are embedded in tools and methods and that are taught to new community members, often implicitly, as the right way to do things. These assumptions concern what is usable information, what is the role of science in practice (or vice versa), what kind of scientific knowledge is applicable and how it can be produced.

In this chapter, the authors conceptualise the gap between safety scientists and practitioners from knowledge and cultural perspectives. The authors then propose a model that integrates these two perspectives and utilise the model to analyse two established safety management methods, the accident analysis tool AcciMap and the concept of safety culture. The purpose of the analysis is to reveal how the differing assumptions of scientists and practitioners have manifested during the historical development of these two methods. Based on the integrative model and the analysis of safety management methods, the authors propose implications for future directions for safety science.

13.1 Conceptualizing the Gap between Safety Scientists and Practitioners

13.1.1 The Gap as a Knowledge Issue

Three distinct views to approaching the gap between theory and practice have been proposed (Van de Ven & Johnson, 2006) that conceptualise the gap as a knowledge issue. First, it can be viewed as a *knowledge transfer problem* (Figure 13.1a). This view conveys the idea that practical knowledge at least partially originates from the knowledge produced by scientists, which

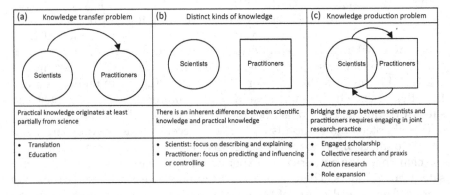

FIGURE 13.1
Three views to theory–practice gap. (Adopted from Van de Ven and Johnson, 2006.)

implies that to remove the gap, the diffusion and translation of scientific knowledge to practice should be ensured. This is a unidirectional approach that does not consider the role of practitioners in reducing (or producing) the gap.

Approaches such as translation and education can be utilised to facilitate dialogue between scientists and practitioners. The *translation* approach involves using third parties as links between science and practice (Ginsburg and Gorostiaga, 2001). The role of these 'link individuals' is to translate the information gained from science into practical use. In the safety field, many consultants act as links between science and practice, with some consultants carrying out research of their own. The *education* approach relies on developing either scientists to better communicate the knowledge they possess to practitioners, or encouraging practitioners to seek out scientific knowledge (Ginsburg and Gorostiaga, 2001). In this process, a translator is not required.

These two approaches with a low amount of dialogue represent a linear and passive interaction between scientists and practitioners (Beier et al., 2017; Cash et al., 2006). Unidirectional interaction ensues when an entity (e.g. a research institute or a private company) funds a scientist to produce scientific knowledge or a product, which, when finished, is delivered in the form of a scientific publication or a finalised product, seldom to be followed-up afterwards.

The gap can also be viewed as resulting from the inherent difference between science and practice: science and practice have different purposes and different methods to achieve the purpose. They are thus complementary, but *distinct kinds of knowledge* (Figure 13.1b). Scientists and practitioners face different challenges in their work and thus it is natural that the knowledge they need (and produce) also differs. Different knowledge is valuable, and usable, to different people. While creating scientific knowledge might be an intrinsic value for a scientist, it might not be so for a practitioner. Practitioners are primarily interested in *influencing* or *controlling* something while balancing other goals in the system. Scientists traditionally seek to *describe* or *explain* phenomena. However, effective control requires description, explanation and prediction. Thus, for the scientific knowledge to be applicable for practical purposes, these goals should be aligned. In reality, we see practitioners struggling with controlling safety-related issues without adequate understanding of the theoretical issues connected to the phenomena to be controlled, and scientists describing phenomena in controlled environments, based on abstracted knowledge and devoid of contextual or relational understanding of how the phenomenon behaves in real life. On the other hand, rigorous investigation by scientists might be the only way to confirm the validity of the knowledge practitioners have created in the field. Such a role may require scientists to reorient themselves or overcome biases to value field experience as a source of scientific discovery.

Finally, the gap can be viewed as a *knowledge production problem* (Figure 13.1c), which implies that to bridge the gap, knowledge should be coproduced together with scientists and practitioners. This requires bidirectional communication approaches or joint reflection between scientists and practitioners, such as role expansion, applied research, action science and collective research and praxis (Ginsburg and Gorostiaga, 2001; see also Anderson et al., 2001).

The *role expansion* approach involves blurring the role definitions between scientists and practitioners, for example, when practitioners engage in scientific work or when scientists do consultancy work (Ginsburg and Gorostiaga, 2001). The *applied research* approach attempts to orient scientists towards creating practically useful scientific knowledge (Ginsburg and Gorostiaga, 2001). This approach includes certain level of joint reflection with practitioners during problem identification or interpretation phases of research. The *action research* approach involves the scientists and practitioners jointly engaging in research activities that have benefits in practice. This approach includes obvious joint reflection; however, there is nevertheless a boundary of roles and responsibilities: while the practitioner has responsibilities in research work, the scientist has little responsibilities in the implementation work.

The approach representing the most extensive dialogue is *collective research and praxis* (Ginsburg and Gorostiaga, 2001). This approach involves scientists and practitioners jointly participating in research and its practical implementation. A similar approach is described by Beier et al. (2017) who describe various routes towards actionable science and emphasise *coproduction* as the vital strategy. Coproduction is the collaboration among the various stakeholders that take interest in scientific work (e.g. scientists, managers and other practitioners), which includes jointly defining research problems, methods and outputs, creating interpretations and developing strategies for the use of the resulting scientific knowledge (Beier et al., 2017). Coproduction is especially important in complex situations such as when situational factors change constantly and there is a need for continuous scientific support, or when problems cannot be pre-specified by neither the scientists nor the practitioners (Beier et al., 2017).

13.1.2 The Gap as an Issue of Differing Cultures

Scientists and practitioners function as a part of their respective communities, simultaneously forming and abiding by sets of rules and norms (Argyris et al., 1985). For scientists, these rules and norms include issues such as what kinds of research problems are of legitimate interest, what are the criteria that define whether a problem has been solved or not, and what methods or paradigms are used to solve the problem. For basic researchers, the characteristic activity is to describe the problem, to understand the contributing phenomena and to formulate these facts into theories. This type of scientific work provides little input to practitioners regarding how to act within the constraints they face in their particular context, and instead puts more emphasis on factors outside of the practitioner's control (Argyris et al., 1985).

However, applied researchers' aims are not only to understand the problem, but also to understand what can be done about it – that is how to achieve a given set of ends. This provides input into how a certain end can be achieved under certain constraints. However, because applied research emphasises specific ends, it often fails to provide much insight into, for example, how practitioners can negotiate conflicting ends. Argyris et al. (1985) remind that the norms that guide research may hinder the application of scientific knowledge to practice, because anything outside the characteristic assumptions underlying each type of research may remain unnoticed.

Safety science is considered an inherently applied science – one that should ultimately result in improved safety (Le Coze et al., 2014). This has often been interpreted as a need to produce tools. A multitude of methods, tools, models and principles exists to tackle the various issues safety practitioners might face in their work. However, the production of tools and methods as such does not make science applicable. Safety science has been hesitant to offer validated models and methods as paradigmatic and widely accepted ways of managing safety (Le Coze et al., 2014). It is often left to the practitioners to decide what methods to use, and how. Safety science has not given much consideration to how the particular methods help solve certain types of problems and what kind of assumptions about safety and management of safety they embed. The assumptions underlying safety science and its artefacts (methods, tools and research findings) often remain implicit and are not communicated to the end users. Subsequently, safety practitioners as the end users discover ways of using these methods that fit their particular context and that help them solve the problems they face in their daily work. This, in turn, results in the formation of new assumptions regarding what are the accepted or plausible ways of using the methods. Safety practitioners then embed the new assumptions in the methods. Gradually, the methods change, sometimes beyond their original intended purpose.

Safety practitioners need to act in context and cannot isolate their work from other technical, human, or organisational issues. They interact with management and other stakeholders in organisations, including the workers who in the end 'do' safety. Thus, their role consists of exerting authority and influence in their respective work communities. The role of a safety professional is socially and organisationally complex (Provan et al., 2017). They need to balance between independence and involvement in daily work and between financial goals and safety goals of the company (Provan et al., 2017; Reiman et al., 2015; Reiman and Pietikäinen, 2014). These contradictions have an effect on how safety science can be utilised in practice. Almklov et al. (2014) discuss what happens when the results of safety science meet the practitioners. They propose that knowledge generated by safety scientists may displace or marginalise existing local or system-specific safety knowledge embedded in operational practices. This is due to the reliance on generic safety management approaches combined with societal trends towards standardisation and bureaucratic control. Overall, the above suggests that emphasis should be put on understanding different practices and their underlying assumptions – the cultures of scientists and practitioners, and their interaction.

13.2 Integrative Model

We propose a multiple-culture model combining the knowledge production perspective with a cultural perspective of different communities of practice. The model illustrates how scientists and practitioners interact and what processes exist that enables this interaction (Figure 13.2). Differences between the two communities emerges from the inherent contextual differences. In this depiction of the model, we will focus on the cultures of scientists and practitioners and their interactions. However, we also acknowledge that there are other stakeholders who are involved in the overall system of safety management. This includes various levels of managers, owners, regulatory and peer authorities, consultants and governments, each having their own 'culture' (Kirwan, 2000; Le Coze, 2013), and thus influencing or providing constraints towards the management of safety, and the interaction between scientists and practitioners. In this sense, the model could be extended with an arbitrary number of cultures (and gaps).

We propose that there are three main cultural elements, which define both science and practice: theories, methods and findings and outcomes. Table 13.1 summarises the definitions of these elements and their relation to science and practice.

One of the messages of this integrative model is that *scientists and practitioners have different assumptions about the world and how to act in it.*

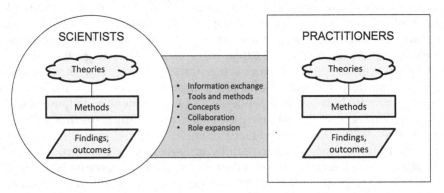

FIGURE 13.2
Integrative science–practice model describing the cultural mechanisms of the two communities and artefacts that may bridge the gap between them.

TABLE 13.1

Descriptions of the Cultural Elements in Communities of Science and Practice

Cultural element	Scientists	Practitioners
Theories *What are the assumptions and the knowledge that steers what is done and why*	*Formal safety science theories*: e.g. Perrow's Normal Accident Theory, High-Reliability Organisations, safety culture, resilience engineering, Rasmussen's sociotechnical model of risk management *Informal or unconscious assumptions*: e.g. what constitutes good or otherwise desirable scientific research in general and in the context of safety, how are problems and research questions specified, how is new knowledge created, what are the criteria for solved problems	*Theories that have been formalised in safety practice*: e.g. behaviour-based safety in industrial settings, CRM in aviation, or safety culture in the nuclear industry *Informal or unconscious assumptions*: e.g. what is acceptable in a particular industrial or organisational context, what phenomena can be influenced and what cannot, what opportunities and constraints are there for interventions
Methods *What is actually done and how*	*What*: e.g. describe or explain a real-life phenomenon, create an application of a theory *How*: e.g. qualitative and quantitative analysis methods and other tools utilised to help find the answer for the research problem, concept and data visualisation, hypothesis formation practices, scientific paper structures, publication practices, etc.	*What*: e.g. assessing safety-related strengths and weaknesses, creating and implementing an intervention strategy and following-up *How*: e.g. safety development, assessment and assurance tools and methods, interventions, risk analysis, organisational structures
Findings, outcomes *What are the results of what is done*	Research results, recommendations, principles, tools or methods resulting from a research study	Specification of the level of safety, organisational strengths and weaknesses with regard to safety, level of safety culture, etc.

These assumptions concern issues such as what counts as valid data, how to gather valid data about safety, what information is needed and how it is gathered, and so on. Some of these assumptions are embedded in the tools and methods that scientists and practitioners utilise. In other words, the developers' assumptions become embedded within the tools and methods. For example, things that scientists perceive as relevant and possible steer the way they develop scientifically based methods and these assumptions define some aspects of the method: questionnaire items are all based on assumptions of what are the relevant things to ask, and how to ask them. When the end user, a safety practitioner, then applies this tool to solve a particular problem, it raises a question: do the scientists and practitioners understand each other the same way? If they do not, there may be a risk that scientists and practitioners either misunderstand each other's intentions, or misunderstand a tool and concept, its purpose and limitations. Eventually, this may result in the misapplication of the tool or concept.

Due to the inherent differences between the contexts of scientists and practitioners, when implementing the methods produced by science in practice, certain level of adaptation to the needs of the particular context is usually required. The practitioners may, for instance, add, drop or reframe questionnaire items, with the purpose of decreasing its length, change its focus, or include something else they consider essential. As a result, the adaptation might no longer share the assumptions with the original method. The second message of the integrative model thus is that as when these tools are taken into use by practitioners, *the assumptions stored in the methods gradually change as practitioners adapt the tools to their particular context*. This can result in a change of the original message embedded in the tool. The need for adaptation may also serve as an indicator that the initial assumptions made the tool either irrelevant or suboptimal to begin with. This is something safety scientists should carefully consider.

The third message of the model is that *there is a risk of information being 'stuck' either within scientists or within practitioners*. This can take place, for example, when scientists introduce a method to practitioners, but do not follow up to understand how the practitioners actually use the method or concept, and whether this is in line with the original intentions of the method or concept. This can be especially problematic if the prevalent assumption concerning bridging the science–practice gap does not view it as an issue of coproduction (Figure 13.1c): in such case, scientists might not consider feedback from the field important.

13.3 Case Examples

We illustrate the application of the integrative model by utilizing it to examine two well-known safety management methods: the accident analysis tool AcciMap and the concept of safety culture. We chose these two as the subjects because of their popularity in both the scientific and practitioner communities, but also because they highlight certain tensions in assumptions between these two communities. In this exercise, we view AcciMap and safety culture from the perspective of science–practice interaction. For the sake of readability and compactness, we simplify the analysis, not incorporating all potential interactions or historical developments. Rather, we focus on what appear to be the most essential elements of the science–practice interaction.

13.3.1 AcciMap

AcciMap was derived from Rasmussen's (1997) seminal work on risk management, which proposed a new, system-oriented approach that views the different levels and disciplines of the sociotechnical system as a whole, with special emphasis on understanding the interactions and adaptations between system components. This approach resulted in the influential sociotechnical risk management framework, which identifies the system levels of work, staff, management, company, regulators/associations and government, and explicated the kinds of interactions that exist between these levels (Rasmussen, 1997). The accident analysis tool AcciMap incorporates the basic idea of this framework by embedding the system levels into the AcciMap analysis process and visualisation with the purpose of modelling the sociotechnical interactions within an adverse event.

Figure 13.3 illustrates the development of AcciMap in scientific domain in the light of the integrative model (steps 1–4). The initial version of the AcciMap method was developed by Rasmussen and Svedung (Rasmussen, 1997; Rasmussen and Svedung, 2000) and consisted of six levels of sociotechnical systems, and a causal tree which spans across the levels, eventually leading to a critical event (loss of control) and the direct adverse outcomes that result from it. AcciMaps were originally intended to be used to model series of accidents to identify the correct risk management strategies for that particular industrial sector (Rasmussen and Svedung, 2000). However, in practice AcciMaps have also been utilised to analyse single accidents. Even though standard practices have been proposed (e.g. Branford et al., 2009), AcciMap is relatively diverse when it comes to its practical application. Waterson et al. (2017) conducted a review of 27 scientific studies where AcciMaps have been utilised for accident analysis to identify how the method has evolved since it was initially introduced. They found that while newer variations do share a lot with the original conceptualisation of AcciMap (e.g. they all share the basic assumption of system levels, and interrelations between them), a large variation between the applications was also found.

Steps 5–8 of Figure 13.3 illustrate the development of AcciMap in the domain of safety practice. In the practitioner community, AcciMap has established itself as an accident analysis tool in many organisations that are involved with accident investigations. For instance, Finnish Safety Investigation Authority (SIA), which investigates all major accidents in Finland, has utilised standardised AcciMaps in most of their investigations since 2013. Finnish Safety and Chemicals Agency (Tukes) also uses AcciMaps in many of their investigations. However, and consistent with the findings of Waterson et al. (2017), the way in which AcciMaps are applied by these practitioners differs from Rasmussen's original conceptualisation. SIA's approach to AcciMap deviates from the original AcciMap by the addition of an event sequence timeline explicating the concrete events that took place during the accident – the phenomena at different levels of the

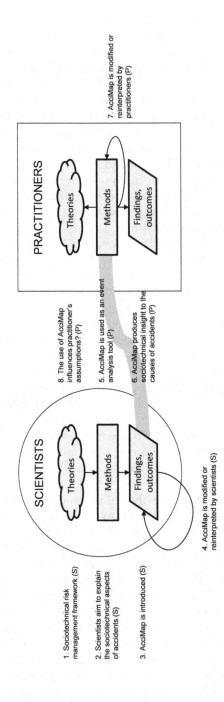

FIGURE 13.3

An illustration of the interaction between scientists and practitioners in the case of AcciMap. The events suffixed with 'S' relate to scientists' cultural elements and the events suffixed with 'P' relate to practitioners' cultural elements. The grey line in between indicates the (probable) connection between the two cultures: AcciMap can be seen as a result of scientific studies for scientists, but an analysis tool for practitioners.

sociotechnical system are then associated with each of the events in timeline without making assumptions of causality (Figure 13.4d). This differs from the standardised AcciMap variation proposed by Branford et al. (2009), who stress the importance of causal, rather than temporal ordering of AcciMap graph elements. Tukes used another interesting variation of AcciMap in 2004 for analysing an explosion at a pulp factory. This variation made a distinction between normal events, contributing events and direct causes (Figure 13.4c). The different types of events were coloured green, yellow and red, respectively. Despite the adaptations, both SIA's and Tukes' variations of AcciMap do include the very essential of AcciMap, which is the analysis of the interactions between the levels of sociotechnical system.

What remains unclear is how the use of AcciMap has influenced the 'theories' of practitioners? Specifically, how do the practitioners' assumptions or meanings given to safety-related phenomena change as they routinely apply AcciMaps to understand accidents or events? One possibility is that the practitioners will embrace a more system-oriented view. Following this line of thought, we can interpret AcciMap as a vehicle for conveying assumptions from scientists to practitioners. Essentially, for the scientific community, AcciMap can be viewed as an output and tangible summarisation of a particular line of research (in this case system safety), which then can be introduced to the practitioners to help make sense of accidents, but also to promote the sociotechnical view of safety. Conversely, the practitioners might view AcciMap as an intuitive, easy to use method of accident analysis that helps structure and visualise the analysis. As a side process, their actual ways of thinking change.

This has several implications in the context of actionable safety science. If AcciMap is a carefully crafted tool by scientists based on previous scientific work on sociotechnical systems, what is then the effect of practitioners' contextual reinterpretation and modification of the tool on the message that AcciMap embeds? For example, some applications abandon the use of causal connections between the elements. This changes the assumptions of the method. How does this change of assumptions affect the assumptions of practitioners? Does AcciMap still effectively promote a sociotechnical view of safety, or does it promote a different, implicit and unknown safety model? Consequently, how can the practitioners' knowledge and assumptions related to how *they* use AcciMap be introduced back to scientists?

13.3.2 Safety Culture

International Atomic Energy Agency (IAEA) coined the concept of safety culture in the aftermath of the Chernobyl nuclear power plant accident in 1986. The IAEA defined safety culture as follows: *Safety Culture is that assembly of characteristics and attitudes in organisations and individuals which establishes that, as an overriding priority, nuclear plant safety issues receive the attention warranted by their significance* (IAEA, 1991, p. 4). The purpose of the concept

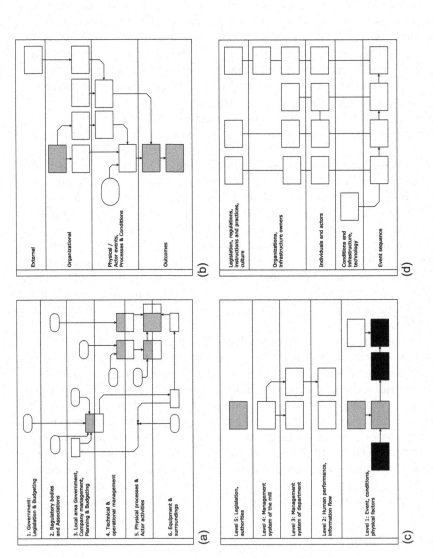

FIGURE 13.4

Examples of different types of AcciMaps: (a) Original type (style adapted from Rasmussen and Svedung, 2000), (b) 'Standardised' AcciMap with strict requirements of causality and vertical flow (style adapted from Branford, 2011), (c) Tukes variation with colours indicating different types of factors and with event sequence (style adapted from Partanen et al., 2004) and (d) SIA variation with event sequence (style adapted from SIA, 2016). (AcciMaps a–b was to highlight that in addition to technical and human factors, social and organisational aspects need to be taken into consideration when assuring nuclear safety. While the concept has originated from the nuclear industry, safety culture has also been utilised in other safety-critical industries, especially offshore, process industries and healthcare.

Figure 13.5 shows the timeline of major safety culture-related events from the nuclear industry and (handpicked) topics of interest in practitioner and scientist communities. We chose nuclear industry as the context because safety culture was first introduced here, and in order to retain readability. The figure also shows the number of safety culture-related journal articles published annually. This number serves as an indicator of how the scientific relevance of the topic has progressed over time.

International Nuclear Safety Group (INSAG), a group of safety experts that provides recommendations on emerging nuclear safety issues, prepared the seminal work (INSAG reports) on safety culture. We interpret this work as originated from practitioner's community, as there were no explicit references to or basis on previous scientific work (cf. Cole et al., 2013). However, their use of cultural concepts was most likely inspired by the trends in management sciences in the 1980s. In the practitioner community, the focus was on how the concept of safety culture can be used for safety management, specifically, how to assess the state of safety culture, and how to practically improve it in operational power plants. Later, there was an expansion of focus in the practitioner community. Application of safety culture expanded from purely operational contexts towards preoperational phases of the nuclear power plant lifecycle triggered by, for example, QA issues found during the construction of Olkiluoto 3 plant (STUK, 2006) and to regulatory regimes, which were highlighted after the Fukushima accident (IAEA, 2015).

Nowadays, safety culture has been heavily institutionalised in the nuclear industry in the form of regulatory harmonisation (WENRA, 2014) and industry-wide standardisation as it is established in IAEA safety standards (IAEA, 2006, 2009, 2016a). This suggests that the concept has become embedded in the practitioner's formal theories. Steps 1–4 of Figure 13.6 illustrate this process.

The way in which safety culture has been viewed in the scientific community has strong connections and parallels to previous scientific work, for instance through cultural concepts in the field of anthropology and organisational management such as Schein's theory of organisational culture (Schein, 2010). Many concepts in safety science that preceded safety culture have similarities which were most likely influential in the conceptualisation and further development of safety culture, namely the man-made disasters theory (Turner, 1976) and safety climate (Zohar, 1980). However, safety culture did not spark major interest among scientists until 1998 and 2000 when the journals *Work and Stress* and *Safety Science* published special issues on safety culture. Only few years after this, we see a clearly rising trend in safety culture publications (Figure 13.5). This indicates that the concept

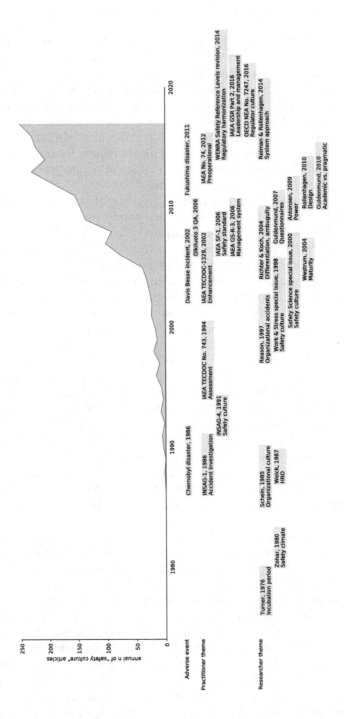

FIGURE 13.5

Safety culture timeline including notable adverse events in the nuclear industry, significant research themes and significant practitioner themes. The line plot illustrates the annual number of scientific journal articles that relate to safety culture until 2017. The statistics for the line plot were retrieved from Scopus using the search term TITLE-ABS-KEY((safety culture)) AND DOCTYPE("ar").

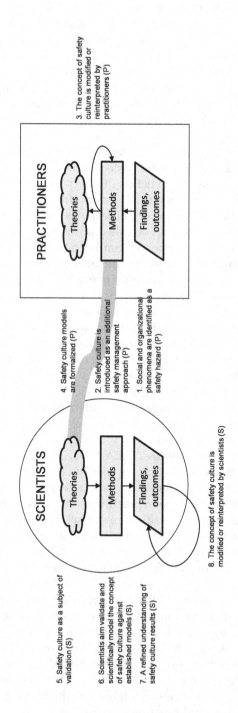

FIGURE 13.6

An illustration of the interaction between science (left) and practice (right) in the case of safety culture. The events suffixed with 'S' relate to scientists' cultural elements and the events suffixed with 'P' relate to practitioners' cultural elements. The grey line indicates the (probable) cultural elements that are connected between the two cultures: safety culture is primarily a safety management method for practitioners but a framework and a subject of research for scientists.

started to establish itself in the research community. A lot of the publications at this point focussed on differentiating safety culture from other related terms (especially safety climate), developing models (e.g. Cooper, 2000; Guldenmund, 2000) and creating an understanding of what are valid ways of assessing it (e.g. Hale, 2000). Afterwards, topics such as questionnaires (e.g. Guldenmund, 2007), power relations (Antonsen, 2009), relation to technical design (e.g. Rollenhagen, 2010) and the relation between safety culture and system safety (Reiman and Rollenhagen, 2014) emerged. Steps 5–8 of Figure 13.6 illustrate this work.

The scientific work on safety culture has resulted in various conceptualisations and models of safety culture. These conceptualisations range from descriptive studies on the social construction of safety to normative models of ideal safety culture/climate dimensions. A basic difference in approaches to organisational culture is whether the organisation is seen as having a culture or being a culture. The former can be called a variable based approach and the latter an interpretative approach (Smircich, 1983). By and large, the models of safety culture bear more resemblance to the variable based approaches than to the interpretative ones (Reiman and Rollenhagen, 2018).

In the early days of its development, safety culture appeared to be an extremely pragmatic concept – one that was used to further any area of safety management not covered by technical and human factors studies. At that point, there was not much research done with a specific focus on safety culture. To some extent, such approach to safety culture is still used by the practitioners. Over time, the scientific work on safety culture certainly shows an influence on the practitioner community. The documentation and tools aimed to practitioner audience are increasingly referring to established scientific work on organisational culture. For instance, IAEA's safety culture materials make heavy use of Schein's theory of organisational culture to illustrate the multilevel nature of culture (IAEA, 2002, 2016b), and some documents even include extensive theoretical supplements (e.g. IAEA, 2016b).

It is, however, unclear, whether the work done in scientific domain converges or diverges from the work done in practitioner domain. If the topics of interest are highly diverging, the practitioners and scientists might not find any common ground. For instance, a lag – and a lack of foresight – between themes of interest can be observed. It was not until major scientific journals published special issues that safety culture slowly became a relevant topic for research in the mid-2000s – 20 years after its first introduction by practitioners. Additionally, scientific community has not been able to provide insight in a timely and proactive manner into topics practitioners have lately shown interest in, including safety culture improvement, pre- and post-operational contexts or regulation. These observations suggest a disconnect between the two communities: is there a fundamental misunderstanding resulting from differing assumptions regarding the nature and use of the concept of safety culture between the two communities which hinders the convergence of the themes that interest scientists and practitioners?

13.4 Discussion

In this chapter, we illustrated how a tool developed by researchers, AcciMap, was adopted and modified by practitioners, and how a concept introduced by practitioners, safety culture, became infused with the theories of culture and safety culture of the scientific community. In both examples, we found little reflection 'across the borders' from practitioners to scientists, or vice versa. This implies that there is a gap between the two communities. Expanding the previous work done on science–practice gap, we suggest that the gap is not only a question of knowledge and its creation but also, and more importantly, a question of differences of assumptions.

Many of these assumptions are embedded in the tools and concepts created by one community and adopted and modified by the other community. The concepts, methods and tools very seldom contain the technical specifications for their use, or the 'safety instruction' about their misuse. The translation of those tools from scientific community to practitioner community, vice versa, or within one community, is very often left to individual members of their respective community. Gradually, new assumptions are formed and embedded into the other community's culture of how the tool is used and understood. Simultaneously, the community where the concept or tool is originated from continues to use and develop their version of the concept, further widening the gap between the two communities of safety research and practice.

How much do safety scientists actually reflect scientific practice and its basic assumptions? Or the assumptions underlying the results, and the subsequent tools and methods, as they are taken into use by practitioners, and transformed and adapted to their community? The same question need be raised for the practitioners. Weick (2001) questions whether practitioners are interested in a nuanced, scientific and accurate understanding of the world and whether this accurate view is even needed in today's complex and rapidly changing environment where being 'roughly right and fast' can mean success over someone who is 'exactly right and slow'. However, this should be an empirical question worthy of research interest. Other potential areas of interest for research raised by this chapter include:

- The roles and practices of scientists and practitioners that would support the formation and utilisation of actionable safety knowledge,
- The ways of coproducing actionable safety knowledge, including ways of feeding back experience on the use of this knowledge in practice,
- The processes of adaptation of tools or concepts to a given context, including the gradual embedding and change of assumptions concerning their tool, and
- The possibility to use some of the existing theories to describe and explain the gap, such as drift (Snook, 2000) and normalisation of deviance (Vaughan, 1996).

To be actionable in the long term, safety science needs diverse paradigms: It needs basic research but also applied research. However, more than anything, it needs to reflect the premises on which it is based. This includes reflecting on the tools and concepts adopted and adapted by practitioners, and studying the process of adaptation. More specifically, there is a need for a deeper understanding on how scientific results gradually are incorporated into daily practice and how they gradually transform as practitioners try to make sense of them based on their own context and its constraints and possibilities. Collective research and coproduction seem viable options for jointly making sense of how safety science and safety practice interact and influence each other.

References

Almklov, P. G., Rosness, R., and Størkersen, K. (2014). When safety science meets the practitioners: Does safety science contribute to marginalization of practical knowledge? *Safety Science*, Volume 67: 25–36.

Anderson, N., Herriot, P., and Hodgkinson, G. P. (2001). The practitioner-researcher divide in Industrial, Work and Organizational (IWO) psychology: Where are we now, and where do we go from here? *Journal of Occupational and Organizational Psychology*, Volume 74: 391–411.

Antonsen, S. (2009). Safety culture and the issue of power. *Safety Science*, Volume 47: 183–191.

Argyris, C., Putnam, R., and Smith, D. M. (1985). *Action Science*, 1st ed. San Francisco: Jossey-Bass.

Beier, P., Hansen, L. J., Helbrecht, L., and Behar, D. (2017). A how-to guide for coproduction of actionable science. *Conservation Letters*, Volume 10: 288–296.

Branford, K. (2011). Seeing the big picture of mishaps: Applying the AcciMap approach to analyze system accidents. *Aviation Psychology and Applied Human Factors*, Volume 1: 31–37.

Branford, K., Hopkins, A., and Naikar, N. (2009). Guidelines for ACCIMAP analysis. In A. Hopkins (Ed.), *Learning from High Reliability Organisations*, pp. 193–212. Sydney, Australia: CCH Australia Ltd.

Cash, D. W., Borck, J. C., and Patt, A. G. (2006). Countering the loading-dock approach to linking science and decision making: Comparative analysis of El Niño/ Southern Oscillation (ENSO) forecasting systems. *Science, Technology, & Human Values*, Volume 31: 465–494.

Cole, K. S., Stevens-Adams, S. M., and Wenner, C. A. (2013). *A Literature Review of Safety Culture* (No. SAND2013-2754). Albuquerque, NM: Sandia National Laboratories.

Cooper, M. D. (2000). Towards a model of safety culture. *Safety Science*, Volume 36: 111–136.

Ginsburg, M. B., and Gorostiaga, J. M. (2001). Relationships between theorists/researchers and policy makers/practitioners: Rethinking the two? Cultures thesis and the possibility of dialogue. *Comparative Education Review*, Volume 45: 173–196.

Guldenmund, F. W. (2000). The nature of safety culture: A review of theory and research. *Safety Science*, Volume 34: 215–257.

Guldenmund, F. W. (2007). The use of questionnaires in safety culture research – an evaluation. *Safety Science*, Volume 45: 723–743.

Hale, A. R. (2000). Culture's confusions. *Safety Science*, Volume 34: 1–14.

IAEA. (1991). *INSAG-4. Safety Culture* (No. 75- INSAG-4). Vienna: International Atomic Energy Agency.

IAEA. (2002). *Safety Culture in Nuclear Installations: Guidance for Use in the Enhancement of Safety Culture*. Vienna: International Atomic Energy Agency.

IAEA. (2006). *Fundamental Safety Principles* (Safety Fundamentals No. SF-1). Vienna: International Atomic Energy Agency.

IAEA. (2009). *Safety Assessment for Facilities and Activities* (General Safety Requirements Part 4 No. GSR Part 4). Vienna: International Atomic Energy Agency.

IAEA. (2015). *The Fukushima Daiichi Accident*. Report by the Director General. Vienna, Austria: International Atomic Energy Agency.

IAEA. (2016a). *Leadership and Management for Safety* (No. GSR Part 2). Vienna: International Atomic Energy Agency.

IAEA. (2016b). *Performing Safety Culture Self-Assessments*. Vienna: International Atomic Energy Agency.

Kirwan, B. (2000). Soft systems, hard lessons. *Applied Ergonomics*, Volume 31: 663–678.

Le Coze, J.-C. (2013). New models for new times. An anti-dualist move. *Safety Science*, Volume 59: 200–218.

Le Coze, J.-C., Pettersen, K., and Reiman, T. (2014). The foundations of safety science. *Safety Science*, Volume 67: 1–5.

Partanen, J., Palmén, M., and Munukka, A. (2004). Räjähdys Sunila Oy:n laimeiden hajukaasujen järjestelmässä 19.10.2004. p. 23.

Provan, D. J., Dekker, S. W. A., and Rae, A. J. (2017). Bureaucracy, influence and beliefs: A literature review of the factors shaping the role of a safety professional. *Safety Science*, Volume 98: 98–112.

Rasmussen, J. (1997). Risk management in a dynamic society: A modelling problem. *Safety Science*, Volume 27: 183–213.

Rasmussen, J., and Svedung, I. (2000). *Proactive Risk Management in a Dynamic Society*. Karlstad, Sweden: Swedish Rescue Services A.

Reiman, T., and Pietikäinen, E. (2014). The role of safety professionals in organizations–developing and testing a framework of competing safety management principles. Presented at the Probabilistic Safety Assessment and Management PSAM, Honolulu, Hawaii.

Reiman, T., and Rollenhagen, C. (2014). Does the concept of safety culture help or hinder systems thinking in safety? *Accident Analysis & Prevention*, Volume 68: 5–15.

Reiman, T., and Rollenhagen, C. (2018). Safety culture. In N. Moller, S. O. Hansson, J.-E. Holmberg, and C. Rollenhagen (Eds.), *Handbook of Safety Principles*, pp. 647–676. Hoboken: John Wiley & Sons.

Reiman, T., Rollenhagen, C., Pietikäinen, E., and Heikkilä, J. (2015). Principles of adaptive management in complex safety–critical organizations. *Safety Science*, Volume 71, Part B: 80–92.

Rollenhagen, C. (2010). Can focus on safety culture become an excuse for not rethinking design of technology? *Safety Science*, Volume 48: 268–278.

Schein, E. H. (2010). *Organizational Culture and Leadership*. San Francisco, CA: John Wiley & Sons.

SIA. (2016). Vakava Vaaratilanne Helsinki-Vantaan Lentoasemalla 28.10.2016. Safety Investigation Authority.

Smircich, L. (1983). Concepts of culture and organizational analysis. *Administrative Science Quarterly*, Volume 28: 339–358.

Snook, S. A. (2000). *Friendly Fire: The Accidental Shootdown of U.S. Black Hawks Over Northern Iraq.* Princeton, NJ: Princeton University Press.

STUK. (2006). *Management of Safety Requirements in Subcontracting During the Olkiluoto 3 Nuclear Power Plant Construction Phase* (Investigation Report No. 1/06). Helsinki, Finland: Säteilyturvakeskus.

Turner, B. A. (1976). The organizational and interorganizational development of disasters. *Administrative Science Quarterly,* Volume 21: 378–397.

Van de Ven, A. H., and Johnson, P. E. (2006). Knowledge for theory and practice. *Academy of Management Review,* Volume 31: 802–821.

Vaughan, D. (1996). *The Challenger Launch Decision: Risky Technology, Culture, and Deviance at NASA.* Chicago: University of Chicago Press.

Waterson, P., Jenkins, D. P., Salmon, P. M., and Underwood, P. (2017). 'Remixing Rasmussen': The evolution of AcciMaps within systemic accident analysis. *Applied Ergonomics,* Volume 59, Part B: 483–503.

Weick, K. E. (2001). Gapping the relevance bridge: Fashions meet fundamentals in management research. *British Journal of Management,* Volume 12: S71–S75.

WENRA. (2014). *Safety Reference Levels for Existing Reactors.* Update in Relation to lesson learned from TEPCO Fukushima Dai-Ichi accident. Western European Nuclear Regulators' Association.

Zohar, D. (1980). Safety climate in industrial organizations: Theoretical and applied implications. *Journal of Applied Psychology,* Volume 65: 96.

14

Safety Research and Safety Practice: Islands in a Common Sea

Steven T. Shorrock

CONTENTS

14.1 Views from Elsewhere..225
    14.1.1 Research and Practice in Safety...226
14.2 Research Practice in Transportation Safety: An Experiential
    Account..227
14.3 Reflections on Research and Practice: A Pact Analysis......................229
    14.3.1 Safety Researchers...229
    14.3.2 Safety Practitioners...230
14.4 Safety Research and Safety Practice: The Current Situation.............233
    14.4.1 The Research Does Not Address Practitioners' Day-to-Day
        Difficulties, Dilemmas, Concerns and Needs...........................234
    14.4.2 Research and Practice Outputs Are Not Consistently
        Accessible..235
    14.4.3 There Is Too Much Research...235
    14.4.4 There Is a Lack of Time and Inclination to Read the
        Products of Each Other's Work..236
    14.4.5 Safety Practitioners and Safety Researchers Spend Little
        Time Together, Especially on Collaborative Activities..............236
14.5 Hope for a Better Future: Four Possibilities.......................................237
    14.5.1 Collaborate – Collaborate at All Stages to Research on
        What Matters in Practice and in a Way That Will Make a
        Difference to Practice...237
    14.5.2 Adapt – Make Research Products More Practice Focussed.....239
    14.5.3 Read – Incorporate More Reading and Awareness of
        Current Research into Practice..240
    14.5.4 Write – Translate Research for Practitioners and Others,
        and Write about Practice..241
14.6 Conclusion...241
References..242

Introduction

As a reader of a book on safety research, it is quite likely that you are a safety researcher, probably working for a university or research institute. Developing a deep understanding of safety research is not a habit for most safety practitioners, just as developing a deep understanding of safety practice is not a habit for most safety researchers.

This is problematic, and it is the reason for this chapter. Safety can be seen as two islands (Figure 14.1). One is the research island, the home of the safety discipline, with safety researchers and professors associated with safety departments in universities, safety research centres and associations, conducting safety research, presenting at safety conferences, writing safety articles and books.

The other is the practice island, the home of the safety profession, with safety practitioners in primary, secondary, tertiary and quaternary industries, and in governmental and intergovernmental organisations.

The degree of interaction between the islands of research and practice would seem especially important for safety; safety is one of very few disciplines and professions that also relates to a value, a feeling, an activity (e.g. a safety check) and an artefact (e.g. a safety barrier); also, it is subject to laws, regulations and standards. One might expect many bridges and ferry crossings so that: (a) safety research informs safety practice (e.g. theory, methods, data), (b) safety practice informs safety research, raising questions so that research approaches the right kinds of problems and opportunities in an ecologically valid way, producing useful and usable products. You might also expect that the research–practice relationship in safety would be a thriving area of inquiry and discussion. There is, however, little research that directly addresses the topic. This will be the theme of this chapter. I contend that researchers and practitioners remain, to use the words of aviator and writer Anne Morrow Lindbergh, *'islands – in a common sea'*. A number of inhabitants from each island might be able to cross this common sea, and visit the other island; other inhabitants might be able to see across to the other island's shore. But what about the rest?

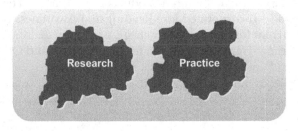

FIGURE 14.1
Research and practice: islands in a common sea.

14.1 Views from Elsewhere

The research–practice relationship has attracted reviews and empirical studies in:

- Healthcare, including nursing, occupational therapy, physical therapy, physiotherapy, speech and language therapy, dietetics and social work
- Library and information science (LIS)
- Human resources (HR)
- Industrial, work and organisational (IWO) psychology
- Human factors (HF) and ergonomics.

Research in these fields has found a significant 'research–practice gap', otherwise referred to as the 'academic-practitioner divide', the 'researcher-practitioner divide' and the 'scientist–practitioner split' (Anderson, Herriot and Hodgkinson, 2001; Gelade, 2006; Rice, 1997). The bottom line is that academics and researchers do not pay much attention to one another's activities and outputs. Two of the fields above that clearly relate to safety are industrial, work and organisational psychology and human factors. These disciplines have researched the issue via questionnaires, focus groups, content analysis and debates (e.g. Anderson et al., 2001; Blanton, 2000; Cascio and Aguinis, 2008; Chapanis, 1967; Chung and Shorrock, 2011; Chung and Williamson, 2018; Green and Jordan, 1999; Nowicki and Rosse, 2002; Meister, 1992, 1999; Sackett and Larson, 1990). To summarise, consistent findings in these fields are as follows:

- *Published research has a university bias:* The vast majority of research published in journals is university-derived research by academics. There has been a gradual decline in the proportion of research by practitioners, and research from industry and government facilities.
- *Published research has limited relevance to practitioners:* Practitioners believe that journal research is not driven by workplace problems and practitioners' local difficulties, dilemmas, concerns and needs. Research is increasingly empirical, methodological and analytical, with a focus on control and explanation, driven by questions, variables, methods and theories that lack relevance to practitioners. Therefore, research only has a limited generalisability to the organisational environment; its applications and implications are unclear.
- *There is a tension between research and practice communities:* There is a disconnect and mutual tension between researchers and

practitioners. Academics regard practitioner approaches as lacking in rigour. Practitioners regard academic approaches as being impractical. Both see the approaches of the other as lacking in validity, but each focuses on a range of different aspects of validity. Researchers and practitioner communities are largely separate, usually only coming together during conferences.

- *Practitioners lack access to and awareness of research and do not integrate it into practice:* Practitioners lack access to research, lack awareness of research, read little research and rarely discuss research. Journal published research is rated low as a source of knowledge for practice, especially relative to discussions with colleagues and personal experiences, and especially for more experienced practitioners. Research is, to a very large degree, not integrated into practice.

Blanton (2000) noted that, 'If we consider practitioners as the customer and research as the product, most research journals would go the way of New Coke' [a failed product with low consumer demand] (p. 239). Researchers appear to be both the suppliers and the customers of research.

14.1.1 Research and Practice in Safety

The safety literature includes few papers on the research–practice relationship. What are the attitudes of safety practitioners towards safety research? To what extent do safety practitioners read safety research? Do they integrate it into their practices? If so, how? The work of safety practitioners does not feature very strongly in safety research, but where it does, the relationship between research and practice is rarely mentioned. For instance, there are surveys of safety practitioners (e.g. Brun and Loiselle, 2002), and deeper qualitative and quantitative analyses of safety practice data collected over several years (e.g. Karanikas, 2015a), but the role of *research* in practice typically does not feature in the studies of practitioners.

As with HF and IWO psychology, the research has a strong university bias. Recent issues of *Safety Science*, *Journal of Safety Research* and *Accident Analysis and Prevention* show that (a) the vast majority of authors are affiliated with universities and research organisations, and (b) a large majority of papers appear to be authored exclusively by researchers. (Also, most are available only to researchers; the articles that I located when writing this chapter – even those on safety practice – are mostly not freely available to practitioners.)

One safety practitioner role (albeit a minority role) that does receive some attention from the research–practice perspective is that of safety investigators. In one of the few interview studies concerning the research–practice gap (focussed on systemic accident models), Underwood and Waterson (2013) reported on interviews with 42 participants (accident investigators, health and safety professionals, HF specialists and researchers). They reported that, 'the majority of practitioners remain unaware of the most frequently cited systemic analysis models, i.e. STAMP, FRAM and AcciMap. ... This is in contrast to the responses of the researchers who were interviewed and indicates that knowledge and use of these models is greater within the scientific community' (p. 156). Underwood and Waterson stated that their study was indicative of a lack of systemic accident model usage within industry and provided evidence of a research–practice gap. According to the authors, a different understanding of systemic accident models exists between the two communities. This study was, however, relatively small and concerns only one aspect of safety.

Other research on safety investigators in the context of safety research has been approached via desktop reviews of accident investigation reports. This research has found a limited impact of research on practice (academic propositions might have not yet affected practice dramatically, Karanikas, 2015b), and suggested wider adoption of research products (use more complex and sophisticated scientific theories and notions, Stoop and Dekker, 2012). Interestingly, research on investigator competency includes competencies such as empathy, imagination, working to (regulatory) standards and effective writing skills, but does not include specific competencies relating to reading, discussing or integrating research into practice (Nixon and Braithwaite, 2018).

On the whole, it appears that day-to-day safety practitioners' *practice* with regard to safety research is largely unexplored, especially via surveys and ethnographic research.

14.2 Research Practice in Transportation Safety: An Experiential Account

I write this chapter having spent almost exactly half of my life working in safety management, as a practitioner (primarily) and as a researcher. In these 22 years, I have worked in a number of industries, but principally transportation (aviation and rail), where I have had involvement:

- as a safety and human factors practitioner in an air navigation service provider (NATS, UK).
- as a safety and human factors consultant in an international consultancy (DNV).
- as an academic in a School of Aviation (University of New South Wales, Australia), researching safety and human factors, and teaching flying and management students.
- as a safety system and human factors practitioner in a train operating and rail infrastructure company (RailCorp, Australia).
- as a part-time sole trader (rail safety and border security, Australia).

- most recently as a safety and human factors practitioner in the safety unit of an intergovernmental air navigation organisation (EUROCONTROL, Europe) – acting as an external consultant to other organisations.

In these transportation roles, I have provided internal and external safety support to commercial, governmental and intergovernmental organisations in over 20 countries. This experience has included: developing safety management systems, writing safety cases, investigating and analysing incidents, doing safety assessments, designing and evaluating both interfaces and working environments, evaluating staffing arrangements, designing and delivering safety training, managing safety for major technical projects, performing safety culture assessments and research, developing safety methods, writing and editing safety articles and editing safety magazines.

Data collection activities have involved interviews, focus groups and ethnographic observation with over two thousand staff in operational, technical, management and support roles. In the last 8 or so years as part of my role in an intergovernmental organisation, I have had regular contact (meeting formally and informally, several times per year) with over 100 aviation safety specialists, safety managers and safety directors who work in around 40 countries.

In these roles, I have also edited, authored and co-authored the following kinds of outputs relating to safety: industry reports (confidential and publicly available), white papers, journal articles, an edited book, edited books of conference proceedings, magazine articles, magazine issues (as editor) and a blog. These activities, particularly as an editor, have provided considerable contact with safety researchers and practitioners in a range of industries (see, for instance, Sharples and Shorrock, 2014, Sharples, Shorrock and Waterson, 2015; Shorrock and Williams, 2016a; EUROCONTROL 2017a,b).

Most of this experience has been in what is often considered an ultra-safe industry – commercial aviation; an industry that is widely held up by other industries as an example of good practice in safety.

Therefore, it seems reasonable to explore the research–practice relationship through the lens of this experience alongside a background understanding of the research–practice relationship in human factors (e.g. Chung and Shorrock, 2011), especially in the absence of wider research on the research-practice relationship in safety. So, in this chapter, I reflect on my experience of over 22 years of:

- being a safety practitioner, safety researcher and safety research-practitioner.
- what others have said to me about research and practice, both informally in conversation at conferences and forums, and formally during safety culture surveys.

- what I have seen when working with safety practitioners and safety researchers.
- what I have read from safety practitioners and researchers.

In the sections that follow, I begin with an analysis of the research and practice context. I then outline what I believe to be the current situation in safety, as well as some opportunities. I have framed many of my reflections specifically within my experience of system safety in aviation. However, from my experience of other industries (certainly rail and chemical manufacturing) the same problems and opportunities apply.

14.3 Reflections on Research and Practice: A Pact Analysis

In this section, I will paint a picture of safety research and practice using a framework of people, activities, contexts and tools (PACT). This is necessarily driven by my experience, but readers will be able to compare this with their own experience of safety researchers and practitioners.

14.3.1 Safety Researchers

By 'researcher', I have in mind someone whose primary role is to conduct research; usually employed by a university, research organisation, or research division. My experience of safety researchers has been via the following:

- being taught and supervised by safety researchers at undergraduate and postgraduate level.
- working as a safety researcher in a university
- supervising postgraduate safety researchers as an academic supervisor.
- working as a practitioner while contracting scientific support from universities and working in partnership with university researchers throughout joint programmes.
- contact with researchers as an editor and author.
- contact via conferences and social media.

Researchers naturally tend to be highly educated, very often to doctoral level. The typical route following doctoral research is post-doctoral training and a series of academic posts. Some primary activities and demands for researchers are in Box 14.1.

BOX 14.1   SOME PRIMARY ACTIVITIES AND DEMANDS FOR SAFETY RESEARCHERS

- Writing research proposals
- Conducting research (reviewing literature, developing aims and hypotheses, using research methods to collect research data, performing analysis, generating and testing theories and methods)
- Reading research publications
- Writing research publications
- Attending and presenting at conferences
- Teaching and supervising
- Working on committees and fulfilling university requirements

Safety research activity is shaped by the academic environment of universities and grant-awarding research bodies. Progression through research posts tends to be driven primarily by research income and (high-ranking) academic publications (for those on a research track, especially in high-ranking, research-intensive universities). Williams and Salmon (2016) sum up that *'The bottom line is that if the work isn't publishable, it has little benefit to the researchers and universities involved'* (p. 145). For researchers, it makes sense from a career perspective to remain employed by universities or research institutes, where the values, goals, demands, incentives and resources (people, environments, equipment, information, money, time) are best suited to winning grants and having research published. Moving to and from other sectors (e.g. commercial) is, broadly speaking, detrimental to publication and therefore career progression.

Safety methods and tools are a key focus of safety research, and those of interest to researchers tend to be more 'advanced', which often means more complicated and requiring more resources (competence, information and time). Some become part of a series of publications, often used by no-one other than developers and close research associates (Shorrock and Williams, 2016b). Others are frequently used and cited in scientific literature but not in commercial organisations (Underwood and Waterson, 2013).

14.3.2 Safety Practitioners

By 'practitioner', I have in mind someone whose primary role is to help identify, understand and manage stakeholder problems and opportunities (in particular, as internal or external consultants, advisors and trainers). My contact with safety practitioners has been via the following:

- working as a safety practitioner.
- working alongside safety practitioners as direct colleagues.
- working with safety practitioners from other organisations (as an external consultant, contractor, contract manager, group/committee member).
- teaching transportation safety practitioners (on short courses and on a distance-learning transportation HF post-graduate degree).
- contact with practitioners as an editor and author.
- contact with practitioners via conferences, social media, etc.

There are also 'research–practitioners' who are involved with both research and practice, for whom one function does not dominate day-to-day activity. This third group is, in my experience, the smallest of the three, but is the group to which I identify. I am therefore not typical of safety practitioners, for whom the research–practice gap is greater, but have spent a lot of time with safety practitioners who have little contact with research or researchers.

In transportation, safety practitioners tend to come from two main backgrounds. One background is as a front-line operational or technical specialist. In air traffic management, many safety specialists did their training as air traffic controllers or technical engineers/technicians. Some retain their licences and continue to perform a few hours of operational or technical duties each month, while some operational and technical specialists perform a few hours of safety duties each month. In most cases, the safety specialists that I have encountered, in around 40 countries, have no university education in safety. Many have no university education at all; it was not required for their original (especially operational) post. Instead, they learned safety management and associated activities on the job, by undertaking safety training courses (usually not provided by universities). This is significant to the research–practice relationship.

The other main background of safety practitioners is as graduates of a variety of programmes. In aviation this tends to mean science, technology, engineering and maths subjects (e.g. avionics, aeronautical engineering, mathematics, physics). Very few enter the profession with a safety-related or a research degree, and very few proceed to obtain either. There are also human factors specialists, who have typically graduated in human factors/ ergonomics, psychology, design or engineering. However, this is a very small group; most transportation organisations (including regulators) have no suitably qualified and experienced human factors specialists.

Strictly speaking, the term 'profession' ordinarily applies where there are certain barriers to entry, such as prolonged study and formal qualifications, membership of professional associations, as well as an adherence to a set of particular professional standards and codes of conduct. However, in many cases this is not required, and people perform roles to fulfil organisational and regulatory requirements, often learned on the job, on a full- or part-time basis.

There are, of course, other stakeholders for safety practice (e.g. operational and technical staff, project managers, middle managers, senior managers, allied specialists and support staff, service users). These other stakeholders are even less likely to read safety research products or interact with safety researchers.

Safety practitioners in aviation, and many other highly regulated industries, spend most of their working time in activities related to safety, or the management system and safety regulation. This work will tend to involve the activities in Box 14.2.

BOX 14.2   SOME PRIMARY ACTIVITIES AND DEMANDS FOR SAFETY PRACTITIONERS

- Conducting safety risk assessment
- Safety occurrence reporting and investigation and analysis
- Conducting safety surveys
- Conducting safety audits
- Doing safety promotion (including training)
- Performing safety monitoring
- Maintaining safety records and documentation
- Writing safety policies
- Responding to internal demands and external demands (e.g. from regulators, airports, air navigation service providers, airlines, suppliers, etc.)

Safety activity, especially in high-hazard industries (e.g. aviation, nuclear, oil and gas, chemical manufacturing), is oriented and shaped by regulation and safety management systems. In air traffic management within the European Union, for instance, safety practice in air navigation service providers is shaped primarily by standards, recommended practices, regulations and work programmes of International Civil Aviation Organisation (ICAO), the European Aviation Safety Agency (EASA) and national regulators. Work is also influenced by intergovernmental organisations (e.g. EUROCONTROL), industry bodies (e.g. CANSO) and public-private partnerships (e.g. SESAR Joint Undertaking). Together, these organisations directly influence most safety work by air navigation service providers.

Most of the work relates to demands from planned and unplanned changes and events inside and outside of the organisation. This work occurs over a timescale of days (e.g. notifying National Supervisory Authorities of occurrences; see EUROCONTROL, 2017c), weeks (e.g. a safety survey or occurrence investigation), months (e.g. a safety assessment), or years (e.g. a safety culture programme or major change).

In terms of reading activity, the practice context directs practitioners to standards and recommended practices (e.g. ICAO), regulations (e.g. European and national regulators), and reports produced internally (e.g. relating to safety investigation and assessment) and externally (e.g. relating to external changes that may affect safety). Depending on the level of interest of safety practitioners, they may also read safety promotion material (e.g. magazines such as EUROCONTROL's *HindSight* magazine). Safety journal research is not, however, a primary source of knowledge for transport safety practitioners. From my discussions, a very small proportion of transport safety practitioners are frequent readers of academic journal articles. Many will not read so much as a single article in a year. Few are involved in safety research that leads to publications in peer-reviewed journals and books. As Kirwan (2000) noted, the order of importance of reading material to practitioners in a company is more-or-less the reverse of the academic ordering.

While researchers are involved to some extent in some safety activities, in general there is little contact between safety practitioners and safety researchers, and little of the day-to-day activity that contributes to safety practice (Box 14.2) is a focus of researchers' activity (Box 14.1).

Safety practitioners use a number of tools and methods in their activities. These are sometimes centralised and used by many users, often situated in different organisations, sometimes internationally, and often embedded into computer software. They are used routinely, change slowly and cannot easily be replaced or adapted. Some are mandated by regulation or are an acceptable means of compliance (e.g. the risk analysis tool [RAT] in European air traffic management).

14.4 Safety Research and Safety Practice: The Current Situation

From the brief analysis above, several important conclusions can be drawn on for safety research and practice in education, experience, values, interests and perspectives.

1. *People*: Safety researchers and practitioners have very different backgrounds (e.g. levels of formal education, industry experience) and aspirations.

2. *Activities*: Safety researchers and practitioners perform very different activities, which are driven by different demands. They routinely read different material, with little overlap.

3. *Context*: Safety research and practice activity is shaped by different goals, values, pressures, incentives, resources, constraints and environments.

4. *Tools*: Safety researchers and practitioners use different tools in order to perform their activities, and value different things in the tools they use.

In this section, I outline five challenges that make it difficult for safety research and practice to align. However, if we are to improve the research-practice relationship, we must attend to them.

14.4.1 The Research Does Not Address Practitioners' Day-to-Day Difficulties, Dilemmas, Concerns and Needs

As shown above, safety practitioners' day-to-day activities, contexts and tools have little overlap with those of safety researchers. Their difficulties, dilemmas, concerns and needs also differ greatly. For safety practitioners these may involve, for example:

- compliance with regulatory demands (e.g. time limits on reporting incidents).
- keeping up with incident investigation (e.g. providing feedback).
- keeping up with changes and managing associated requirements (e.g. safety assessments).
- getting management commitment to safety.
- obtaining adequate resources (e.g. full time staff, operational and technical expertise).
- communicating and collaborating with internal and external parties.
- involving operational staff in safety activities.
- making effective recommendations.
- handling difficult relationships (e.g. with senior managers or operational staff).
- understanding relatively new safety issues and domains (e.g. drones, digitisation, cybersecurity).
- understanding repeating incidents.

Without formally studying the people, activities, contexts and tools of safety practice, and without having acted in the role of a safety practitioner for a meaningful period of time, it would be difficult for researchers to understand safety practice. But when research ignores safety practitioners,

practitioners will struggle to see value in research (especially when research is not accessible).

14.4.2 Research and Practice Outputs Are Not Consistently Accessible

While safety researchers may have wide access to academic journals and databases, safety practitioners often have no more access than the general public. This is one of the most significant barriers to research application in HF/E (Chung and Shorrock, 2011). In recent years, this situation has improved. Some authors now ensure that open access is included in funding, and so some articles and even whole journal issues, are open access. Practitioners may be unaware of the gradual shift. Still, for many mainstream journals, open access remains rare. For the four issues of *Safety Science* that were published while this chapter was being written (March, April, June and July 2018), only 7 out of 110 articles were open access.

Purchasing individual articles is rarely an option. The current cost per article for *Safety Science* is US$39.95. This makes it unlikely that safety practitioners would request permission from their organisations to purchase articles, which can require more administrative effort than is worthwhile for an unknown return on investment. As a practitioner, seeing that *most* articles are available exclusively for purchase can be a major disincentive to even go looking for research.

It is also true that researchers have very limited access to practitioner outputs; most safety data and reports are not publicly accessible (exceptions include major accident reports and information on fatalities). This means that safety researchers cannot easily ascertain how practitioners collect and analyse data, write reports; and therefore cannot easily understand the current state of affairs in safety practice.

14.4.3 There Is Too Much Research

Can there be too much research? Today, we have hundreds of journals that are relevant to safety, publishing hundreds of safety-related articles per month. For practitioners (and, I suspect, researchers), this is overwhelming. Safety practitioners cannot hope to even keep up with research developments within their own specific industry, even where access exists. Many wouldn't know which journals might be relevant. In the context of HF/E, practitioners rated the dispersal of research in multiple locations as a major barrier to reading (Chung and Shorrock, 2011).

The increasing size and number of journals appears to be related not to the needs of practitioners, but to the academic incentive system and its associated pressure to publish, the ambition to be respected, and academic tenure and job opportunities (Meister, 1999; Norman, 1995; Singleton, 1994; Wilson, 2000).

14.4.4 There Is a Lack of Time and Inclination to Read the Products of Each Other's Work

Even if all of the outputs of safety researchers and practitioners were freely available and accessible, I am not convinced that this would make a fundamental difference to safety researchers and practitioners, whose (mismatched) backgrounds, goals, activities and contexts act as a deterrent to reading each other's work. The contextual issues here are:

- *Temporal*: For practitioners, there is often insufficient time to read lengthy research papers (or even to meet day-to-day demands with the thoroughness that practitioners would like to afford the activities). In the context of HF, practitioners cited lack of time to read research as one of the top five biggest barriers to research application (Chung and Shorrock, 2011).
- *Educational*: Research does not lend itself to easy reading by practitioners, especially under time pressure, and written in a language other than one's own. Technical industry reports may have little meaning to researchers.
- *Social*: Because people read the same sorts of things at work as their peers, there is little social incentive to read or discuss anything else.
- *Economic*: Reading research publications carries direct and indirect financial and opportunity costs.
- *Regulatory*: There is no regulatory requirement to read, participate in, or conduct research, or to write publications.

Safety researchers and practitioners have little need to read the products of each other's work in order to perform their activities and meet their goals.

14.4.5 Safety Practitioners and Safety Researchers Spend Little Time Together, Especially on Collaborative Activities

As noted above, safety researchers and safety practitioners have little need to meet, spend time together, or stay in frequent contact with one another in order to perform their activities and meet their goals. Neither perceives the other to be a primary stakeholder. They also have little contact with each other's stakeholders, activities, contexts or tools.

There also appears to be a lack of organisational interest in research. This has been evidenced by the closure of many research divisions in organisations. A recent example was the closure of the Liberty Mutual Research Institute for Safety, which opened in 1954 and remained a significant contributor to applied non-proprietary research in occupational safety and ergonomics until it closed in 2017.

14.5 Hope for a Better Future: Four Possibilities

This section outlines four possibilities for researchers and practitioners to improve research application in practice. In effect, this means trying to increase the overlap between research and practice, so that research is meaningful and relevant to the practitioners who help to shape and propagate research.

14.5.1 Collaborate – Collaborate at All Stages to Research on What Matters in Practice and in a Way That Will Make a Difference to Practice

a. More contact and communication. The most frequent suggestions made by practitioners and researchers for improved research application in Chung and Shorrock's (2011) study were increased collaboration, communication and networking between researchers and practitioners, especially in-person communication and networking between practitioners and researchers (in person and online) to allow for a greater sharing of experience, knowledge, research findings, approaches and implementation strategies. Many respondents suggested more practice-oriented and research–practice conferences. Conference programmes need to move towards more interactive and participative exchanges in order to understand the needs of safety practitioners and industry.

b. Focus on real world safety problems and opportunities and the needs of safety practitioners. To be relevant and useful safety research should:

- stem from real-world problems that have been identified in the real environment together with safety practitioners.
- involve workplace-based scenarios and situations, and workforce members.
- generalise to real-world performance.
- be meaningful and relevant to safety practitioners, and help them in their work.
- where relevant, produce outputs (e.g. tools and methods) using a human-centred design approach, to ensure that they are useful and usable by safety practitioners in their context (with reasonable adaptations).

Safety research might answer a specific applied question and help meet an expressed need, while at the same time contributing to broader knowledge.

c. More industry experience for researchers. Expertise in safety requires expertise in safety theories and methods and expertise in the context of industry. How much can we understand of safety in highly regulated industries – with the associated demands, pressures, personnel, equipment, competencies, regulations, policies, procedures, incentives, punishments and activities – without immersive experience in those industries? Without this experience, we may think we know more than we do.

The academic context is not supportive of job changes that reduce research publication output. But researchers must have access to organisations to be able to understand and investigate problems in real-world contexts (Kirwan, 2000; Singleton, 1994). Some possibilities are as follows.

First, experienced safety researchers could take sabbaticals, internships or part time roles in industry. Hyatt et al. (1997) thought that this could uncover what really matters to organisations. This information could be taken back to the university, informing research projects to address the issues faced by the specific organisation. The organisation would then benefit from having the latest knowledge as well as an external perspective.

Second, organisations could integrate experienced safety researchers into formal company processes in the longer term, as an adjunct or in a part time role, in the same way that practitioners are sometimes adjunct researchers. This could involve offering strategic advice or oversight or acting as a critical friend and link to the research.

Third, formal links could be formed between academic institutions and industry organisations to (a) conduct joint research–practice safety project, and (b) supervise and mentor postgraduate or post-doctoral students working on research projects within the industry. Co-supervision of postgraduates can improve access to research for organisations, while academics can learn more about the field of practice with which they are involved (Jarvis, 1999). This was my entry to air traffic management.

I have seen the latter two suggestions work well in transport organisations (see also Sharples, 2016). More industry experience helps to build stronger connections, alliances and partnerships between researchers and practitioners. It helps researchers to understand the context, current challenges and needs of the industry; and, working together with practitioners, it helps co-create practical solutions to problems and opportunities. It also facilitates publication of applied, collaborative research, including case studies. However, it does require that practitioners raise the awareness of the importance of research and new thinking within organisations and governments and gain support for applied research.

d. Increase practitioner participation in research. Evidence from industrial, work and organisational psychology suggests that practitioner involvement in the research process leads to findings that practitioners are more likely to find useful and implement, and other academics are more likely to cite (Rynes, 2007).

Safety practitioners could be involved in each phase of research:

- understanding the need for research real-world practical problems.
- planning and designing the research.

- conducting the research.
- writing, tailoring and publishing the research.
- integrating the research into practice.

The benefits of researcher–practitioner collaboration might include:

- experiencing professional growth.
- obtaining high-quality, ecologically valid organisational data.
- sustained impact on organisational effectiveness.
- communicating research to others.

Of course, for practitioners to become real partners in research requires that they actually want this role, which may require resources (e.g. time, travel budget), support and incentives.

14.5.2 Adapt – Make Research Products More Practice Focussed

a. Make research freely accessible and widely promoted. In order to enable practitioner access, relevant and readable research information must be openly shared and distributed throughout a range of accessible channels at no cost. There is no impediment to this. Even where open access fees cannot be built into research funding (these fees will rarely be paid for by practitioner organisations), drafts can be placed on university, personal or social media websites. Articles that are made freely available online have a far greater impact than publications that require payment (Kingsley, 2007). At the time of writing, eight of the ten most downloaded *Safety Science* articles are open access. Open access safety journal articles are increasing, but slowly, and the percentage of open access articles remains low. Syntheses, summaries and reviews are likely to be of particular interest to practitioners and should be provided freely, either in the journal itself or via widely publicised and easily available drafts.

It should be noted, however, that journal publications have been identified as one of the most passive and ineffective forms of communication of research information to practitioners in the allied disciplines (Schlogl and Stock, 2008; Sitzia, 2001). Researchers need to use a broader range of media (e.g. blogs) and adapt outputs to meet practitioner needs.

b. Report text and statistical analyses in a more understandable, clear and readable manner. As Singleton (1994) noted, for practitioners and others in industry, trying to use research can prove to be baffling and frustrating. It is curious that writing in academic journals and books is often so far removed from other writing in everyday life, e.g. in newspapers and magazines. If researchers wish that practitioners read their outputs, then writing must be more inclusive. This requires a clear, concise writing style, using plain language and less academic or scientific jargon; using clear headlines and summaries.

c. Make implications clear. Salas (2008), editor of *Human Factors* from 2000 to 2004, stated that authors do not take the time to provide precise implications for practice or system design. He recalled that reviewers of rejected articles commonly said, 'Who cares about this issue?' and 'Why are the author(s) examining this problem?' or 'Where is the need?' As editor, Salas recalled that, as much as he pushed some authors, 'they were resistant to the problem or apathetic or lacked know-how' (p. 353).

Meister (1999) suggested that articles should include at least one paragraph specifically outlining the applications of the research. This would require an editorial policy that encourages the inclusion of practical implications in research articles (Gelade, 2006). In Chung and Shorrock's (2011) study, the need for more tangible conclusions and practical recommendations was the second most frequent suggestion aimed at researchers. This requires a 'translation' of theoretical constructs and research results into real practice (e.g. implications for the safety practitioner activities), perhaps in a standardised 'practical applications and implications' section. This activity would be best done via collaborating with safety practitioners.

14.5.3 Read – Incorporate More Reading and Awareness of Current Research into Practice

While it might be unhelpful to suggest that safety practitioners simply 'read more', maintaining competence via continuous professional development is one of six criteria generally recognised as being necessary of professionals (Rice, Duncan and Deere, 2006).

How can we incorporate this into practice, while still being mindful of the barriers outlined above? Possibilities include:

- incorporating reading research into development plans.
- building literature-reviewing into project planning time, and as part of the billed activity.
- discussion groups based on key articles or methods.
- changing work practices based on specific research findings to enhance performance.

Better understanding of theory and methods is likely to correlate with effectiveness (Breedveld and Dul, 2005). This requires an open minded and reflexive attitude from safety practitioners, who could usefully question their safety theories, approaches and methods, and reflect on personal and system barriers to new thinking and good practice (Shorrock, 2013). Self-reflection might prompt practitioners to assess the value of research (as well as their reading and use of that research), including the time they spend trying to understand implications arising from safety research.

14.5.4 Write – Translate Research for Practitioners and Others, and Write about Practice

a. Develop research translation expertise. Safety practitioners and researchers must summarise and translate research into the language and context of industry (as well as other practitioners), alongside with their application ideas. This can be done for particular concepts, theories, methods, or individual articles that are of particular relevance, perhaps in safety practitioner magazines, blogs, social networking websites and newsletters. Expanding the publication of research into a wider range of media would allow for quicker and easier access, increase awareness of research findings, create greater interest in research findings and encourage more integration of research into practice.

b. More practitioner research and publication. If researchers are to better understand the context of safety practice, practitioners have to write; yet lack of time, inclination, or reward means that practitioners seldom conduct research or publish papers (Norman, 1995). Inspection of some of the recent volumes of *Safety Science* and *Journal of Safety Research* shows that relatively few papers are written by practitioner authors.

Practitioners – or practitioner-researchers – could share information about:

- ordinary work and system behaviour, perhaps using basic ethnographic and systems thinking approaches.
- industry and practitioner problems, opportunities and needs.
- how organisations apply theory, methods and research findings in the real world.

This would help researchers to understand the realities of practice (people, activities, context and tools), and current applied difficulties and dilemmas. It would also give researchers feedback on the relevance, acceptance, use and ecological validity of their work.

Attracting articles from practitioners can take a lot of encouragement, coaching and patience, however. Thus, better support, incentives and outlets are required to encourage more practitioner writing, from researchers, publishers, practitioners and practitioner organisations.

14.6 Conclusion

This chapter has attempted to explore the research–practice relationship in safety, from the perspectives of experience and research from other fields. I have highlighted fundamental differences in the people, activities, context and tools in safety research and practice. From this analysis, I outlined some challenges facing safety research and practice, as well as some ideas from the literature (and from experience) that, if implemented by individuals and organisations, could possibly improve the current situation. Researchers and practitioners need to find ways to collaborate at all stages to research what matters in practice and in a way that will make a difference to practice. Researchers must adapt to make research products more practice-focussed; practitioners must find ways to incorporate more reading and awareness of current research into practice. Practitioners must also try to translate research for practitioners and others and write about practice. Some of these are factors over which individual researchers and practitioners, along with their organisations, have significant control over. Therefore, if it matters to us, we should take action.

But we have to be realistic. The idea that practice is the mechanical application of research products is a misguided notion of 'technical rationality' (Schon, 1993) – 'practice-as-imagined'. Schon remarked that, *'when research-based theories and techniques are inapplicable, the professional cannot legitimately claim to be expert but only to be especially well prepared to reflect-in-action'* (p. 345). We are yet to understand the day-to-day practice and reflection-in-action of safety practitioners, nor their needs from safety research. What is clear is that for practitioners, research is just one element that informs and influences 'the craft of practice'. The problem is that, currently, research seems to be a very weak influence. At best, this means a huge amount of wasted funding on research that does not inform practice. At worst, it may contribute to losses to life, assets and the environment. The research–practice gap is a practical and ethical problem in safety that we have ignored for too long.

<hr>

References

Anderson, N., Herriot, P. and Hodgkinson, G. P. (2001). The practitioner–researcher divide in industrial, work and organisational (IWO) psychology: Where are we now, and where do we go from here? *Journal of Occupational and Organisational Psychology*, Volume 74, 391–411.

Blanton, J. S. (2000). Why consultants don't apply psychological research. *Consulting Psychology Journal: Practice and Research*, Volume 52(4), 235–247.

Breedveld, P. and Dul, J. (2005). *The position and success of certified European ergonomists*. Rotterdam, The Netherlands: RSM Erasmus University, in collaboration with the Centre for Registration of European Ergonomists (CREE).

Brun, J.-P. and Loiselle, C. D. (2002). The roles, functions and activities of safety practitioners: The current situation in Québec. *Safety Science*, Volume 40(6), 519–536.

Cascio, W. F. and Aguinis, H. (2008). Research in industrial and organisational psychology from 1963 to 2007: Changes, choices, and trends. *Journal of Applied Psychology*, Volume 93(5), 1062–1081.

Chapanis, A. (1967). The relevance of laboratory studies to practical situations. *Ergonomics*, Volume 10(5), 557–577.

Chung, A. Z. Q. and Shorrock, S. T. (2011). The research–practice relationship in ergonomics and human factors – surveying and bridging the gap. *Ergonomics*, Volume 54(5), 413–429.

Chung, A. Z. Q. and Williamson, A. (2018). Theory versus practice in the human factors and ergonomics discipline: Trends in journal publications from 1960 to 2010. *Applied Ergonomics*, Volume 66, 41–51.

EUROCONTROL. (2017a). *HindSight. Work-as-imagined and work-as-done*. Issue 25, Summer. Brussels: EUROCONTROL. Retrieved from https://www.skybrary.aero/index.php/Hindsight_25

EUROCONTROL. (2017b). *HindSight.: Safety at the interfaces: Collaboration at work*. Issue 26, Winter. Brussels: EUROCONTROL. Retrieved from https://www.skybrary.aero/index.php/Hindsight_26

EUROCONTROL. (2017c). Special section on regulation 376/2014 and the law of unintended consequences. *HindSight. Work-as-imagined and work-as-done*. Issue 25, Summer. Brussels: EUROCONTROL. [Online Supplement.] Retrieved from https://www.skybrary.aero/index.php/Hindsight_25

Gelade, G. A. (2006). But what does it mean in practice? The *Journal of Occupational and Organisational Psychology* from a practitioner perspective. *Journal of Occupational and Organisational Psychology*, Volume 79, 153–160.

Green, B. and Jordan, P. W. (1999). The future of ergonomics. In M. A. Hanson, E. J. Lovesey, and S. A. Robertson (Eds.), *Contemporary ergonomics 1999*. London: Taylor & Francis, 110–114.

Hyatt, D., Cropanzano, R., Finfer, L. A., Levy, P., Ruddy, T. M., Vandaveer, V. and Walker, S. (1997). Bridging the gap between academics and practice: Suggestions from the field. *The Industrial–Organisational Psychologist*, Volume 35(1), 29–32.

Jarvis, P. (1999). *The practitioner-researcher: Developing theory from practice*. San Francisco, CA: Jossey-Bass Publishers.

Karanikas, N. (2015a). Human error views: A framework for benchmarking organisations and measuring the distance between academia and industry. 49th ESREDA seminar: Innovation through human factors in risk assessment and maintenance, Brussels, Belgium.

Karanikas, N. (2015b). Evaluating the horizontal alignment of safety management activities through cross-reference of data from safety audits, meetings and investigations. *Safety Science*, Volume 98, 37–49.

Kingsley, D. (2007). The journal is dead, long live the journal. *On the Horizon*, Volume 15(4), 211–221.

Kirwan, B. (2000). Soft systems, hard lessons. *Applied Ergonomics*, Volume 31, 663–678.

Meister, D. (1992). Views on the future of ergonomics: Some comments on the future of ergonomics. *International Journal of Industrial Ergonomics*, Volume 10, 257–260.

Meister, D. (1999). *The history of human factors and ergonomics*. Mahwah, NJ: Lawrence Erlbaun Associates.

Nixon, J. and Braithwaite, G. R. (2018). What do aircraft accident investigators do and what makes them good at it? Developing a competency framework for investigators using grounded theory. *Safety Science*, Volume 103, 153–161.

Norman, D. A. (1995). On differences between research and practice. *Ergonomics in Design*, April, Volume 3(2), 35–36.

Nowicki, M. D. and Rosse, J. G. (2002). Managers' views of how to hire: Building bridges between science and practice. *Journal of Business and Psychology*, Volume 17(3), 157–170.

Rice, C. E. (1997). The scientist–practitioner split and the future of psychology. *American Psychologist*, Volume 52(11), 1173–1181.

Rice, V. J., Duncan, J. R. and Deere, J. (2006). What does it mean to be a "professional" … and what does it mean to be an ergonomics professional? [FPE Position Paper: Professionalism]. 1–11. Accessed 21 June 2019, https://ergofoundation. org/images/professional.pdf

Rynes, S. L. (2007). Editor's afterword: Let's create a tipping point: What academics and practitioners can do, alone and together. *Academy of Management Journal*, Volume 50(5), 1046–1054.

Sackett, P. R. and Larson, J. R. (1990). Research strategies and tactics in industrial and organisational psychology. In M. D. Dunnette and L. M. Hough (Eds.), *Handbook of industrial and organisational psychology* (2nd ed., Volume 1, pp. 419–489). Palo Alto, CA: Consulting Psychologists Press.

Salas, E. (2008). At the turn of the 21st century: Reflections on our science. *Human Factors*, Volume 50(3), 351–353.

Schlogl, C. and Stock, W. G. (2008). Practitioners and academics as authors and readers: The case of LIS journals. *Journal of Documentation*, Volume 64(5), 643–666.

Schon, D. A. (1993). The reflective practitioner. New York: Basic Books.

Sharples, S. (2016). Ergonomics and human factors research and practice: Lessons learned from Professor John R. Wilson. In S. Shorrock and C. Williams (Eds.), *Human factors and ergonomics in practice: Improving system performance and human well-being in the real world*. CRC Press 405–413.

Sharples, S. and Shorrock, S. (2014). *Contemporary ergonomics and human factors 2014*. London: Taylor & Francis.

Sharples, S., Shorrock, S. and Waterson, P. (2015). *Contemporary ergonomics and human factors 2015*. London: Taylor & Francis.

Shorrock, S. (2013). Why do we resist new thinking about safety and systems? [Blog post]. Retrieved from https://humanisticsystems.com/2013/04/12/why-do-we -resist-new-thinking-about-safety-and-systems/

Shorrock, S. and Williams, C. (2016a). *Human factors and ergonomics in practice: Improving system performance and human well-being in the real world*. Boca Raton, FL, CRC Press.

Shorrock, S. and Williams, C. (2016b). Human factors and ergonomics methods in practice: Three fundamental constraints. *Theoretical Issues in Ergonomics Science*, Volume 17, 5–6.

Singleton, W. T. (1994). From research to practice. *Ergonomics in Design*, July, Volume 2(3), 30–34.

Sitzia, J. (2001). Barriers to research utilisation: The clinical setting and nurses themselves. *European Journal of Oncology*, Volume 5(3), 154–164.

Stoop, J. and Dekker, S. (2012). Are safety investigations pro-active? *Safety Science*, Volume 50(6), 1422–1430.

Underwood, P., and Waterson, P. (2013). Systemic accident analysis: Examining the gap between research and practice. *Accident Analysis and Prevention*, Volume 55, 154–164.

Williams, C. and Salmon, P. (2016). The challenges of practice-oriented research. In S. Shorrock and C. Williams (Eds.), *Human factors and ergonomics in practice: Improving system performance and human well-being in the real world*. Boca Raton, FL: CRC Press.

Wilson, J. (2000). Fundamentals of ergonomics in theory and practice. *Applied Ergonomics*, Volume 31(2000), 557–567.

Section II

Pioneers

15

Safety Research: 2020 Visions

Rhona Flin

CONTENTS

15.1 Evolution 250
15.2 Challenges... 251
    15.2.1 The Gulf.. 251
    15.2.2 A Replication Crisis in Safety Science?................................. 252
    15.2.3 In Defence of Individual Analysis.. 254
15.3 New Directions 255
    15.3.1 Organisational Drift.. 256
    15.3.2 Safety Culture.. 256
    15.3.3 The Impact of New Technologies on Safety Management 258
15.4 Conclusion 259
References... 259

Introduction

As a cognitive psychologist who meandered into safety research (after studying face recognition and eyewitness testimony), I feel ill-qualified to offer pronouncements on the state of safety science or its evolutionary progression. Nonetheless, the temptation to preview the current perspectives coming out from the newer blood of safety researchers proved irresistible and the price was to provide some reflections. Undoubtedly, these are myopic and ingrained with the theoretical and methodological preferences of my own discipline.

The chapters presented an impressive array of theoretical approaches and enticing avenues for the safety researchers of the 2020s. It would be impossible to do the contributions justice in such a short commentary, given the depth and range of conceptualisation involved, but their accounts stimulated my own ruminations on where safety research could be heading, in relation to evolution, challenges and new directions.

15.1 Evolution

The progression in safety science from early normative models, through a preoccupation with deviance from rules and errors, to a range of descriptive approaches examining actual work activities, has been charted by Rasmussen (1997a), who noted that this progression reflected parallel trends in other fields, such as decision science (Rasmussen, 1997b). The extent to which the theoretical basis has evolved further in the decades since Perrow's (1984) normal accidents; the high-reliability organisation (HRO) case reports (Rochlin et al., 1989); Weick's (1995) sensemaking and Reason's (1997) organisational accidents is debatable. Certainly, when I began to work in safety science by the 1990s, there was a distinct focus on the organisation, as well as much talk of leading indicators rather than simply attending to accident data and lagging indicators. The HRO literature had shown the value of studying organisations that have low accident rates while operating in high risk settings. Similarly, Reason's (1990) work on human error emphasised latent conditions in the organisation – as well as sharp end failures – and his four safety paradoxes (Reason, 2000) remain relevant today. Taking this volume as an indicator of current research interests, significant attention is now being paid to more nuanced contextual factors and there are theoretical refinements, but these developments do not appear to constitute a paradigmatic shift in scientific thinking on safety.

According to Shorrock (this volume, Chapter 14), safety practitioners rarely read scientific journal articles; whilst I suspect this to be true, I believe that they do read accessible books on safety, such as the writings of Hollnagel (2014) and Dekker (2015). These authors have been advocating a revised approach to safety management similar to the conceptual shift outlined above. This advises trying to understand what goes well in organisational operations, rather than overly relying on the analysis of accident data and a concomitant obsession with safety incident statistics. Analysing events that went wrong does not provide a full understanding of why tasks can be executed successfully without the occurrence of adverse events. Similar views on the importance of studying the reality of work activities are made by Haavik (this volume, Chapter 7), although I am not entirely convinced that more neologisms are beneficial.

The recent versions of the 'new view' of safety (as I have heard it called by safety practitioners) emerged under the guise of resilience engineering (Hollnagel et al., 2006), which proposed flexibility, learning, adaptation, anticipation; echoing principles espoused by the HRO theorists and others. This has evolved into what Hollnagel (2014) now calls 'Safety-II', understanding the ability to succeed under varying conditions, whereas Dekker (2015) talks of 'Safety Differently' arguing, amongst other things, against the over-bureaucratisation of safety. Both approaches are characterised by a positive orientation. Within psychology, a similar shift in focus took place 20 years ago. Seligman, originally famous for his learned helplessness theory (Maier and Seligman, 1976), instigated a movement which became known as positive psychology (Seligman and Csikszentmihalyi, 2000), in this case advocating that psychologists should investigate well-being, happiness and life enhancing behaviours, as well as mental illness, maladaptive behaviours and negative thinking.

Practitioners seem to find positive safety an attractive prospect, and the logic of trying to understand effective and safe task execution, as well as appreciating the expertise of the workforce seems irrefutable, although the evidential basis for some of the component ideas is not extensive. Having recently listened to some practitioners espouse the wonders of 'the new view', I suspect that some consideration may need to be given as to whether this is being understood as an augmentation to existing practice, or as an alternative. Hayes (this volume, Chapter 12) argues convincingly that the risk of devaluing accident analysis could mean the loss of valuable information on system protection. Likewise, Downer (this volume, Chapter 5) commends the aviation industry for *fastidiously deconstructing accidents*. There definitely appears to be a risk of some significant safety tenets being thrown out with the 'old view' bathwater.

15.2 Challenges

There are considerable challenges in conducting applied research; safety science is no exception. I discuss three particular areas: the science/practice divide, standards of scientific investigation and the risks of demonising individual factors.

15.2.1 The Gulf

Shorrock's chapter, 'Safety Research and Safety Practice: Islands in a Common Sea' capitalises on his dual experience as a safety practitioner and a safety scientist in order to delineate the gulf that exists between the two camps. Reiman and Vitanen (this volume, Chapter 13), also address the same topic and comment that, *'practitioners seem to view safety science as disconnected from everyday problems'*, portraying the issue as one of assumptions. This divide is not peculiar to safety science, but it is undoubtedly a problem that can limit the performance of both parties. For industrial psychology research, it is almost impossible to collect data without the cooperation of managers and workforce members of the domain under investigation. Planning discussions with them can help to reduce the gap, especially if the project has a steering group from the practitioner community, able to help shape study designs and convey the findings that they could use in their daily practice.

My recent experiences studying aspects of patient safety, for example, into non-technical skills in operating theatres (Flin et al., 2015), has required a high level of collaboration, but with practitioners rather than safety managers. This may be closer to the type of coproduction of knowledge that Reiman and Vitanen describe. Working as co-investigators with surgeons, nurses and anaesthetists was challenging, as we learnt to recognise their concerns, priorities and styles of reporting. I preferred to publish this type of research in the scientific journals from my collaborators' areas of professional practice, as this was the audience we wished to inform and influence, rather than safety scientists or academic psychologists. I am aware that junior researchers in universities can be cautioned against this practice, as it might not be career-enhancing within one's own discipline. In our case, it also required some adaption of the presentation style we would use for psychology or safety journals. When we wrote the draft of our first empirical paper for a surgical journal, our surgeon co-author said, 'Very good but you'll need to lose 50% of it. Surgeons only have time to read concise papers'. The knock-on risks of unintended dilution and conceptual over-simplification when translating safety theory for audiences of practitioners is discussed by Bergström (this volume, Chapter 11) in his analysis of discursive effects. However, if safety scientists wish to influence workplace practices, they need to appreciate which scientific journals the technical practitioners read (and these are not typically safety journals).

Since 2014 in the United Kingdom, universities now have to prove on a regular basis (e.g. every six years) that a proportion of their research output has had an impact on policy or practice, in order to obtain government funding for their research. This carrot has encouraged academics to not only design their studies with applications in mind, but to make sure that they disseminate their findings to practitioners and policymakers. They have to be able to provide the evidence that their work has been applied by practitioners, and, ideally, to demonstrate that it has made a positive difference in some way. Safety researchers typically work closely with practitioner communities, but this type of academic incentive encourages us to gather the evidence on how our theories, tools or findings are being used in practice, and to find out whether or not they are having any effect. Having a better insight into the reality of safety practitioners' daily work (see Guennoc et al., 2019) is very helpful in this regard.

15.2.2 A Replication Crisis in Safety Science?

In the academic world, salutary lessons have recently been learned in psychological science that have implications for other disciplines, and therefore, possibly also for safety science. What is now known as 'the replication crisis' in psychology (Schmidt and Oh, 2016; Shrout and Rodgers, 2018) started with concerns regarding the robustness of findings from laboratory-based social psychologists, before developing into a wider state of alarm. The principal issue was the revelation of failures to replicate notable findings that had become established in the literature, principally on the basis of single reports. Setting aside dramatic cases of outright scientific fraud (see Bhattacharjee, 2013), the ramifications of scientific standards being questioned have forced psychology researchers to carefully examine not only their own scientific practices (which are primarily quantitative), but also the systems in place for reviewing and publishing research findings.

Many of the technical concerns relate to the favoured methods of statistical analysis, with over-reliance on significance testing to judge the presence of real effects, at the expense of paying appropriate attention to other parameters, such as context, effect sizes and confidence intervals. Other problems, such as the use of small samples, may be relevant to safety scientists who conduct experimental research on human performance in laboratory or simulator settings. Researchers who conduct surveys to collect their data generally gather large samples, but self-report questionnaires have their own weaknesses and almost all empirical findings require independent confirmation before they can be relied upon. Survey researchers are usually more inclined to devise their own questionnaire rather than use one designed by another team, unless the latter has become the established standard. Michael Frese, a psychology professor, once said at a conference that psychologists would rather use other people's toothbrushes than their questionnaires. The attraction in producing a new version is the kudos of authoring an apparently novel instrument. This means, as we have seen in quantitative studies of safety climate and culture, that a proliferation of different instruments exist, many of them ostensibly measuring the same concepts. This is not useful for safety managers looking for reliable measurement tools, and it is not conducive to proper scientific development of the field. Moreover, as Reader (this volume, Chapter 2) points out, survey data only provides an incomplete picture of culture and should be complemented by qualitative assessments.

Part of the cause of the replication crisis was that journal editors and reviewers – across disciplines – tended to favour novel findings over repeated studies using the same parameters, especially where non-significant effects had been found. Thus, we have the 'file drawer' problem in science (Rosenthal, 1979), also known as publication bias (Sterling, 1959), where none of the studies that failed to replicate an effect were published, even though there may have been several of them. Also, reviewers prefer smooth, coherent narratives; as such, they can sometimes advise that sub-studies and components of results that appear to add little (or detract from) the central argument, are dropped from a paper. While this might improve readability, it could actually be better for scientists to see the rough edges of a particular study. Moreover, the question might be asked if the specialist safety journals are doing enough to enhance research standards, for example: encouraging replication, pre-registration of study methods and hypotheses (to reduce data mining, post hoc theorising) and making data sets available for independent scrutiny.

Whilst initially uncomfortable, recent navel gazing in psychological science has proved to be extremely useful in opening up an overdue debate on method drift, publication practices and the evolution of our knowledge base. Safety science has a broad base of disciplines, not all of which analyse quantitative data. Nevertheless, the underlying message of the self-scrutiny of our investigation and publication practices is applicable, and does extend to qualitative methods. We could also ask if we have sufficiently tested the plethora of existing (and emerging) theories, or if we just have an endless proliferation of slightly different models and newish theories resulting in a tangle of neologisms. Perhaps more work needs to be done in order to develop a common language; to engage in an exercise of reviewing, evaluating and reaching an expert consensus on the value of our existing theories and measurement methods. This is already being achieved in other disciplines. For instance, in health psychology, an attempt was made to identify the main theories of behaviour change relevant to health improvements, and to systematically review the evidence on their efficacy and conditional constraints. The resulting consensus has created an extremely useful framework called the behavioural change wheel (Michie et al., 2013, 2014). Whilst it was initially designed for health researchers, the work has had clear applications for those interested in behaviour changes relevant to safety outcomes (see Chadwick, 2017).

15.2.3 In Defence of Individual Analysis

I am aware that focussing on worker behaviour, or on other aspects of the individual within safety research, now risks being labelled as 'Lutheran', 'bad apple theory', 'blaming the worker', 'shifting responsibility to the worker', and so on. Certainly, behaviour-based safety approaches (which are based on conditioning principles) are overly narrow as a workplace intervention, but they can include components, such as monitoring and reinforcement that have proven effects on altering behaviour patterns (see Gottcheva, Aaltonen and Kujala, this volume, Chapter 3). While psychologists specialise in studying the mental, emotional and behavioural life of individuals, this does not mean that they are ignorant of, or indeed disregard, the technical, environmental, political, organisational and social influences that pervade the workplace, acting as constraints or facilitators of human behaviour. Also, it does not imply they believe that workers should be deemed primarily accountable for the state of safety at a worksite. Studying human factors does not equate to advocating for the blaming of workers as the prime cause of accidents – what Perin (1995) called, 'the human factors, gotcha' technique of investigation.

While this might be unfashionable to state, there is a level of personal accountability for worksite safety in almost every occupation. As Rasmussen (1997a) pointed out, humans are not automatons entirely at the mercy of systems, rules, bureaucratic constraints, managerial foibles, manipulative incentive schemes and badly designed work environments. There can be choices about how work is done, and the celebration of human ingenuity and engagement in the resilience movement pays testimony to this autonomy. Nonetheless, humans do make errors, even if some now seem to find this fact unpalatable. Therefore, safety scientists need to understand how people think and feel about their work activities; so, at the risk of being called a 'behaviourist', I believe they also need to investigate the factors that influence behaviours during the execution of tasks. Bergström (this volume, Chapter 11) notes that the European aviation regulator has added resilience skills [which may or may not be achievable] to the latest Crew Resource Management training requirements, and he suggests that this reveals a belief *'that operational resilience is located at the level of pilot mental processes'*. Having just ploughed through these indigestible EASA documents (Flin, 2019), as far as I am aware, nowhere in these new regulations is it stated or implied that operational resilience is only located at an individual level. These are regulations focussing on the training requirements for pilots, so it is hardly surprising that they concentrate on the individual rather than the responsibilities of the operator which are covered elsewhere.

While it is perennially pointed out that we require multidimensional approaches to studying safety, the reality is that we have a multidisciplinary subject composed of various breeds of social scientist specialising at different levels of analysis. It is not wrong for some safety researchers to focus on trying to understand the cognitive, emotional, physiological and social components of the individual worker; nor is it inappropriate for others to analyse organisational power structures, hierarchies, systems, engineering practices or regulatory philosophy. Establishing false dichotomies does not help, either. It is not a question of examining power or culture; individual or collective phenomena; the workplace or the worker; safety I or safety II, safety the same or safety differently, error or resilience. Safety science has to encompass all of the above. What benefits safety science (see Le Coze, 2016) are concerted efforts to synthesise these perspectives, to create hybrid models and to 'merge our paradigms' as Rasmussen (1997a,b) illustrated in a series of diagrams thirty years ago. Le Coze (this volume, Chapter 10) persuasively shows how this type of theoretical visualisation might help to reconceptualise our thinking about safety paradigms.

15.3 New Directions

The chapters in this volume provide a host of new ideas for future safety research and I consider three of them in this final section.

15.3.1 Organisational Drift

The metaphor of organisations sliding towards danger zones has been used for some time: Rasmussen (1997) postulated a process of 'systematic migration', envisaged in the analysis of rail accidents as the 'unrocked boat' (Perin, 1992 cited in Reason, 1997), and of the space shuttle accidents under the guise of a normalisation of deviance (Vaughan, 1996) and organisational drift (Woods, 2005). This concept has been extended and linked to risk attenuation by Pettersen Gould and Fjæren (this volume, Chapter 8), who examine three types of drift: practical, organisational and societal. Failing to detect changes in the culture relating to risk tolerance underlies the work on organisational drift, which has broad implications for practice, and to some extent, could even explain what might have happened in the pockets of psychological science outlined above. They also argue that safety scientists should pay more attention to interest groups, media and NGOs who can influence safety agendas.

Translating to an individual level, drift can also occur as personal rule compliance and risk-taking starts to slip towards more hazardous practices or riskier behaviours. This can occur during the execution of a single episode of task execution, as studies of plan continuation error and goal seduction have demonstrated (Bearman et al., 2009). One antidote to the underlying build-up of complacency or risk attenuation which often underpins drift is to foster the state of 'chronic unease' – a tendency of wariness towards risks described by Reason (1997, p. 37) who called it, 'the price of safety'. This idea of a requisite level of anxiety for safety tends to feature strongly in the HRO literature, for instance Weick (1997, p. 119) refers to 'chronic suspicions'. Whilst the terminology might be a touch pathological for many in the industry, the central message itself is valid, and we have now begun to identify core components of the chronic unease concept (Fruhen and Flin, 2016), in addition to starting to develop some methods for its measurement, which we are currently testing.

15.3.2 Safety Culture

The study of safety culture continues to form a major topic of inquiry (Reiman and Rollnhagen, 2018) and is still being implicated in major accidents (e.g. NAS, 2016). Reiman and Vitanen (this volume, Chapter 13) point out that the scientists may be failing to provide insights into what really concerns practitioners, such as safety culture improvement, although new studies are starting to examine this aspect (Lee et al., 2019). New developments in this domain include somewhat overdue attention to the role of other types of culture present in organisations, such as the influence of national cultural differences. Reader (this volume, Chapter 2), who has been involved in large-scale studies from European air traffic control, demonstrates important consistencies in safety culture across national boundaries, but he also finds some intriguing interactions between common facets of national cultures (identified by Hofstede in IBM many years ago), such as uncertainty avoidance and safety culture dimensions. He argues for a SIGN. Antonsen and Almakov (this volume, Chapter 6) also discuss national culture in relation to the power distance variable. I would argue that their concern about bold pronouncements on national characteristics correlating with differences in safety from studies using Hofstede's model is justified. Not only are correlational data tricky to interpret; great care must be taken to not place an over-reliance on indicative psychometric measures. Again, with relevance to culture, Engen and Lindøe (this volume, Chapter 4) outline a risk regulation regime for a dominant and unionised industry (offshore oil and gas) in a small country (Norway). They consider the extent to which a tripartite system that works in this setting would work just as well within a larger jurisdiction (or in a different national or political culture entirely).

Gotcheva et al. (this volume, Chapter 3) exploring the risks in large, complex networked projects, which can span several countries, mention a performance culture, a powerful component in many organisations and rarely examined by safety scientists, even though it may be clashing with the safety culture. On similar ground, Waring and Bishop (this volume, Chapter 9) tackle the role of professional culture – something which is also infrequently considered within safety research, and, from my experience of working within healthcare, an extremely powerful force of constraint and enablement. Waring and Bishop provide a fascinating account of the norms of risk acceptance and self-regulation within the established professions, and how this may be threatened by the increasing professionalism of safety specialists. The clash of the professional culture with the safety system is certainly under-researched and while this is probably most noticeable in healthcare, their work alerts us to its possible presence elsewhere.

Another cultural trend beginning to emerge in safety research conducted in industry is a relatively new focus on product safety culture, which is concerned with the organisational culture that prioritises the safety of the users of produced goods. There is an established literature looking at organisational culture influencing the safety of service users, for example: patients in healthcare (Waterson, 2013) or passengers in aircraft being guided by air traffic controllers (Reader, this volume, Chapter 2). Compared to service users, consumers purchasing and using manufactured goods can be very distanced in time and space from the organisation that has a legal responsibility for the safety of the product (Conway, 2018). A continuing series of product failure events, including fatalities relating to faulty cars or domestic goods, shows the severity of the consequences (Suhanyiova et al., under review). The review of the crash of an RAF Nimrod aircraft in 2006 with 14 deaths, allegedly due to weaknesses in its redesign is subtitled: 'A Failure in Leadership, Culture and Priorities' (Haddon-Cave, 2009); contributing causes which regularly feature in industrial process accidents. However, it appears that workers in manufacturing who are demonstrating some concern for remote, unknown consumers might have distinct ethical and motivational attributes (De Boeck et al., 2017; Suhayionova et al., 2017). The ethical component has not been surfaced to the same degree in studies of worker and plant safety culture, although Hayes (this volume, Chapter 12) touches on a similar notion of a lack of 'professionalism'. Commenting on Pacific Gas and Electric staff interviews (after an accident) she notes, *'there was little sense from them that they feel a sense of professional responsibility for the results of their actions'*. This case study of the San Bruno pipeline rupture has a focus on public safety, a topic which Downer (this volume, Chapter 5) also considers in his discussion of epistemic accidents, drawing on cases of biohazards in agriculture and nuclear power plant events. The current legal cases on both public and worker safety involving DuPont and the manufacture of non-stick coatings (Richter et al., 2018) are now raising serious questions on both corporate and regulatory safety cultures. One promising approach to gathering external risk data concerning an organisation involves the systematic analysis of complaints of service users in search of safety failings (Gillespie and Reader, 2019).

15.3.3 The Impact of New Technologies on Safety Management

I was recently at an offshore drilling conference, where a safety manager showed me that on his phone, he could monitor a whole set of real-time parameters on the drilling rigs his company was currently operating across the globe. At the same meeting, there were presentations on the latest developments in protective monitoring, where sensors on large pieces of moving equipment could detect humans in collision range. The introduction of this type of device and other kinds of visual recording on worksites means that workers' behaviours are now increasingly being recorded during their task operations. While this could have considerable benefits for safety interventions and accident analysis, the employment implications for the workers might need to catch up with these developments. The airlines now routinely download data on technical parameters throughout a flight, and where these exceed a given margin, the pilots are then liable to be interviewed at a later time, in order to explain their actions. Naturally, airline pilots are used to having their voices recorded during flights, but in this industry there are powerful unions which have secured the necessary individual protections for the subsequent analysis of this type of material. For other occupations, the increasing use of on-site monitoring may not have been so thoroughly debated prior to implementation, and the safety benefits, data protection and employment law issues may still need to be established.

In a similar vein, Almklovv and Antonsen (this volume, Chapter 1) discuss the safety implications of the increasing use of standardisation of work procedures, which can be a forerunner to various forms of digitalisation, and, in some cases, automation. They acknowledge the benefits and attractions of this development, but argue that the downside can be a loss of 'situational adaptation' caused by a managerial failure to appreciate the uncaptured tacit knowledge (technical, social, organisational) that facilitates normal work activities, and can be critical for non-normal event management. This presents a particular aspect of the discrepancy that can exist between work as it is imagined, versus work as it is actually done; a recurring theme in the search for organisational resilience (Hollnagel et al., 2006). For a volume on new directions in safety research, there were surprisingly few mentions of innovative technologies, especially the shift to digitalisation and the overwhelming reliance on organisations' information technology systems. In their second chapter (Antonsen and Allmklov, this volume, Chapter 6), whilst arguing that there has been a neglect of the study of power in safety research, again make passing reference to the risks of cyber-attacks, remote control and monitoring of processes, as well as system dependencies, but do not expand any further. Le Coze (this volume, Chapter 10) touches on a suite of other, less familiar risks, which could have an impact on workplace safety and beyond, such as nanotechnology, synthetic biology and transhumanism. The safety scientists of the 2020s and beyond are likely to have to broaden their horizons and investigate a much wider range of emerging threats to human well-being and security.

15.4 Conclusion

The chapters in this volume serve to indicate that safety research as it is imagined by academics has greater differentiation than safety research as it is applied in practice. Behind idiosyncratic labelling and differences in theoretic positioning, considerable conceptual resonance exists across the disciplines of this field. This unusual volume makes a distinct contribution to the continuing search for common ground, thus strengthening the fundamental knowledge base of safety science.

References

Bearman, C., Paletz, S. and Orasanu, J. (2009). Situational pressures on aviation decision making: Goal seduction and situation aversion. *Aviation, Space and Environmental Medicine,* Volume 80, 556–560.

Bhattacharjee, Y. (2013). The mind of a con man. *New York Times,* April 26, 28.

Chadwick, P. (2017). Safety and behaviour change. In C. Bieder, C. Gilbert, B. Journe and H. Laroche (Eds.), *Beyond safety training: Embedding safety in professional skills.* FonCSI. Springer Briefs in Safety Management. Accessed 17.11.18. Retrieved from https://www.foncsi.org/en/publications/collections/springerbriefs.

DeBoeck, E., Mortier, A., Jacxsens, L., Dequidt, L. and Vlerick, P. (2017). Towards an extended food safety culture model: Studying the moderating role of burnout and job stress, the mediating role of food safety knowledge and motivation in the relation between food safety climate and food safety behaviour. *Trends in Food Science and Technology,* Volume 62, 202–214.

Dekker, S. (2015). *Safety differently: Human factors for a new era.* Boca Raton, FL: CRC Press.

Flin, R. (2019). Non-technical skills for European pilots. In B. Kanki, J. Anca, and T. Chidester (Eds.), *Crew resource management* (3rd ed.). New York: Elsevier.

Flin, R., Youngson, G. and Yule, S. (2015). *Enhancing surgical performance: A primer on non-technical skills.* London: CRC Press.

Fruhen, L. and Flin, R. (2016) 'Chronic unease' for safety in senior managers. *Journal of Risk Research,* Volume 19, 645–663.

Gillespie, A. and Reader, T. (2018). Patient-centered insights: Using health care complaints to reveal hot spots and blind spots in quality and safety. *The Milbank Quarterly,* Volume 96(3), 530–567.

Guennoc, F., Chauvin, C. and Le Coze, J. C. (2019). The activities of occupational health and safety specialists in a high-risk industry. *Safety Science,* Volume 112, 71–80.

Haddon-Cave, C. (2009). *The Nimrod review: An independent review into the broader issues surrounding the loss of the RAF Nimrod MR2 Aircraft XV230 in Afghanistan in 2006.* London: Stationery Office.

Hollnagel, E. (2014). *Safety-I and Safety-II: The past and future of safety management.* Farnham: Ashgate.

Hollnagel, E. Woods, D. and Leveson, N. (Eds.) (2006). *Resilience engineering. Concepts and precepts.* Farnham: Ashgate.

Le Coze, J. C. (2016). Vive la diversité! High reliability organisation (HRO) and resilience engineering (RE). *Safety Science,* online.

Lee, J., Huang, Y., Cheng, J., Chen, Z. and Shaw, W. (2019). A systematic review of the safety climate interventions literature. Past trends and future directions. *Journal of Occupational Health Psychology* (online first).

Maier, S. and Seligman, M. (1976). Learned helplessness: Theory and evidence. *Journal of Experimental Psychology: General,* Volume 105, 3–46.

Michie, S., Atkins, L. and West, R. (2014). *The behaviour change wheel: A guide to designing interventions.* Great Britain: Silverback Publishing.

Michie, S., Richardson, M., Johnston, M., Abraham, C., Francis, J., Hardeman, W. and Wood, C. (2013). The behavior change technique taxonomy (v1) of 93 hierarchically clustered techniques: Building an international consensus for the reporting of behavior change interventions. *Annals of Behavioral Medicine,* Volume 46(1), 81–95.

NAS (2016). *Strengthening the safety culture of the offshore oil and gas industry.* Washington: National Academy of Sciences.

Perin, C. (1995). Organisations as contexts: Implications for safety science and practice. *Industrial and Environmental Crisis Quarterly,* Volume 9, 152–174.

Perrow, C. (1984). *Normal accidents.* New York: Basic Books.

Rasmussen, J. (1997a) Risk management in a dynamic society: A modelling problem. *Safety Science,* Volume 27, 183–213.

Rasmussen, J. (1997b) Merging paradigms. In R. Flin, E. Salas, M. Strub, and L. Martin (Eds.), *Decision making under Stress.* Aldershot: Ashgate.

Reason, J. (1990). *Human error.* New York: Cambridge University Press.

Reason, J. (1997). *Managing the risks of organisational accidents.* Aldershot: Ashgate.

Reason, J. (2000). Safety paradoxes and safety culture. *Injury Control and Safety Promotion,* Volume 7(1), 3–14.

Reiman, T. and Rollenhagen, C. (2018). Safety culture. In N. Moller, O. Hansson, J. Holmberg, and C. Rollenhagen (Eds.), *Handbook of safety principles.* Chichester: Wiley.

Richter, L., Cordner, A. and Brown, P. (2018). Non-stick science: Sixty years of research and (in) action on fluorinated compounds. *Social Studies of Science,* Volume 48(5), 691–714.

Rochlin, G., La Porte, T. and Roberts, K. (1987). The self-designing high-reliability organisation: Aircraft carrier flight operations at sea. *Naval War College Review,* Volume 40(4), 76–90.

Rosenthal, R. (1979). File drawer problem and tolerance for null results. *Psychological Bulletin,* Volume 86, 638–641.

Seligman, M. and Csikszentmihalyi, M. (2000). Positive psychology: An introduction. *American Psychologist,* Volume 55, 5–14.

Schmidt, F. and Oh, I. (2016). The crisis of confidence in research findings in psychology: Is lack of replication the real problem? Or is it something else? *Archives of Scientific Psychology,* Volume 4, 32–37.

Shrout, P. and Rodgers, J. (2018). Psychology, science and knowledge construction: Broadening perspectives from the replication crisis. *Annual Review of Psychology,* Volume 69, 487–510.

Sterling, T. (1959). Publication decisions and their possible effects on inferences drawn from tests of significance—Or vice versa. *Journal of the American Statistical Association,* Volume 54, 30–34.

Suhanyiova, L., Flin, R. and Irwin, A. (2017). Safety systems in product safety culture. In L. Walls, M. Revie, and T. Bedford (Eds.), *Risk, reliability and safety: Innovating theory and practice.* London: Taylor & Francis Group.

Vaughan, D. (1996). *The Challenger launch decision. Risky technology, culture and deviance at NASA.* Chicago: University of Chicago Press.

Waterson, P. (Ed.) (2014). *Patient safety culture. Theory, methods, application.* Aldershot: Ashgate.

Weick, K. (1987). Organisational culture as a source of high reliability. *California Management Review,* Volume 29, 112–127.

Weick, K. (1995). *Sensemaking in organisations.* Thousand Oaks, CA: Sage.

Woods, D. (2005). Creating foresight: Lessons for enhancing resilience from *Columbia.* In W. Starbuck, and M. Farjoun (Eds.), *Organisation at the limit: Lessons from the Columbia disaster.* Oxford: Blackwell.

16

The Gilded Age?

Erik Hollnagel

CONTENTS

16.1 The Allure of Unitary Solutions 265
16.2 Confirmation Bias 266
16.3 Conclusion 267
References 268

Introduction

The stated aim of this book is to show how a new generation of safety researchers look at – or have looked at – safety; what main themes or problems they see, and how they pursue them in words and deeds. It is proposed they do this by applying, developing, expanding, adapting or challenging the existing concepts that were developed during the 1980s–1990s (often called the 'golden age' of safety research) with new disciplinary insights from a variety of fields, such as the sociology of science and technology; computer supported coordinated work; actor network theory; distributed, embodied or extended cognition; globalisation studies; strategic decision making and so on. To complement this, several 'key authors' who, in one way or another, are supposed to have contributed to the 'golden age', have been invited to comment and share their reflections on the new generation's approach to safety research from their own perspective and experience. Accordingly, this is what I shall endeavour to do on the following pages.

The chapters of this book certainly represent a number of aspects of safety that are accepted as worthy of further thought or attention by a vaguely defined 'safety community'. This sweeping interest in safety is possibly the reason why they show a diversity of views and positions rather than a coherent or unified approach. This lack of coherence is unfortunately symptomatic of the state of safety science at the present time (the end of the second decade of the 21st century), and of the many diverse concerns and interests that have been assembled under the safety umbrella. The question is whether the diversity should be seen as a strength or a weakness.

For nearly a century, safety has been a central concern for organisations and societies, initially with an emphasis on production and transportation, but gradually expanding to include the services that are deemed necessary to maintain a reasonably comfortable way of living in the industrialised world. In the beginning the concern was mainly related to accidents and injuries at work (Heinrich, 1931), but as everyday life became increasingly dependent on technology (transportation and communication, for instance), the concern was widened. Indeed, the 'golden age' of safety research seems to coincide with the rapid and often uncritical uptake of prolific technological innovation across societies. The unintended consequence of this was the occurrence of accidents that challenged the established understanding of safety and defeated the available methodologies (Hale and Hovden, 1998). The 'golden age' was thus more due to a need to think in new ways and to develop more powerful analytical tools – perhaps even new paradigms – than to a sudden spur of creativity and insights. Since the 1990s, however, other fields such as healthcare have been added (IOM, 1999), and most recently the amalgamation of safety and security concerns in the guise of cyber-risks have demanded attention (Masys, 2018). Despite significant technological progress the importance of safety has not diminished; it does not look likely to diminish. On the contrary, the eager and sometimes inconsiderate application of new technology is seen by many as being the primary reason for why safety is now a growing concern again. Indeed, most safety professionals will probably agree that safety is more of an issue today than it was in the 1930s, and that there are few (if any) areas of life and activity where we can feel truly safe and relaxed because we know that nothing can go wrong.

When multiple stakeholders – companies, authorities, politicians and public opinion – impatiently demand that the safety problem is solved once and for all, safety professionals, researchers and practitioners alike inevitably become more concerned with solving or eliminating the manifest problems – accidents, incidents, and so on – than with improving the understanding of what safety actually is. It is thus hardly surprising that an unspoken agreement of what safety is can be found throughout this book. Indeed, it is probably significant that few (if any) of the chapters bother to put forward a definition – or even their own definitions – of safety. Nowadays, to not do so is almost the norm; and, in fact, it is unusual to do otherwise. We usually take for granted that the people we interact with, both professionally and in everyday situations, understand what safety is and that it therefore is a meaningful concept for them. Whereas terms such as 'resilience' or 'actor network theory' may prompt people to ask for some explanation, the term 'safety' does not. Not only do people recognise the term 'safety' and feel that they understand it; they assume that others do so, too – moreover, they feel that others understand it the same way they do. Therefore, practically any treatise on safety tacitly assumes the existence of a common and shared understanding that discussions and analyses can be based on, and this book is no exception to that.

To some extent, this assumption is warranted. Although there may be variations in the definitions of safety, the variations are insignificant compared to the near universal agreement on the fundamental meaning of the term, namely that safety is a state where as little as possible goes wrong. This can be seen in the ALARP principle, which dates back to 1949, in ANSI's definition of safety as the freedom from unacceptable risk, and in the current concerns about the unintended and harmful behaviour that may emerge from poor design of real-world AI systems. There seems to be universal agreement among academics and practitioners from every field that the essence of safety is to be without something – harm, risk, injury, losses, and so on, in good accordance with the etymology of safety which comes from the Latin *salvus,* meaning 'uninjured'. Indeed, no one in their right mind could possibly disagree with the need to be without harm or without injury (except perhaps a masochist). But, as Reason (2000, p. 3) pointed out, 'safety is defined and measured more by its absence than by its presence'. The common understanding of safety thus refers to conditions where there is a (partial) lack of safety, rather than to conditions where safety is present. The unspoken assumption is that we can achieve a condition of perfect safety, of zero accidents and zero harm, by focussing on how harm occurs and by eliminating or attenuating the 'causes' of harm. It is therefore hardly surprising that this assumption can also be found in all of the chapters in this book. Yet if we heed Reason's warning, we should look to the presence of safety, to the situations where work goes well and where 'nothing' happens, corresponding to Weick's influential notion of dynamic non-events (Weick, 1987).

16.1 The Allure of Unitary Solutions

It is quite common that book chapters and academic papers present or represent a specific view on the chosen topic; that they choose depth rather than breadth. And the chapters of this book generally – with a few exceptions – do that. But given the nature of the subject matter – safety – it is not very helpful to imply, or even worse, to wholly advocate, that these problems could be solved by a single perspective or approach, or that there is one perspective that is more important than others.

The book provides examples of this in its discussions of standardisation; of the issue of power; of the role of epistemic accidents; of international regulation; of governance; of sensework; of accident investigation; of safety culture, national culture and of drift. Although the chapters mostly refrain from offering up their chosen perspective as a universal solution, they nevertheless neglect to acknowledge that safety – regardless of how it is interpreted – must look at how systems function from a broad perspective. As a whole, the book does represent the many different views on safety, but it is done chapter by chapter, without a commensurate attempt to pull things together. The editors, of course, feel the obligation to do that. But it would have been preferable if the new generation of safety researchers had considered it natural to do so themselves. It would seem appropriate in this day and age to embrace multifaceted solutions. The main issue with safety – and the reason why it is an ever-growing problem – is that the lack of safety is neither due to a single factor, nor can it be comprehended by a single view. Safety is not a unitary problem and there are no unitary solutions. A contemporary approach to safety should recognise that; it should try to embrace multiple perspectives in a comprehensible way. A few of the chapters attempt that, but it would have seemed natural if they had all done so.

As humans, we do have a preference for thinking of the world in terms of single themes or issues because it is easier for us to focus on a single concept or idea. Examples can easily be found in politics and science – and in business, as well. We tend to describe the world as black or white and use that as the basis of our explanations and communication. This reinforces the mistake that it is easier to control and communicate something that is binary, rather than something that is not.

A strategy of depth-before-breadth makes it possible to pursue a single idea until the objective has been achieved, while at the same time limiting the mental effort required. A strategy of breadth-before-depth means that several foci must be considered at the same time, and so it is therefore generally less appealing. Depth-before-breadth allows us to queue up our problems and consider them one at a time, whereas breadth-before-depth does not. A depth-before-breadth mindset is sufficient as long as one can afford to have a single focus; it may be adequate as long as the world is tractable. Unfortunately, this is no longer the case, and it has not been so for many years (cf., Perrow, 1984). But this essential lesson from the 'golden age' seems to have been only partly absorbed by the new generation of safety researchers represented in this book.

16.2 Confirmation Bias

Since the focus of safety is on the negative, on what goes wrong or fails, the efforts focus on how to eliminate the causes (which in turn implies a belief in causality rather than emergence). Yet it is not possible to improve something (work) by only preventing it from failing. To prevent that something does *not* work still leaves us with the problem of how to make it work. When something goes wrong or when some accident occurs, we analyse it in order to understand how it happened, since this understanding is seen as a necessary foundation for the changes and interventions that will prevent future occurrences. But by the same logic we also need to understand how something works when it goes well, since it will otherwise be impossible to manage or control it, hence increase the number of acceptable outcomes.

When something goes wrong, when an accident happens, there is an obvious conflict with our expectations and our intentions. Accidents are therefore taken as evidence that our understanding is incomplete or deficient. The consequence of that is a need to improve our understanding by trying to find explanations for what has happened. The situation is quite different when outcomes are acceptable: when work goes well. In these cases, our expectations and intentions correspond to what happens. The acceptable outcomes are therefore seen as proof or evidence that our understanding and actions are (or were) correct. That being the case, there is clearly no need to look at them more closely.

The confirmation bias, or positive view, means that we think we understand why work goes well, and therefore never find it necessary to check our underlying assumptions. Even worse, we habituate to the constant occurrence of acceptable outcomes so that we no longer pay any attention to them. This problem was recognised by Karl Weick who pointed out that 'reliable outcomes are constant, which means there is nothing to pay attention to' (Weick, 1987, p. 118). This lack of visibility is the reason why safety is traditionally concerned with what has gone wrong, and why little or no efforts are spent on understanding that which has gone well. The problems are exacerbated by a concomitant lack of terminology, methods and models. The confirmation bias is ubiquitous, characterising every chapter of this book.

16.3 Conclusion

The chapters of this book show how the established understanding of safety can be addressed by thoughts and ideas that have their origins in other sciences and schools of thinking. This may, in some cases, open up the possibility of finding new solutions to problems that continue to plague safety management and safety research – primarily, the occurrence of unexpected and unpredictable adverse outcomes. But the chapters also – and perhaps unintentionally – demonstrate that there have been few (if any) changes in the fundamental understanding of what safety means and what the real problems of safety actually are. The urgency of finding solutions to well-known manifestations – losses, injuries and the like – prevent us from reflecting on whether the problems really are the same today as they were in the 'golden age' – whether the aetiology is the same, just because the phenomenology is. New perspectives and new approaches are brought on to solve the old problems, but the problems are, on the whole, understood in the same way, at times with a sprinkle of complexity added. Perhaps it would be appropriate to step aside for a moment; to look carefully at what we have been doing – what we *are* doing – in order to ensure we are indeed trying to solve the right problems.

References

Hale, A. R. and Hovden, J. (1998). Perspectives on safety management and change. In A. R. Hale and M. Baram (Eds.), *Safety Management*. Oxford: Pergamon, 1–18.

Institute of Medicine (1999). *To Err Is Human*. Washington, DC: National Academy Press.

Masys, A. J. (Ed.) (2018). *Security by Design. Innovative Perspectives on Complex Problems*, Cham, CH: Springer International Publishing AG.

Perrow, C. (1984). *Normal Accidents: Living with High-Risk Technologies*. New York: Basic Books, Inc.

Reason, J. T. (2000). Safety paradoxes and safety culture. *Injury Control and Safety Promotion*, Volume 7(1), 3–14.

Weick, K. E. (1987). Organisational culture as a source of high reliability. *California Management Review*, Volume 29(2), 112–128.

17

Observing the English Weather: A Personal Journey from Safety I to IV

Nick Pidgeon

The musician Terry Riley once remarked that the experience of listening to his music, in common with much of the minimalist output of his composer peers in the 1960s and 70s, was akin to watching the clouds drift by in a gentle summer breeze. If you observed them intently enough it appeared that nothing much was changing. Look away for a moment, however, and when you glanced back the apparent uniformity had given way to subtle new patterns and relationships: and so it was with much of his music. Being asked to return to reflect upon the safety domain some 30 years on from my close collaborations with Barry Turner, David Blockley and Brian Toft feels just a little bit like this. As the many excellent contributions to this volume demonstrate, many things that we took for granted then remain fundamental within the safety field today: the 'surprise' and intellectual reframing but also the opportunities for deep learning that all genuine disasters bring ('how could we have been so sightless?'); the multi-causal complexity and ambiguity of risky systems that can serve to defeat even the very best risk assessment and management; culture, people and organisations as the key actors in major accident sequences; and above all the inevitable fact that somebody, somewhere, sometime, probably held all or parts of the knowledge that could have interrupted an incubating disaster. And yet for all of that, glancing back now few things appear exactly as they were.

I myself came to the topic of major accidents around about 1970, while still studying at school. We had been assigned the task of presenting to our English class a research project on a topic of our own choice. For this I had used an account of the 1967 Stockport Air Disaster, in which 72 people (both in the aircraft and on the ground) had died. My school itself was in Windsor and located directly under one of the western landing paths to London's busy Heathrow airport. This was much to the outrage of the English teacher because, even in those days, the noise of arriving jets would often disrupt his lessons. I saw it somewhat differently, of course. Like many of my school friends living locally, I had developed a keen interest in aircraft and aircraft recognition. So an aviation disaster looked just about right for the English teacher and just about right for my own project, too. The Stockport accident itself had occurred at the end of a flight from Europe, when two out of the four engines of a Canadair Argonaut airliner became starved of fuel just seven miles on the approach to Manchester Airport. A poorly understood and flawed system of multiple fuel tanks, badly designed controls which had allowed fuel to bleed between tanks unnoticed during the flight, and a degree of pilot fatigue had resulted in the two engines suddenly shutting down (Stockport Accident, 1967). This immediate accident chain had however been compounded by overlooked warnings from earlier fuel-loss incidents with the Argonaut aircraft type. As my reading about these events progressed, I was left asking myself repeatedly how a perfectly serviceable aircraft, one with a fully trained crew and sufficient fuel on board, could be brought to such a terrible end by such an obviously avoidable circumstance? In an insight only too familiar to most practitioners of safety research, I also remember being struck at the time by what appeared to be the sheer bad luck underlying the pre-accident combination of events.

My project presentation was favourably received, and that might well have been that but for a final-year undergraduate class on the psychology of international relations in 1979 – the year coincidentally also that of the Three Mile Island accident. There I was introduced to Graham Allison's multi-layered account of the Cuban Missile Crisis *Essence of Decision*, as well as Irving Janis' classic text *Victims of Groupthink*. Both of these books demonstrated that there might well be more to learn about failures (and successes) in risky situations, and that they could be understood and analysed in organisational and social psychological terms. One year later, a close friend who had by coincidence shared that psychology class with me, came to visit waving the first edition of *Man-Made Disasters*. It was crystal clear to both of us that here was a work of profound intelligence, one that by working systematically across multiple different case studies, revealed in its own quiet language a very new way of thinking about systems, hazards and risk (Turner, 1978). In doing so, it also offered answers to many of our own questions about major failures. Barry Turner (1995) himself later admitted that, at the time, he felt his book and its message – marketed abysmally by his then publishers – had disappeared without any trace whatsoever. This was not entirely true, as my own experience (and that of others) was to subsequently prove to him. We now know that a very small group of readers – academics around the globe, some safety professionals, and the simply curious – had read and digested the key messages in the book. I still keep my own, heavily annotated first edition on the shelf in my office; and others who read it at the time also recognised the genuine step-change in thinking that it had opened up. For me in particular, glancing back to Stockport, it seemed that the weather had indeed changed for the better.

I then had the very good fortune to be offered my first post-doctoral position in 1983, working with both Turner and the civil engineer David Blockley on a project looking at the organisational causes of construction accidents

(Pidgeon, Blockley and Turner, 1986, 1988). In part, they had met because Barry had been looking for an engineering domain in which to further apply Man-Made Disasters theory, while Blockley recognised that he needed new social sciences insights from outside the engineering sphere. We made an unlikely team – sociologist, social psychologist and engineer. But Blockley had been concerned with fundamental problems of uncertainty, reliability and their relationships with failures in engineering systems for some time. He had also published his own careful analysis of some classic construction failures – concluding, independently of Turner, that human and management error had been a primary cause (Blockley, 1980). Heavily influenced by the writings on scientific knowledge by Karl Popper, Blockley also believed that failure was one of the most important routes by which engineers could progress their knowledge, and that, along with the help of social scientists, he might also be able to improve safety by pushing forward his own understanding of what 'human error' really meant.

As a number of authors in this volume remark, the 1980s were also marked by the publication of Charles Perrow's far better-known book, *Normal Accidents* (1984). Its appearance was accompanied by a string of major industrial and transport disasters, which Turner's Disaster Incubation Model and to some extent Perrow's account helped to explain. As I have argued elsewhere (Pidgeon, 2011, 2012), a close reading of the contrast between Perrow and Turner is constructive and yields an essential tension still present in safety research and practice today. Perrow's somewhat deterministic view was that certain types of increasing sociotechnical system complexity, uncertainty and ambiguity would inevitably lead us to encounter major system accidents. By contrast, Turner believed that failures in such risky systems, while organisationally messy and complex, held much in common with other sociological phenomena; a stance which led him to the idea of 'sloppy management' (Turner, 1994a). If this was indeed the case, then disaster incubation in complex organisations could in principle be diagnosed, understood and potentially shaped. In the United States, this idea became associated with the research conducted at UC Berkeley, to understand how high-reliability organisations operate (Weick, 1987; Roberts, 1993; La Porte, 1996); a research program which yielded the important idea of 'organisational mindfulness' as a core strategy for detecting weak signals of incubating failure (Weick and Sutcliffe, 2001). In Europe, following the discourses applied to Chernobyl, much of the focus switched to the somewhat more ephemeral idea of a good 'safety culture' (Pidgeon, 1991). Something that we all recognise now, but was not entirely evident at the time, is that both developments represented a move away from simply diagnosing vulnerability to one of understanding system resilience (Pidgeon, 1997).

In all of these arguments one can discern the emerging outlines of the subsequent debate about Safety I and Safety II, and, in particular, that of resilience engineering thinking – was it better to anticipate a class of broadly 'known' failures through conducting risk assessments and specifying better operating procedures, or, in the face of ultimately unpredictable hazards, was it better to foster locally embedded resilience and flexibility (Turner, 1994b; Pidgeon, 1998)? In truth, safe operations have probably always needed a degree of both, but striking the correct balance for any particular circumstance remains the safety practitioner's goal. The notion that key trade-offs might exist between anticipation and resilience had already been addressed by the political scientist Aaron Wildavsky in his book *Searching for Safety* (Wildavsky, 1988). It would be wrong, however, to mischaracterise contemporary thinking about Safety I and II solely in his terms. Essentially, a neo-liberal argument about the negative consequences of over-burdensome safety regulations, and a polemic against precautionary thinking in particular, Wildavsky argued that anticipatory risk management approaches might fail tests of cost effectiveness, as well as result in unanticipated hazards. As many readers of this volume will know, contemporary resilience engineering, by contrast, in principle stresses the search for excellence in *both* anticipation and resilience, alongside the negotiation of any trade-offs that do occur between the two.

The subtle similarities and distinctions between the different schools have subsequently been debated by others, and I will not repeat the details of those discussions here. Hopkins (2014) for example, argues that Safety II is simply the high-reliability organisation in another guise (or perhaps even a sub-set of what high-reliability organisations purport to do). Le Coze (2016) disagrees, highlighting the different disciplinary origins of the two – specifically, human factors and cognitive science underpinning resilience engineering, as opposed to sociology, social psychology and politics in relation to high-reliability organisations. Such distinctions matter – whatever we might think about the desirability of achieving a requisite synthesis – since disciplines socialise their members into particular understandings of how the world works and how to study it. In my experience of working on interdisciplinary risk research in many contexts, it is clear that mathematicians, sociologists, psychologists and engineers, by the very nature of their profession, think with different concepts (probability, human error, power), address different phenomena (people, society, systems) and sometimes even espouse different epistemologies. Not only does that bring with it the long-standing problem of correctly communicating between diverse disciplines (where terms differ or the same term means somewhat different things), but at times it brings heated border disputes over whose approach is to be considered the most appropriate or 'best' way forward. Equally, as Hopkins (2014) rightly points out, such labels are as much about the search for an academic or professional identity as they are about fundamental disagreements about ontology or epistemology. Looking back to 1985, my own hope at the time was always that a synthesis of disciplines could indeed be attained, given what was at stake in preventing future disasters. If Le Coze (2016) is correct, however, we may now in reality be facing three distinctive perspectives on systems and risk: namely Safety I (conventional risk rules and standards), Safety II (resilience engineering and human factors) and Safety III (high-reliability organisations and culture). I am aware that, for some, this will seem a controversial distinction to propose, but the reader should bear with me a little longer for the sake of this argument. It seems to me that, here, the clouds on the horizon have not so much cleared as parted.

What is also evident, from the varied contributions to this volume and elsewhere, is that more fine-grained social sciences methods and theories have indeed been systematically applied to the problem of studying disas- . ters and safety over the last 30 years, thereby greatly enriching and expanding the evidence base alongside our suite of methodological tools. To take just a small personal example, during the 1990s I spent much of the time discussing safety with Mike O'Leary, who was both a British Airways pilot and a trained psychologist (O'Leary and Pidgeon, 1995; Pidgeon and O'Leary, 1994, 2000). I had met O'Leary quite by accident, when he was completing his cognitive psychology PhD part time at Birkbeck College, where I had just started lecturing in applied psychology. When not flying an A320 he was responsible for human factors at British Airways Safety Services, and also for developing and managing some of the incident reporting systems that the airline had begun using. It was O'Leary who first introduced me to the ideas underpinning what later became known as a component of a 'just culture' (Dekker, 2007): one which aims to minimise individual blame in the attempt to gain maximum levels of incident reporting across an organisation – and through this – to otherwise hidden incident and safety data. In this regard, British Airways had already realised that they needed very careful internal procedures and tacit agreements with their staff, to ensure that their safety investigators could foster and maintain trust as part of the reporting culture. This was achieved through a very careful separation of the function of organisational learning from the function of instigating sanctions where significant procedural violations had occurred (Pidgeon and O'Leary, 2000; Le Coze, 2013).

Shorrock (this volume) argues that academics and practitioners of safety have rarely worked well together – a claim I would contest; of all the areas of research that I have worked in, safety sees some of the smoothest academic-practitioner collaborations. James Reason, for example, quickly translated the latent/incubating failures model directly for industry (e.g. for Royal Dutch Shell and British Airways, to name just two) to such good effect that one rarely sits down today with a senior manager with any safety interest who does not know all about 'Swiss cheese'. Of course, when pressed over their dessert, they rarely know exactly where this particular brand of cheese initially came from, but they definitely know what it means for their safety systems and bottom line. Likewise, at British Airways, Mike O'Leary was concerned that the UK Civil Aviation Authority's recommended method for categorising aviation risk at the time – the well-known, consequence x severity matrix from risk assessment – was not fit for purpose when thinking about prioritising for action the many incident reports that the Safety

Services department routinely received. Where was the really important information about safety that they needed to know and follow up? If the matrix was not working, how were the safety investigators in his team, all experienced pilots and engineers, actually assessing incident risk for each report? And while Safety Services currently seemed to be doing a good job in filtering and dealing with these reports, could we learn more about how they did this so as to improve their future practice?

This difficult set of issues that Mike had posed subsequently became Carl Macrae's PhD project (Macrae, Pidgeon and O'Leary, 2002; Macrae, 2006). After unsuccessful attempts to apply traditional human factors methods to the problem (Safety II), we reasoned that a period of close ethnographic field-work embedded with the Safety Services department (viz. Safety III) might do the trick, which it ultimately did. The research yielded the insight that, for any given incident report, judgements about the level and acceptability of risk depended upon the detailed interpretive work and sensemaking of the safety investigators – labelled by Macrae as 'interpretive vigilance', this can be thought of as a concrete example of Weick and Sutcliff's (2001) idea of organisational mindfulness. One cussswe to identify where an event was high in risk was where it could be shown to have breached several of the organisational and technical safety barriers, with few or no barriers remaining to prevent a catastrophe. Hence, risk became less about abstracted probability and consequence than one of degraded organisational capability. Of more interest to us was what happened next, after a suspected significant risk had been identified – here a process of intense social negotiation and organisational networking ensued, as an investigator went out into the airline to discuss extensively with those directly involved, and others within the airline who might shed light on the issue at hand, the reasons for the events reported, as well as some possible solutions. In effect a much wider, distributed process of mindfulness was being triggered. The role of the safety team was then not to 'do safety' themselves, but to reflexively prompt relevant parts of the organisation to take ownership of the problem, and through this, seek requisite solutions. Always the goal was to return any potential incubating hazard to a situation of defence in depth. In this way, Safety III was approached within the airline through an intensely *social process* of interpretation and networking, negotiation between teams and collective sensemaking. Once again, or so it seemed to me at the time, safety science was capable of making a radical step-change in thinking and practice.

To conclude this short commentary I want to return to the subject of Wildavsky's neo-liberal philosophy. In addressing what might have changed for contemporary safety research in the interim, we should not forget that safety practice does not stand in isolation from the wider trends that are constantly occurring within and across societies. In particular, since the late 1970s, neo-liberal thinking has driven a pattern of globalisation in many countries: characterised by extensive deregulation, privatisation of state assets, outsourcing of responsibility for standards and audits and a retreat by the state from the direct provision of many social goods formerly taken for granted by citizens (such things as welfare, healthcare, employment, water and energy, financial protection). One would be highly surprised to learn that safety had been completely insulated from such intense global pressures.

Much has been written by contemporary sociologists, often with more heat than light, about the changing conditions of late modernity alongside that of globalisation and the place of risk within it. In searching for evidence of the risk society (Beck, 1992, Giddens, 1990) one can point to the increasingly complex and interconnected nature of some global technological, financial and environmental risks, alongside novel emerging technologies and systemic existential risks. Hopkins (2006, Ch. 13) rightly comments that this has not necessarily led to all of us becoming the eternally anxious 'risk subject' that Beck foresees. However, we may well now face a risk society where individualised responsibility for securing safety from harm sits alongside an erosion of trust in institutions and the unwinding of the social certainties that traditional identities (such as being a family member, employee, trade unionist, home owner, even safety engineer) always brought us (Henwood and Pidgeon, 2014). Beck himself did not entirely foresee that the first genuine empirical test of his thesis would come in the United States and Britain, amidst the near systemic collapse in 2008 of global banking systems because of the global proliferation of opaque, and ultimately worthless, financialisation instruments.

Le Coze (2017) also raises this important question of globalisation, and in so doing points to what I believe to be a genuine deficit in our contemporary safety thinking – and for once a cloudless sky appears. As befits mainstream sociology, the work of Beck, Giddens and commentators (e.g. Mythen, 2004; Lash, 2018) tends to lend its focus solely to social critique, and through this to ways of characterising the conditions of contemporary society. We might classify the yawning gap for the safety sciences here as one requiring a Safety IV (macro-global oriented) approach – although a more precise definition may be more challenging to specify. Mainstream sociology pays scant attention to how material (un)safety might well be the logical consequence of a risk society, since this was never their problem to resolve in the first place. But what is clear for contemporary accident research and practice is that the trends embedded within globalisation may well be bringing with them unanticipated, unwelcome forms of societal and organisational 'drift' (Pettersen Gould and Fjæran, this volume, Chapter 8), ones that serve to attenuate our perceptions of risk while simultaneously undercutting the established norms and processes for dealing with hazards. For example, globalisation of finance and the financial systems has meant that companies and other organisations are being transformed into simple 'financialised' entities within global balance sheets, with the one sole overriding goal of the identification and extraction of short-term value (so-called 'value chasing'). Almost by definition this subordinates all other useful objectives or societal functions that an organisation might well have delivered for society in the past: such things as, making products, providing useful services, protecting the environment, industrial capacity and corporate longevity, providing (safe and stable) employment and contributing to regional or national economies. To repeat, it seems unlikely that industrial safety practice can remain completely insulated from these major global developments. This in turn sets a research agenda to fully understand the ways in which safety is indeed being impacted, and an immediate challenge to respond with new approaches and methods of risk governance (e.g. Renn, 2008; Engen and Lindøe, this volume, Chapter 4).

In his own glance back at former times, Le Coze (2017) perceptively observes that the fingerprint of globalisation may now be unfolding in an uncanny re-run of the disasters which were so common in the 1980s, and which served as the main stimulus to the first phase of sustained organisational safety research. These include the *Challenger* (ocurring in 1986)/*Columbia* (in 2003) accidents, the level 7 nuclear failures at Chernobyl (1986)/Fukushima Daiichi (2011) and Piper Alpha (1988)/Deep Water Horizon (2010) from the oil and gas industry. My personal addition to his list would be the catastrophic fires at the Summerland Leisure Complex in 1973 and at Grenfell Tower Block in 2017 which resulted in the deaths of 72 Londoners. Summerland in particular formed one of Turner's ideal-typical cases in the original *Man-Made Disasters* account, and occurred at a new leisure facility on the Isle of Man. The architects had specified a novel – but highly flammable – acrylic sheeting for the building, alongside other panel materials with limited fire resistance. When a small external fire set an external panel ablaze, the fire quickly spread through the building, leaving many families struggling to locate friends and children or to evacuate the building.

Although at the time of writing this contribution the Grenfell inquiry is still ongoing, what is clear is that the proximal cause of this fire was the re-cladding of the entire building in insulation panels with low fire resistance. Accordingly, when the fire broke out in one flat because of an electrical fault in an appliance, it was able to quickly jump over to the other flats via the burning external walls. This form of fire behaviour in exterior cladding had already been described in an earlier inquiry into similar events at Lakanal House (2009) in South London; an event that had resulted in six deaths. Not only did the fire behaviour at Lakanal and Grenfell run counter to all conventional assumptions about high-rise safety (that a fire would, by design, be contained in any single flat 'cell' until it was extinguished by the fire services), it also fatally undermined the rescue effort. The fire service procedures and norms for tackling a major tower block fire at that time dictated that the safest advice for residents in such circumstances would be to stay put, protected in what were assumed to be fully fire-resistant flats. As a result the emergency services failed to recognise quickly enough that, as the fire rapidly grew out of control up the sides of the building, prompt evacuation might have been a wiser counsel.

What seems most important about Grenfell for our purposes is not the proximate events of the night, horrific as they were, but some of the underlying reasons behind the building's vulnerability. In the United Kingdom, following the initial privatisations of the 1980s and the further acceleration of fiscal austerity measures from 2010 onwards, local authorities had sought to aggressively cut their costs, whilst outsourcing many building and maintenance contracts (e.g. for refurbishment). At the same time many of their internal building engineering departments were downsizing in capacity or disappearing entirely (Blockley, 2018), and in ways that may well have compromised their ability to fully evaluate and assess construction and other building risks. One can hypothesise here, and with good reason, that the local authority involved with Grenfell did not particularly want to scrutinise, or were not even in a position to scrutinise properly, the small cost savings achieved by adopting more flammable cladding as against the increased risks of uncontained fire spreading quickly.

As my final comment on this question of the globalisation of risks and safety we can only hope that the argument here is misplaced, although like Le Coze (2017), deep down I suspect that it is not. And what that suggests for safety and safety sciences only the future will show us. One implication concerns the changing nature of the trade-off to be made between an organisation's (short-run) profitability and its efforts to improve long-run safety performance. In the worlds of Safety II and III the hope was always that an organisation's striving for excellence would ultimately be to the benefit of both profitability and safety performance. In the globalised, deregulated, financialised and digitised world of Safety IV, these assumption may no longer hold true. Since the 1980s, we have made significant and steady progress in our understanding of organisational safety and risks, but Safety IV may well set the toughest challenges yet for the coming generation of researchers and practitioners. If the current safety outlook has indeed changed radically while we were looking elsewhere, the clouds may now be taking on a more turbulent hue.

References

Beck, U. (1992) *Risk Society. Towards a New Modernity*. London: Sage.

Blockley, D.I. (2018) personal communication.

Blockley, D.I. (1980) *The Nature of Structural Design and Safety*. Chichester: Ellis-Horwood.

Dekker, S. (2007) *Just Culture: Balancing Accountability and Safety*. London: Ashgate.

Engen, O.A. and Lindøe, P.H. (this volume) Coping with globalisation: Robust regulation and safety in high-risk industries.

Giddens, A. (1990) *Modernity and Self-Identity*. Cambridge: CUP.

Henwood, K.L. and Pidgeon, N.F. (2014) Risk and identity futures. Future Identities Programme, London: Government Office of Science.

Hopkins, A. (2014) Issues in safety science. *Safety Science*, Volume 67, 6–14.

Hopkins, A. (2006) *Safety, Culture and Risk*. Australia: CCH Books.

La Porte, T.R. (1996) High reliability organisations: Unlikely, demanding and at risk. *Journal of Contingencies and Crisis Management*, Volume 4, 60–71.

Lash, S. (2018) Introduction: Ulrich Beck: Risk as indeterminate modernity. *Theory, Culture and Society*, Volume 35(7–8), 117–129.

Le Coze, J.-C. (2017) Globalisation and high-risk systems. *Policy and Practice in Health and Safety*, Volume 15(1), 57–81.

Le Coze, J.-C. (2016) Vive la diversité! High reliability organisations (HRO) and resilience engineering (RE). *Safety Science*. doi:10.1016/j.ssci.2016.04.006.

Le Coze, J.-C. (2013) What have we learned about learning from accidents? Post-disasters reflections. *Safety Science*, Volume 51, 441–453.

Macrae, C. (2006) *Assuring organisational risk resilience: Assessing, managing and learning from flight safety incident reports*. PhD Thesis, University of East Anglia.

Macrae, C., Pidgeon, N.F. and O'Leary, M. (2002) Assessing the risk of flight safety incident reports. In C. Johnston (Ed.), *Proceedings of Workshop on Investigating and Reporting of Incidents and Accidents (IRIA 2002)*, GIST Technical Report 2002-2, Computer Sciences, University of Glasgow, Scotland, 99–106.

Mythen, G. (2004) *Ulrich Beck: A Critical Introduction to the Risk Society*. London: Pluto Press.

O'Leary, M. and Pidgeon, N.F. (1995) "Too bad we have to have confidential reporting systems": Some reflections on safety culture. *Flight Deck*, No 16, Summer 1995, 11–16.

Perrow, C. (1984) *Normal Accidents: Living with High-Risk Technologies*. New York: Basic Books.

Pettersen Gould, K. and Fjæran, L. (this volume) Drift and the social attenuation of risk.

Pidgeon, N.F. (2012) Complex organisational failures: Culture, high reliability, and lessons from Fukushima. *The Bridge*, Volume 42(3), 17–22.

Pidgeon, N.F. (2011) In retrospect: Normal accidents. *Nature*, Volume 477, 404–405.

Pidgeon, N.F. (1998) Safety culture: Key theoretical issues. *Work and Stress*, Volume 12(3), 202–216.

Pidgeon, N.F. (1997) The limits to safety? Culture, politics, learning and man-made disasters. *Journal of Contingencies and Crisis Management*, Volume 5(1), 1–14.

Pidgeon, N.F. (1991) Safety culture and risk management in organisations. *The Journal of Cross Cultural Psychology*, Volume 22(1), 129–140.

Pidgeon, N.F., Blockley, D.I. and Turner, B.A. (1988) Site investigations: Lessons from a late discovery of hazardous waste. *The Structural Engineer*, Volume 66(19), 311–315.

Pidgeon, N.F., Blockley, D.I. and Turner, B.A. (1986) Design practice and snow loading: Lessons from a roof collapse. *The Structural Engineer*, Volume 64(A), 67–71.

Pidgeon, N.F. and O'Leary, M. (2000) Man-made disasters: Why technology and organisations (sometimes) fail. *Safety Science*, Volume 34, 15–30.

Pidgeon, N.F. and O'Leary, M. (1994) Organisational safety culture: Implications for aviation practice. In N. Johnston, N. McDonald and R. Fuller (Eds.), *Aviation Psychology in Practice*, Aldershot, Avebury Technical, 21–43.

Roberts, K.H. (1993) Cultural characteristics of reliability enhancing organisations. *Journal of Managerial Issues*, Volume 5, 165–181.

Renn, O. (2008) *Risk Governance: Coping with Uncertainty in a Complex World*. London: Earthscan.

Shorrock, S.T. (this volume) Safety research and safety practice: Islands in a common sea.

Stockport Accident Inquiry (1967) *Flight International*, 7th December, p. 935.

Turner, B.A. (1995) A personal trajectory through organisation studies. *Research in the Sociology of Organisations*, Volume 13, 275–301.

Turner, B.A. (1994a) Causes of disaster: Sloppy management. *British Journal of Management*, Volume 5, 215–219.

Turner, B.A. (1994b) Flexibility and improvisation in emergency response. *Disaster Management*, Volume 6(2), 84–90.

Turner, B.A. (1978) *Man-Made Disasters*. London: Wykeham Press.

Weick, K.E. (1987) Organisational culture as a source of high reliability. *California Management Review*, Volume 29(2), 112–127.

Weick, K.E. and Sutcliffe, K.M. (2001) *Managing the Unexpected: Assuring High Performance in an Age of Complexity*. San Francisco: Jossey-Bass.

Wildavsky, A. (1988) *Searching for Safety*. New Brunswick: Transaction Books.

18

A Conundrum for Safety Science

Karlene H. Roberts

CONTENTS

18.1 The State of Safety Science 282
18.2 Some Issues Addressed in This Volume and What Needs to Be
Done... 283
18.3 Who Should Do It?... 285
References... 286

Introduction

On a worldwide basis, organisations are becoming interdependent within their units and interdependent with each other (e.g. Tjosvold, 1986). This trend is not likely to be reversed in a world where communication mechanisms grow steadily more sophisticated. Increased organisational interdependence provides increasing opportunities for errors within and across organisations, and, with that, increasing opportunities for errors that can have catastrophic consequences. In fact, the reinsurer, Swiss Re reported that global insured losses for insured disastrous events in 2017 were the highest that they had ever been (Swiss Re, 2018).

While many practitioners and researchers no longer simply focus on individual causes of errors and accidents (leading to efforts to find the culprits, and then either fire or train them) most of us are not very good at digging deeper into the situation in search of what are often less obvious and multiple influences. This paper discusses the general state of safety science and suggests several strategies that might reduce confusion about what safety science is.

18.1 The State of Safety Science

Both Hopkins (2014) and Schulman (this volume, Chapter 19) argue that there is no such thing as safety science. Hopkins focusses on three issues: negotiable boundaries; the problematic nature of the theories used to support the existence of the field, and the paradox of major accident inquiries in that they cannot identify preventative strategies. Schulman emphasises the fact that the engineering and social science approaches to the problem are not integrated or additive, and that the key terms used are ambiguous and even sometimes contradictory. Both of these arguments have merit. They might be usefully amended by considering two additional pieces of information. First, most recent accident investigations do include suggestions for avoidance strategies, whether they are drawn directly from the accident or from other situations (e.g. BP U.S. Independent Safety Review Panel, 2007; National Transportation Safety Board, 2011). Second, many would argue that the key terms used across the social sciences are ambiguous and sometimes contradictory. Some social scientists work to reduce these problems, others simply add to the cachet of ambiguous and contradictory terms.

One definition of safety science is that it is the science and technology of human safety, and that it is multidisciplinary (Aven, 2014). There is probably a good deal of tension among members of the communities who call themselves safety science practitioners and those who identify as researchers, as to the appropriate ways in which to better evaluate existing concepts, strategies and methodologies. This is probably linked historically. One progenitor of what is today called safety science is human factors research. That research is multidisciplinary and draws some of its findings from engineering, which is closer to being a canonical science than are any of the social sciences. Researchers and practitioners coming from this perspective might view the many concepts, strategies, and so on, coming from the social sciences as just more 'fads, fashion and folderol' (Dunnette, 1966). This would lead to the often-cited perception that psychologists and other social scientists are engaged in physics envy (e.g. Howell, Collison and King, 2014).

While several authors in this book (e.g. Reiman and Vitanen, this volume, Chapter 13; Shorrock, this volume, Chapter 14; Waring and Bishop, this volume, Chapter 9) focus on the different values held by practitioners and researchers, none discusses the major difference. Practitioners go into the field to solve problems and make money, and their reputations would be sullied if they did not do that. Researchers go into the field to publish in journals whose values include using acceptable methodologies and theoretical approaches. Moreover, they would not be promoted if they did not do that. These two approaches lead to the growing gap between researchers and practitioners discussed by Shorrock (this volume, Chapter 14).

18.2 Some Issues Addressed in This Volume and What Needs to Be Done

A variety of issues are addressed in this volume, ranging from individual differences (such as shared power distributions) to more macro concerns such as networking, globalisation, and so on. What they share in common is that each one of them is taking a nibble from a different part of the cheese wheel called safety science. A useful activity might be to catalogue the most commonly discussed issues in this general area (perhaps arraying them along a micro to macro dimension). Other dimensions may add greater clarity in thinking about safety. For example, putting similar subjects into the same broad categories might aid a second task of trying to reduce or blend the number of subjects in any one category. To that, one could add a listing of practices emanating from those categories. Alternatively, it might be helpful to have a listing of safety practices that then lead to more general discussions and analyses of their components.

These activities might benefit from thinking about an issue discussed in this book. It might aid our understanding of safety to further the Le-Coze discussion of visualising safety (this volume, Chapter 10). To do this, one might bring to the table as helpmates people in professions who regularly use visualisation in their work activities (such as architects, artists, etc.). Those people might be able to help social scientists extend their visualising capabilities beyond Swiss cheese slices and geometric figures.

With the advent of more sophisticated technologies, it is now possible to go beyond two-dimensional representations of physical and cognitive properties. Not only does that allow for the enrichment of variables, but the inclusion of background characteristics in both theoretical and practical approaches to safety. For example, one might be interested in how individuals are motivated to engage in safety-enhancing behaviours. Hypothesising that, training, time-of-day, cultural background and pay level all potentially affect behaviour is interesting, and easily treated as a regression problem. But, actually, designing a 3-D picture of all this might have more of a significant impact on managers' decisions about assignments; workers doing those jobs may have a more immediate and higher-level impact on the people operating within the organisation than any statistical model.

Going beyond the two-dimensional models, Le Coze cites allows us to represent the world as the complicated environment it really is, rather than as a 2 × 2 box or a piece of Swiss cheese. The researcher has at his disposal lines (straight and curved), geometrical and other shapes, colours, distances, depth, height, surface characteristics, and so on.

In addition to clarifying the language of safety research and practice, it is imperative that researchers and practitioners recognise and deal with the fact that the organisational world is becoming ever more global and interdependent. These issues are discussed by several chapters in this book (Engen and Lindøe, this volume, Chapter 4; Gotcheva, Aaltonen and Kujala, this volume, Chapter 3; Reader, this volume, Chapter 2). Reader offers an interesting discussion of the overlap of safety culture and national culture using Hofstede's model (2003). Workforces are becoming more mobile and more eclectic in large-scale organisations. However, many research questions are still unanswered about this development. For example, how do organisations acquire cultures to improve safety across units? Some organisations adopt standardised safety performance rules. Such rules can get in the way of solving a safety problem that falls outside the confines of the rules.

Perhaps further insight about how to extend our knowledge about safety processes can be obtained from Gotcheva, Altona and Kujala's (this volume, Chapter 3) discussion of safety within inter-organisational project networks. They discuss how key elements of governance and safety are intertwined. Gotcheva et al., remind us that the key challenges in complex projects are organisational. They devote some discussion to noting that key elements, such as communication and decision-making (where in the organisation and how) can influence safety outcomes. Firm project networks should not be viewed as hierarchical structures, but as complex network relationships. The key elements in the aforementioned governance discussion should also be investigated as potential key elements in safety culture; goal setting, monitoring, roles and decision-making, incentives, coordination and capability building.

Engen and Lindøe (this volume, Chapter 4) discuss how national risk-regulating regimes deal with the growing complexities and uncertainties associated with global challenges, such as new technologies, outsourcing, mergers and acquisitions. They offer sociotechnical theory as a potential lens for identifying factors that influence complexities and uncertainty.

> ...companies can be seen as innovative production systems accompanied with hazards, incidents, and accidents that represent threats to employees, the environment, and third parties. However, an integrated element of a resilient production system is the development and maintenance of activities, processes best-practices procedures, and enterprise risk management (ERM). Such processes are compatible with programs of 'quality management', 'continuous improvement', and 'quality circles', 'Plan-Do-Check-Act, etc'.

These authors also focus on trust and distrust as they manifest themselves in the global regulatory world. They note that in the maritime industry Norway had developed a system in which trust operates at a high level. Flexibility and resilience are parts of a regime that is an organisational learning system.

18.3 Who Should Do It?

Let us take the first issue, cataloguing the various definitions and approaches to safety science. Since the one thing most students of safety science agree on is that it is a multidisciplinary approach, it might be advantageous to draw on the many disciplines involved. *The American Society of Safety Professionals* is the oldest group of safety professionals in the world.* Begun in 1911, it now has over 38,000 members in 80 countries. Members of this organisation might be able to take the lead in putting together a panel or committee for the task of sorting through the various concepts, thoughts, ideas, notions, approaches, and so on, held (and written about) by safety professionals to sort out those that seem most useful.†

Members of other professional societies who also address safety issues might best populate such a committee or panel. Some of those professional associations are:

- *The Academy of Management*, founded in 1936 has 19,000 members and 25 divisions and interest groups. Members of several of these groups might be interested in this work, including *Critical Management Studies, Health Care Management and Organisational Behaviour*.

- *The American Psychological Association*, founded in 1892, is the largest professional organisation for psychologists with 117.500 members The APA has 54 divisions, 2 of which are *The Society for Military Psychology* and *The Society for Industrial and Organisational Psychology*.

- *The American Political Science Association*, founded in 1905, has 22,000 members and 49 sections. Members of the *Public Policy* or *International Security* sections might be interested in safety issues.

- *The American Sociological Association*, founded in 1905, has 52 sections with 26,347 members. Members of the *Organisations, Occupations and Work* section or the *Social Psychology* section might be interested in this task.

- *The Human Factors and Ergonomics Society*, founded in 1957, has 4500 members, 67 chapters around the world, 46 are student chapters and 24 are technical groups. The Society is federated with the *International Ergonomics Association* and it has other societal members who might be helpful in completing this task

---

* The society was founded after the Triangle Shirtwaist Factory fire.
† They can also define 'useful'.

The combined issues of increased globalisation and interdependence among organisations requires different expertise. These examinations need to recognise that using the idea of national culture is too simplistic. Culture needs to be seen as contested, temporal and emergent (Myers and Tan, 2002). Engen and Lindøe (this volume, Chapter 4) remind us that these issues should be examined in the context of trust.

Who needs to do this? The most obvious candidates are research teams in global organisations or the academic teams with whom they are associated (this is not a one-man job). Second candidates are the international organisations that represent various global organisation interests (e.g. *International Institute for Applied System Analysis – IIASA, Union of International Associations – UIA*).

Bringing science back into organisational safety is not going to be easy. It requires both stamina and resources. However, it should not require trying to convince people that organisational safety is important. All one has to do is peruse newspaper articles and the growing number of accident investigations to know how important it is and how many lives and other assets are lost because of inactivity in this area.

References

Aven, T. (2014). What is safety science? *Safety Science*, Volume 67, 15–20.

Dunnette, M.D. (1966). Fads, fashions and folderol in psychology. *American Psychologist*, Volume 14, 313–352.

Engen, O.A. and Lindøe, P.H. (in press). Coping with globalisation: Robust regulation and safety in high risk industries. In J.C. Le Coze (Ed.), *Safety Science Research: New Evolutions, Challenges and Directions*. Abingdon, UK: Routledge.

Gotcheva, N., Aaltonen, K. and Kujula, J. (in press). Governance for safety in interorganisational project networks. In J.C. Le Coze (Ed.), *Safety Science Research: New Evolutions, Challenges and Directions*. Abingdon, UK: Routledge.

Hofstede, G. (2003). *Culture's Consequences: Comparing Values, Behaviors, Institutions and Organisations across Nations*. Newbury Park, CA: Sage.

Hopkins, A. (2014). Issues in safety science, *Safety Science*, Volume 67, 6–14.

Howell, J.L., Collison, B. and King, K.M. (2014). Physics envy: Psychologists perceptions of psychology and agreement about core concepts. *Teaching of Psychology*, Volume 41, 330–334.

Myers, M.D. and Tan, F.B. (2002). Beyond models of national culture in information systems. In E.J. Szenezat and C.R. Snodgrass (Eds.), *Human Factors in Information Systems*, Myers and Tan, London: IRM Press, 1–19.

National Transportation Research Board. (2011). Pacific Gas and Electric Company Natural Gas Transmission Pipeline Rupture and Fire, San Bruno, California, 9 September 2010. Pipeline Accident Report NTSB/PAR 11-01, Washington, DC.

Reader, T. (in press). The interaction between safety culture and national culture. In J.C. Le Coze (Ed.), *Safety Science Research: New Evolutions, Challenges and Directions*. Abingdon, UK: Routledge.

Reiman, T. and Vitanen, K. (in press). Towards an actionable safety science. In J.C. Le Coze (Ed.), *Safety Science Research: New Evolutions, Challenges and Directions*. Abingdon, UK: Routledge.

Schulman, P. (in press). Some thoughts on future directions in safety research. In J.C. Le Coze (Ed.), *Safety Science Research: New Evolutions, Challenges and Directions*. Abingdon, UK: Routledge.

Shorrock, S.T. (in press). Safety research and safety practice: Islands in a common sea. In J.C. Le Coze (Ed.), *Safety Science Research: New Evolutions, Challenges and Directions*. Abingdon, UK: Routledge.

Swiss Re: Swiss Re Institute. (2018). Natural catastrophes and man-made disasters in 2017: A year of record breaking losses.

The Report of the BP U.S. Refineries Independent Safety Review Panel. (2007). Commonly referred to as the Baker Report.

Tjosvold, D. (1986). The dynamics of interdependence in organisations. *Human Relations*, Volume 79(6), 517–540.

Waring, J. and Bishop, S. (in press). Safety and the professions: Natural or strange bedfellows. In J.C. Le Coze (Ed.), *Safety Science Research: New Evolutions, Challenges and Directions*. Abingdon, UK: Routledge.

19

Some Thoughts on Future Directions in Safety Research

Paul R. Schulman

CONTENTS

19.1 A Starting Premise 290
19.2 Limiting Models and the Future of Safety Research? 292
19.3 The Future Scale, Scope and Time Frame of Safety
       Science Research 293
19.4 The Future of Safety Research as a Cognitive Process....................... 295
19.5 Conclusion 297
References... 297

Introduction

This volume offers an impressive set of chapters – covering a wide range of topics pertaining to safety research and its challenges. The chapters range in topics from more general perspectives on safety as a research and practical undertaking (as well as the limitations of its knowledge), to an examination of safety research as a profession, and, from there, to more specific safety issues such as complexity and globalisation and their likely impacts on safety; the impact and analysis of national culture and various other subcultures on safety culture; safety management challenges in large-scale projects, and, ultimately, to the gap between safety research and safety practice, and to some practical suggestions for reducing this gap, including modifications of the roles of researchers and practitioners.

Reading the chapters has taught me much about these topics, and I have been inspired to think more carefully about future directions for safety research. I will not even attempt to undertake any reviews of particular chapters. Rather, I will discuss their insights, as well as some of my own research experiences over the past four decades, to suggest some possible future directions for safety research.

19.1 A Starting Premise

I start with a premise that will not be shared by everyone. I believe that the future success and impact of safety research depends on addressing two current failures in our research process itself:

> A. The failure to develop truly integrated and additive sociotechnical systems research, and the resulting separation of social science research on sociotechnical systems from the engineering research perspectives on these systems. Safety science has been called by some an interdisciplinary research process. But right now, it isn't: it is not inter-disciplinary or even additive between science, engineering and social science research of sociotechnical systems. Much of safety science research is simply just that: social science research. Science and engineering research continues to conceive of 'sociotechnical' systems as technical systems. Technical designs remain largely uninfluenced by social science findings (save for some of the physiological findings of human factors research).

> B. The failure to express key social science concepts, definitions and models pertaining to safety in more formal terms, rather than the ambiguous and sometimes contradictory natural language expressions currently applied to them. Presently, the dominance of natural language expression of major concepts makes it difficult to forge agreement on their meaning and permit comparative measurements across organisations and technologies (Gerring, 1999; Gerring and Baressi, 2003). I believe this to be a major barrier to the development of integrated social science and engineering safety research.

As an example, we still do not have a clear definition of 'safety'. Is it a variable or a discrete property? Can safety be described as a continuous variable: can we describe *more* or *less* safety? And if so, how would we measure it? Can safety be prospectively determined, or only indirectly measured retrospectively, by lagging indicators of 'unsafety' within incident and accident rates? Alternatively, is safety only a nominal category from an organisational standpoint and therefore indeterminate from a measurement standpoint?

What is the relationship between safety and risk? Are they opposites? Is more safety achieved proportionately through the mitigation of defined risks? An international study group composed of aviation regulators has asserted that 'safety is more than the absence of risk' (SMICG, 2013, p. 2). Risk is about loss, but safety seems to be about assurance. While a number of accidents or failures can occur within a given period without invalidating the probability component of a risk assessment, a single failure can invalidate the assumption of safety. And yet we routinely talk about 'safety management systems' and 'risk management'. Are these the same thing? How exactly are they different? Make no mistake: these ambiguities of concept translate into deficiencies in measurement.

These questions surround the central concept we are trying to address in safety science. Similar ambiguity surrounds another currently important concept – that of 'resilience'. David Woods notes at least four distinctive meanings for the concept currently in use (Woods, 2016). My colleague Emery Roe and I have identified at least five different types of resilience in our research on interconnected infrastructures (Roe and Schulman, 2016). Yet we do not formally differentiate resilience into types or categories and the special properties and requirements for each. These are definitional and conceptual uncertainties we have scarcely tried to resolve.

The Bergström chapter, 'The Discursive Effects of Safety Science', raises the additional spectre that concepts within the safety sciences, such as 'crew resource management' or 'safety culture' can be transformed by others from our analytic ambiguity into moral categories, and then into legal standards for accountability. Ambiguities in our concepts can then be 'clarified' not by researchers, but by law. In this process, Bergström suggests, 'law [will] now claim its legal meaning to govern its legal use'.

Couldn't more careful specification of the meaning of our basic concepts at least help resist their appropriation by lawyers or consultants for normative purposes well beyond their analytic roots? Couldn't this clarification also allow for more measurement in relation to key concepts. Without this, how can we hope to build up a cumulative knowledge base? The clarification and standardisation of meaning and commitment to cumulative knowledge-building are defining properties of science and engineering research. These are not yet prominent features of safety research.

The Hayes chapter, 'Investigating Accidents' offers an excellent account of the San Bruno gas explosion in the San Francisco Bay area, along with descriptions of the many organisational and managerial failings associated with it. But Hayes also points out critiques and recent arguments suggesting that accident analysis is 'obsolete', and that the field of safety science should try to move beyond accident analysis.

Isn't one reason for this critique that, despite the quality of individual accident analyses, so little in the way of cumulative learning has come out of these case analyses? Isn't this is at least partly because organisational properties are expressed in unstandardised natural language descriptions? The concepts described (such as the centralisation or rigidification of decision-making; overconfidence in models or normalisation of deviance, for example) are in fact only nominal categories – there is no way to measure and scale them in common units that would allow for comparison.

I readily concede that the same has been true for a concept at the heart of much of my research over many years – that of 'high-reliability organisations'. This was a nominal category we offered to describe a distinctive set of organisations we observed. But it was never a continuous variable. It never allowed us to define and differentiate 'higher' from 'lower' reliability organisations. Nor, without this scaling ability, did we have any model of how organisations could become HROs. And, indeed, the category HRO has since been appropriated by consultants as a normative standard to be used in evaluating organisations, many of which, given the character of their technical cores, can never hope to be high reliability organisations.

Because many of our basic organisational concepts used in accident analysis such as centralisation, complexity, safety culture or resilience cannot be expressed as continuous (or even formal) ordinal variables, we do not have the metrics for them that would allow, for example, regression analyses across cases in relation to the contribution these organisational features make to incident or accident variables, such as frequency, fatalities, cascade processes, response or recovery times, and so on.

While it may be difficult to formalise definitions and descriptions of our concepts to allow their translation into metrics, I do not believe it to be impossible. For example, with more formalisation, useful comparative 'maturity models' could be developed for a variety of organisational properties, such as safety culture or even some forms of resilience. Maturity models have been used in economics and psychology as a way to convert nominal properties into ordinal categories. (For a look at the strengths and weaknesses of such efforts on safety culture so far, see Filho and Waterson, 2018.)

19.2 Limiting Models and the Future of Safety Research?

Four chapters in this volume offer arguments on the limits of safety research and safety management in preventing accidents. The chapter 'On Ignorance and Apocalypse' by Downer (this volume, Chapter 5) offers a philosophical speculation on the role of inevitable epistemic uncertainty in limiting our understanding of complex technical systems, and the impact of this (despite efforts of safety management) on the likelihood – if not the inevitability – of major accidents. The Waring and Bishop chapter 'Safety and the Professions' (this volume, Chapter 9) argues that complex specialised organisations are problematic settings for effective safety management, and that safety science (itself emerging as a profession) is a form of human organisation, bringing its own culture and roles within other organisations (as 'experts') that will limit the successful application of its knowledge within them. Both the Shorrock chapter 'Safety Research and Safety Practice' (this volume, Chapter 14) and the Reiman chapter 'Towards Actionable Safety Science' (this volume, Chapter 13) specifically detail some of the practical difficulties suggested by the problem noted by Waring and Bishop (this volume, Chapter 9).

Of course, a limiting model, even if it were accurate, does not mean that there are not significant additions and improvements that can be made in organisations by safety science and its analytic coverage, with many life-saving results. Further, an analytic model of limits is still subject to its own limits. Conceding inevitable uncertainty in knowledge, for instance, does not mean that all current unknowns can only have a negative impact on safety. It has recently been argued that complex networked systems can actually have hidden stabilisers – 'weak links' or latent elements not recognised beforehand – and when a system is under stress these can turn out to have beneficial effects to restore equilibrium (Csermely, 2006). Complexity can increase latency (and therefore uncertainty), but it can also increase robustness (just as a crisis can increase stress but also drive unforeseen creativity).

It is a useful exercise to develop limiting models for safety management. For one thing, it can serve as an antidote to hubris within organisations – a belief by leaders and managers that their safety and risk management frameworks have 'solved' the safety problem for good. Or the confidence among regulators that compliance with safety rules and regulations is sufficient for a technical system to be deemed 'safe'. (As a safety inspector at a large public utility regulatory commission remarked: 'If they're [the utilities] in compliance; they're safe'.) It can also moderate expansive statements of potentially misleading safety goals such as 'zero accidents'.

Probing for potential limiting conditions for safety is also important in more fully analysing uncertainty. Ironically, uncertainty can itself be information. There are different forms of epistemic uncertainty. For instance, the U.S. Nuclear Regulatory Agency has differentiated parametric, modelling and incompleteness uncertainty (USNRC, 2009). Each form requires a different analytic, as well as managerial, strategy. There is a role for philosophers, cognitive psychologists and social scientists in formulating differentiated models of uncertainty and their impacts on safety. Sensitivity analysis can help determine just how important specific types of uncertainty might be to safety management and safety outcomes. This would seem to be a promising area of exploration for future safety science research.

19.3 The Future Scale, Scope and Time Frame of Safety Science Research

Several chapters in this volume argue for increasing the scale and scope of safety research. Take, for instance, the Engen and Lindøe chapter, 'Coping with Globalisation: Robust Regulation and Safety in High-Risk Industries' (this volume, Chapter 4). It suggests that globalisation will have important effects on safety in regulated organisations. Regulation will be more complex with multiple players, including international regulatory agencies. This implies that safety science will have to enlarge the scale of its analytic concepts and its research. The 'balance of trust and distrust' – both important elements in successful global coordination according to Engen and Lindøe – will have to be analysed and managed on a global scale.

Another increase in analytic scale for safety science lies in the area of inter-organisational – even global – project networks. As Gotcheva, Aaltonen and Kujala point out in 'Governance for Safety in Inter-Organisational Project Networks' (this volume, Chapter 3), the scope of much of safety science research, including basic concepts such as high-reliability organisations, safety culture and resilience engineering have been based on single organisations, not inter-organisational networks or networked projects. They suggest that project governance must evolve different frameworks from the dominant options of hierarchy or markets, and that governance arrangements have to shift across different phases of a project from design to completion.

In addition, the Reader chapter 'The Interaction between Safety Culture and National Culture' (this volume, Chapter 2), highlights the potential impact of national culture in relation to safety cultures and practices within organisations. Given the diversity of cultural backgrounds in the organisational workforce and the interaction among organisations in global projects, products and processes, the impact of differing national cultures on safety will also be an important research issue. Each of these chapters suggests that future safety research, in order to be effective, will have to be wider in scope and larger in geographic scale than has previously been the case.

Future safety research will have to achieve what I have come to term a 'higher-resolution' focus at both ends of the spectrums of scale and scope and time. On one end, being able to analyse multiple organisations networked in a project, or in assessing safety and reliability risks in long-linked and geographically dispersed supply chains. Concepts and models will have to be enlarged and elaborated on to guide this analysis. For example, a concept of inter-organisational safety culture might be necessary in order to assess and promote in practice safety behaviour across inter-organisational dependencies, workflows and communication processes.

Also, higher resolution models on longer timescales will be necessary to analyse and assess interconnected infrastructure risks, including identifying those latent connections between infrastructures that can emerge over time as infrastructures shift in their technical conditions or states. These shifts can run from normal operations to disruption, failure and then to recovery processes, before finally arriving at a set of 'new normal' relationships encompassing post-recovery changes that may have occurred in technology, regulation and policy (Roe and Schulman, 2016).

In addition, safety science analysis will have to enlarge its time-frame of analysis to encompass long-term shifts or drifts in policy, practice and culture. The Pettersen and Fjæran chapter 'Drift and the Social Attenuation of Risk' (this volume, Chapter 8) suggests that it is important to analyse organisations over extended periods of time, because safety could well be a perishable property in human organisations subject to erosion in the face of competing organisational values and shifting perceptions.

Higher-analytical resolution is also needed to drill down to cover smaller-scale and shorter-time safety-related phenomena. The chapter 'Standardisation and Digitalisation' by Almklov and Antonsen (this volume, Chapter 1), asserts an important 'role of digital technologies, both as enablers of detailed control, carriers of a standardising discourse and the possibilities they present for new forms of situated action'. Here is a good example of how diverse technologies can have their own special (and micro-) impacts on organisational structure and practices and, ultimately, safety. Safety science research will have to encompass detailed, perhaps micro-ethnographic, descriptions in order to fully account (analytically) for these differences and their impacts.

As high-reliability research has discovered, a great deal of reliability and safety is derived from judgements, actions and responses made at the 'sharp end' of organisations by control operators and other individuals in real time (Roe and Schulman, 2008). The chapter 'Revisiting the Issue of Power in Safety Research' also by Antonsen and Almklov (this volume, Chapter 6), suggests that there is 'safety in power', such as the power to define the shape and scope of accident investigations and 'power in safety' – the power associated with the roles of doing 'safety work' in organisations. A similar argument is also made in the Haavik chapter 'Sensework' (this volume, Chapter 7) from the 'actor network theory' perspective for researching the impact of 'sensework' in organisations. In their effects on safety behaviour, both safety and sensework are worth studying in micro-analyses within organisations and work sub-groups.

Higher-resolution analysis in both macro and micro directions requires models that can include more variables and higher amounts of information. This leads to another future requirement for effective safety research: new approaches to the gathering, processing and display of information.

19.4 The Future of Safety Research as a Cognitive Process

In his chapter 'Visualising Safety' (this volume, Chapter 10) Le Coze argues for the role of graphical images in enhancing the understanding of safety models and findings. What is the role of visualisations in future safety science research? While Le Coze asserts their previous role in stabilising meanings within the network of safety researchers, it's important to recognise that images can also generate misinterpretation and misunderstanding among observers. Consider, for example, maps that misrepresent the scale and proportion of the distances they portray. Images can be just as error inducing as a series of vague concepts described in ambiguous words.

But images and graphic representations could also generate powerful advances in the field of safety science. Right now, images used in safety modelling are primarily two-dimensional representations with a few (generally straight) lines. Does a two-by-two table with four boxes filled with words really constitute a graphical representation? Safety research can do much better than this. To use images effectively as elements in both communication *and* discovery – and to realise Le Coze's promising vision for them in the field of safety research – more research and development should be devoted to graphical modeling itself among safety researchers.

Consider, for example, the possibility for researching risks associated with a gas pipeline infrastructure. Imagine that the analysis starts with a two-dimensional map of pipelines in a defined geographical area. Instead of a series of simple lines designating the pipelines, the lines could vary in thickness or colour to differentiate their age, diameter, flow rates and/or operating pressure. Then a 3-D overlapping layer could superimpose population densities surrounding different line segments could be superimposed onto the line diagram. Another layer could demonstrate the spatial proximity of physical structures, including other infrastructures, such as water and sewage lines as well as electric transmission lines and rowers. Yet another layer might describe potential flood plains overlapping with line segments; another layer still could map and describe organisations with management and operational responsibility for the segments. Finally, another level could depict regulatory and emergency response jurisdictions that cover the pipeline.

This visual model could allow any two or more layers to be seen simultaneously through image transparency. Models could then be formulated and displayed, describing the various interactions between levels. Furthermore, imagine that this model could also drill down to individualised line segments, even down to work and inspection groups, thereby analysing safety risks segment-by-segment.

The use of 3-D visual images and models such as this could help cope with the need for higher resolution analysis, by incorporating more information and allowing an analytical expansion of scale and scope. Creative visualisations could also allow the integration of social science and engineering information, facilitating their working together in safety analysis for a truly inter-disciplinary, sociotechnical mode of research.

Visualisations could also be an important way to represent and communicate uncertainty in forecasts and risk assessments. One approach, used in storm trajectory forecast models, draws a cone of uncertainty over a map to indicate the range of possible prediction error in storm trajectory models over a defined time period. In the same way similar visualisations (with dotted lines or fuzzy borders) can be used to portray the range of uncertainty in risk models and in risk mitigation calculations. This could be an important way to signal to the public an uncertainty which is currently often masked in seemingly precise 'risk/spend' calculations that utilities present to public regulators.

19.5 Conclusion

This brief sketch of future challenges and possibilities for safety research has been inspired by the diverse and provocative works contained in this volume. Others will certainly have different takes on the field – its accomplishments, strengths and weaknesses. My own belief is that the time is ripe for new advances into integrative and additive connections between science, engineering and social science research in a concerted effort to yield cumulative advances in our sociotechnical understanding of safety, reliability, risk and error. Perhaps there has never been as pressing a practical need for such advances. These chapters signal to me that a great deal of imagination and research talent exists among younger writers set to fuel a bright and exciting future for safety science.

References

Csermely, P. (2006). *Weak Links: The Universal Key to the Stability of Networks and Complex Systems*. New York: Springer.

Filho, A. and Waterson, P. (2018). "Maturity Models and Safety Culture: A Critical Review." *Safety Science*, Volume 105(June), 192–211.

Gerring, J. (1999). "What Makes A Concept Good?: A Criterial Framework for Understanding Concept Formation." *Polity*, Volume 31(3), 357–393.

Gerring, J. and Baressi, P. (2003). "Putting Ordinary Language to Work: A Mini-Max Strategy for Concept Formation in the Social Sciences." *Journal of Theoretical Politics*, Volume 15(2), 201–232.

Roe, E. and Schulman, P. (2008). *High Reliability Management*. Stanford, CA: Stanford University Press.

Roe, E. and Schulman, P. (2016). *Reliability and Risk: The Challenge of Managing Interconnected Critical Infrastructures*. Stanford, CA: Stanford University Press.

SMICG (2013). Measuring Safety Performance. Safety Management International Collaboration Group. https://www.skybrary.aero/bookshelf/books/2395.pdf.

U.S. Nuclear Regulatory Commission (2009). *Guidance on the Treatment of Uncertainties Associated with PRAs in Risk-Informed Decision-Making*. Washington, DC: NRC.

Woods, D. (2016). "Four Concepts for Resilience and the Implications for the Future of Resilience Engineering." *Reliability Engineering and System Safety*, Volume 141(2015), 5–9.

20

Redescriptions of High-Risk Organisational Life

Karl E. Weick

CONTENTS

20.1 The Carrier *Carl Vinson* (CVN 70) as a Continuing
Representative Example..300
20.2 Compound Abstracting 302
20.3 Impermanence and Regularising...304
20.4 Sensework and Sensemaking 306
20.5 Conclusion 308
Notes 309
References..309

Introduction

'Objective truth is no more and no less than the best idea we currently have about how to explain what is going on' (Rorty, 1979, p. 385). The preceding chapters provide optional vocabularies that can unsettle 'normal' discussions about safety and send them off in a new direction. Those alternative vocabularies impose different meanings that may lead us to deal with reality differently. To deal with reality differently is to describe, explain, predict and modify it, 'all of which are things we do under descriptions' (Rorty, 1979, p. 375). In other words, the 'things' of safety do not exist independent of the words that call them into being. Those words take their meaning from other words, not from the accuracy with which they represent essences (Rorty, 1979, p. 368). These chapters, in other words, are less about converging to a consensus than about redescriptions which have the potential to enlarge perceptions and options.

A good example of what I mean by redescription is Haavik's[*3] replacement of the earlier, more micro and more sharp-end preoccupation with

---

* To conserve space, references to the preceding chapters are indicated by a number rather than by authors. The numbers reflect the order in which the present author studied the chapters while they were in manuscript form.

'airmanship', by substituting the more system-sensitive vocabulary of 'airlineship'. In his words, 'While pilots of earlier days were highly dependent on individual skills and experience, the safety of commercial passenger flights today results from increasingly tight couplings between individual pilots, computers, organisational regimes, standardised aircrafts and procedures, aircraft manufacturers and international legislation....To account for these changes, we introduced a systemic variant of airmanship: airlineship, where de-identification and interchangeability of personnel and aircrafts are central system parameters'. Another example is in Downer's[1] argument that what looked to Diane Vaughan like the normalisation of deviance might just as well have been normal performance within a complex technical system where results match predictions imperfectly in ways that would not be apparent to engineers in advance of the failure.

As I studied the preceding chapters, a handful of ideas that I had found useful in studies of organising and sensemaking seemed to suggest both an overlap with the themes in this book, as well as a common ground between older and newer formulations. In the following, I describe four components of this common ground: the aircraft carrier *Carl Vinson* (CVN 70) as a continuing representative example of error mitigation, compound abstracting, impermanence and regularising, as well as sensework and sensemaking.

20.1 The Carrier *Carl Vinson* (CVN 70) as a Continuing Representative Example

An early study (Rochlin, La Porte and Roberts, 1987) of flight operations on the aircraft carrier *Carl Vinson* (CVN 70) foreshadows several issues in this volume. For example, the carrier 'became' a representative example of high-reliability organising while it was led by its *second* captain, not by the first one. Under the second captain, operators undid an excess of digitalisation in the ZOG system, which had previously concealed and put out-of-reach procedures that could be adapted to cope with unanticipated surprises. As stated in footnote 9 in (Rochlin et al., p. 89), efforts to rely on the first-generation ZOG computer system: 'would have required that almost complete knowledge about all details of ship operations be known and entered if the system were to function as originally intended. In retrospect, this can be seen as a near-impossible requirement without the mounting of a considerable special effort to collect and organise the data' (p. 89). Notice that here, the fear is not focussed on normal accidents with their random, one-in-a-billion coincidence of events, but on epistemic accidents caused by specific, consequential events which challenge the conventional understandings of those events.[1]

The first-generation system was vulnerable to the 'Tyranny of the drop-down menu'[2] (e.g. you just can't go to the next step before the proceeding one is filled out). Work as imagined was initially inscribed into carrier operations which meant that 'digital compliance and accountability compromised situational adaptability'.[2] Thus, the unwinding of the original system under the second Captain led to the modification of procedures so that now they were more visible, controllable and adaptable. These modifications, together with an increase in reliable performance, suggested one reason why a million accidents waiting to happen on the deck simply never did.

The *Carl Vinson* also continues to be an exemplary case because of its more generalised caution and wariness. Personnel on carriers were as worried about anticipating and making sense of surprises as they were about making decisions. Their sensemaking seemed to fall into a pattern where they were guided to pay more attention to failure than success; avoid simplicity rather than cultivate it; focus on operations as much as on strategy; organise for resilience rather than anticipation, and allow decisions to migrate to experts, regardless of where they happened to be located within the hierarchy (Weick, Sutcliffe and Obstfeld, 1999). Personnel were worried about making sense of the ongoing flux, and enacting order back into what they experienced. Within that context, their attention to failure, simplification, operations, resilience and migrating expertise makes perfectly good sense. All five register more of the flux and provide structure for continuing engagement, as well as animating that engagement. These five shifts in attention serve to connect local words and practices back to concrete experiences, perceptions and flux. Chronic questions on the deck seemed to be: 'what's failing', 'am I oversimplifying this', 'what am I doing', 'what do I have to work with' and 'who knows more than I do?' All five questions constitute a culture of wariness. By increasing an organisation-wide sense of vulnerability, emerging answers to these questions mitigated the tendencies to normalise or overlook any discrepancies in the face of production pressures.

A further continuity between carrier operations and the current chapters is that reliable functioning was maintained despite a markedly high potential for accidents. The original question in the carrier study was not, 'how could reliability be improved', but 'how is reliability maintained'. Rochlin put it this way: 'we were not seeking to design ways to improve the reliability of these organisations, but in the spirit of organisational research, to try to understand how they maintained high reliability under time pressure in risk-laden environments' (Rochlin, 2011, p. 17). Reliability on carriers seemed to be maintained relationally by senior chiefs, incoming officers who shared new experiences with the deck crew and uninterrupted on-board training and retraining (Rochlin et al., 1987, p. 80).

The importance of activities done repeatedly and continually is that there will be fewer unexpected moments and/or those that occur will be slower to develop, caught earlier in their development and contained more fully. To maintain reliability is to keep the system running, adapt to situational variability and make sense of events that are unexpected. An early observation that anticipated mechanisms for reliability maintenance was that the carrier CVN 70 'appears to us as one gigantic school, not in the sense of rote learning, but in the positive sense of a genuine search for acquisition and improvement of skills. One of the great enemies of high reliability is the usual "civilian" combination of stability, routinisation, and lack of challenge and variety that predispose an organisation to relax vigilance and sink into a dangerous complacency that can lead to carelessness and error' (Rochlin et al., 1987, pp. 82–83).

The mention of an ongoing fear of 'dangerous complacency' is substantively fleshed out by Gould and Fjæran's[9] discussions of drift and risk attenuation. Drift in relation to accidents has been described as a gradual behaviour change on the part of those individuals at the sharp end of systems; when rules did not match current task demands, individuals adjusted their behaviour accordingly (Snook 2002). This practical drift, where the baseline performance of a system subtly moves away from its original design, potentially increases the possibility of an incident or accident as performance moves too far away from the expected baseline.[9]

20.2 Compound Abstracting

'Language is the quintessential organising technology that enables us to selectively abstract from the otherwise intractable flux of raw experiences' (Chia, 2003, p. 123). In the context of safety, the key phrase is 'selectively abstract'. Abstractions are general concepts that suppress the complex details in order to highlight a subset of apparent similarities. Safety is about limited conceptions of unlimited interdependencies. Abstracting, therefore, is inherent in any discussion of safety. It involves a progression rather than a duality, a progression that extends from perceptual acquaintance through conceptualisation on to reification and formalisation.

Robert Irwin (1977) describes this progression as being one of 'compounded abstraction'. He proposes that the progression begins with *perception* (synthesis of undifferentiated sensations), followed by *conception* (initial crude differentiation into unnamed zones of focus), *form* (zones of focus are named, transformed into entities and communicated), *formful* (named entities are related, compared, preserved as patterns), *formal* (patterns get reified, for example, the pattern of above and below is reified into superior/ subordinate) and finally *formalised* (reifications are treated as true and dictate behaviour thereby maximising their distance from direct perceptual experience and their elevation to the status of facts. The progression 'is not so much temporal progression as a phenomenological one'.

At any given moment, for any given individual, all six stages are operating simultaneously, and yet the earlier phases [perception, conception] exist prior to the later ones in the sense that they ground them, they constitute the source out of which the more compound stages emerge' (Weschler, 1982, p. 183). But each time a whole is represented, it is less than the parts which were present in the original perception. We gain efficiency. We lose information. We increase risk.

One way to reverse this progression and move back towards 'earlier phases' is to cultivate the ability to anomalise (Weick and Sutcliffe, 2006, p. 518). To anomalise is to take proactive steps to become alert to discrepancies, to understand them more completely, and to be less encumbered by history and familiarity. In other words, to anomalise is both to notice discrepancies and to work actively to understand them without simplifying them into familiar categories (Sutcliffe, 2011). Ongoing attention to differences, nuances, discrepancies, the unexpected, as well as outliers weakens the tendency to simplify events into familiar events and strengthens the tendency to differentiate events into unfamiliar events. The more people hold on to differences, the more slowly they normalise the details and the more nuanced and fine-grained their understanding of context ultimately is.

Anomalising is also important because it corrects the property that reliable performance is a dynamic non-event (Weick, 1987, p. 118), in which adjustments rarely call attention to themselves. This lack of visibility can increase vulnerability to threats from complacency, inertia and drift. In Gould and Fjæran[9] words, 'In relation to risk attenuation, the fact that "nothing happens" represents a warning signal that the organisation needs to pay attention to developing perceptions of risk and the everyday behavioural changes that may follow'.[9]

A related way to slow down compounding is to practice Irwin's maxim that, 'seeing is forgetting the name of the thing seen' (Weschler, 1982, p. 180). As naming and abstracting unfold, the conceptions that accomplish this soon 'mean something wholly independent of their origins' (Irwin, 1977, p. 25). It is this potential for meanings to become wholly independent of their origins that leads to misconstruals and misjudgements. Weak signals of danger are ignored, mislabelled, or converted into familiar labels that activate familiar routines. To forget the names of the things seen is to move back toward conception and perception.

Compounding is not just an individual progression. Compounding is social, with the potential to make sharing adaptive but costly. Perceptions have to be described in order to be shared, and this means that early impressions which are more stimulus-driven become more simplified and schema-driven when they are described to others using commonly shared abstractions (Baron and Misovich, 1999). Schemas tend to normalise discrepancies and foster movements farther away from initial perceptions. That's why sharing is 'costly' and also why redescriptions and anomalising can recover crucial details.

Previous chapters also discuss ideas intended to reverse the progression of abstracting. For example, an accessible yet complex visualisation[4] of systemic interactions, such as Figure 10.7 moves back towards perception and conception when it activates a greater variety of labels, preserves wholes despite subsequent abstracting, conveys motion and cycling, suggests anomalies embedded within sequences of normal functioning, and reminds observers that any depiction, regardless of its face validity, is still a relation between map and territory.

Abstracting is also slowed when investigators feel the sting in Collins (1985) compact observation that, 'Distance lends enchantment'.[1] 'Outsiders [and the present author is just such a person] tend to be less sceptical of facts and artefacts than the experts who produce them, simply because outsiders are not privy to the epistemological uncertainties of knowledge production'. For example, from afar safety looks to be 'straightforward' and seems subsumed under five principles that involve failure, simplification, operations, resilience and expertise. Metal fatigue also looks straightforward from afar, but 'from the perspective of the engineers closest to it, it is striking in its formidable complexity and ambiguity: a reflection, as we will see, of deep epistemic constraints'.[1]

20.3 Impermanence and Regularising

The term 'safety' is mischievous in the sense that it is static. The preceding chapters suggest otherwise. As Antonsen and Almklov[8] mentions, safety is not so much an output variable as it is an 'evolving, political, negotiated order in organisations' (Henriqson et al., 2014, p. 474).[8]

If safety is evolving and constantly in need of being negotiated, then we need to describe the organisational context for safety differently. Organisations are 'nothing more than islands of relatively stabilised relational order in a sea of ceaseless change' (Chia, 2003, p. 131). Organising now becomes a stance rather than a place. Chia, for example, argues that: 'organising could be more productively thought of as a generic existential strategy for subjugating the immanent forces of change [and] that organisation is really a loosely coordinated but precarious 'world-making' attempt to regularise human exchanges' (p. 123). Chia's reference to 'a loosely coordinated attempt to regularise human exchanges' alerts us to the reality of dealing with distributed temporary resources. And his reference to 'world-making' alerts us to the difficulty of sensemaking and finding common ground when resources are distributed.

The resulting picture of impermanence and organisation looks something like this: 'we perceive the processes of organisation to be a restless searching to fix its structure through the generation of texts, written and spoken, that reflexively map the organisation and its preoccupations back into its discourse and so, for the moment, produce regularity... It is the existence of such texts and the text-worlds they constitute that makes the organisation visible and tangible to people' (Taylor and Van Every, 2000, p. 325).

Chia's 'immanent forces of change' are 'subjugated' in part by communication embedded in sites and surfaces. Conversation is the *site* for organisational emergence, and language is the textual *surface* from which organisation is read. Thus, organisations are talked into existence locally; they are read from the language produced there. This means that, for an organisation to act safely, its knowledge must undergo two transformations: (1) it has to be textualised so that it becomes a unique representation of the otherwise multiple distributed understandings; (2) it has to be voiced by someone who speaks on behalf of the network and its knowledge (Taylor and Van Every, 2000, p. 243).

Those who speak on behalf of safety usually do so based on their power to edit and silence texts. The expression 'safety in power' refers to the fact that some actors are more powerful than others in defining what has gone wrong in accidents investigations. There is a lot of safety in having the power to say just what happened; to be able to identify risk; to write history, or to assign causes and consequences after an accident.[8] For example, as Waring and Bishop[5] makes clear, the power of the medical profession is not only found in its ability to heal the sick, but, more broadly, in its ability to define the nature of illness and to shape the organisation of services concerned with healing the sick. Thus, the management of professional safety and risk by non-professionals raises the possibility that they will re-organise professional work in the name of safety, and that they could call into question the foundations of professional legitimacy or jurisdiction.

A recurring image in this book, such as in Gotcheval et al.'s[12] detailed picture of project governance, or Reader's[11] air traffic management system, or Hayes's[7] pipeline explosion, is that of safety formed within a network of multiple, overlapping, loosely connected conversations, spread across time and distance. The network preserves patterns of understanding that are more complicated than any one node can reproduce. The distributed organisation literally does not know what it knows until powerful macro-actors articulate it (Taylor and Van Every, 2000, p. 243). This ongoing articulation gives partial voice to the collectivity, enabling interconnected conversations and conversationalists to see an edited, dominant version of what they have said; what it might mean, as well as who they might be. The importance of macro actors is made clear when Antonsen and Almklon[8] remind us that Geertz's notion of culture consisting of webs of significance that man 'himself has spun', was criticised by Scholte, who observed that 'very few do the actual spinning, while the majority is simply caught'.[8]

Edited meanings are not just a product of power. They are also a fixture in conversing itself, conversing that can inadvertently create the illusion of safety. Conversations about risk can become sites that serve as 'stations of attenuation'. An original message reporting that no serious incidents have occurred can be altered into a message which suggests that the risk of major accidents is low (see Campbell 1958 for mechanisms that are involved in this alteration). Networks of conversations have mixed effects on safety. As Haavik[3] reminds us, 'facts and truths are states of affairs that are stabilised by networks of humans and non-humans – the denser network, the stronger fact'.

20.4 Sensework and Sensemaking

Throughout the chapters there is the tacit message that we need to take a closer look at what people are doing at the time when they single out portions of streaming inputs for closer attention. The important questions are how they size up and label what they think they face, and how continuing activity shapes (and is shaped by) this activity. This shaping has been called 'sensemaking', an explanatory process built out of the cyclical entanglement of actions and interpretations. Maitlis and Christianson (2014) formally describe sensemaking as: 'a process, prompted by violated expectations, that involves attending to and bracketing cues in the environment, creating intersubjective meaning through cycles of interpretation and action, and thereby enacting a more ordered environment from which further cues can be drawn' (p. 67).

Safety emerges out of a reality where 'significance and valence do not pre-exist "out there", but are enacted, brought forth, and constituted by living beings' (Thompson, 2007, p. 158). Typically, this constituting occurs relationally, by interactive practices that stabilise interpretations and facilitate swift, coordinated action (Barton et al., 2015, p. 75.)

Two basic interaction practices are respectful interaction and heedful interrelating. They generate shared interpretation and shared action. Respectful interaction, the combination of trust, honesty and self-respect, increases the likelihood that people will speak up about issues of concern, share their perspectives and ask other questions about their interpretations. Heedful interrelating, the combination of contributing, representing and subordinating, increases the likelihood that people are able to make sense of how their work fits in with others (Weick and Roberts, 1993). Whenever one of these practices is weak or missing, an adverse event is more likely to occur.

But the prevention of adverse events requires more than just trust and heed. Part of what is missing from an overly social view are technological systems that include technical devices, organisational routines and procedures, legislative artefacts and scientific or other knowledge elements, such as skills, rules of thumb and established norms for the proper handling of technology.[10]

Relational infrastructures that blend the social and technical also blend the actions of supervisors and frontline staff (Barton et al., 2015, p. 75). The frontline has access to concrete situational details, what Baron and Misovich

(1999) call stimulus-driven knowledge by acquaintance. Supervisors work with schema-driven knowledge, which converts acquaintance into knowledge through the use of description that can then be disseminated (Weick, 2011, p. 23). Description is essential for sensemaking and organising, but acquaintance is critical for the perception of details and cues that begin to clarify ambiguity.

The blending of the social and technical is at the centre of Haavik's[3] argument that, 'sense connotes not only to meaning that is developed by social collectives, but as much to the indispensable sensors through which phenomena not readily available to human senses speak and produce meaning'. When these sensors are incorporated into sensemaking, the result is 'sensework', which combines heterogeneous tools and practices, such as real-time information from sensor instrumentation, formalised models of causal relationships, operational experience, formal procedures and ad hoc workarounds. Sensework is pragmatic by definition.[3]

Sense, then, is built from circumstances that are turned into a situation (sensework) that is comprehended explicitly in words (sensemaking), serving as a springboard for action (a pragmatic outcome) (adapted from Taylor and Van Every, 2000, p. 40). In the earlier discussion of flight operations on carriers, sensemaking took the form of thinking while acting and thinking through acting. These fusions of thought and action stir up details and meanings that people had simply missed or forgotten, as they tried to both adapt and limit the expansion of the unexpected. The practical takeaways of this perspective have been, 'don't stop the action too quickly,' or, as John Dewey put it, 'so act as to increase the meaning of present experience' (Dewey, 1922, p. 283).

One way to describe sensemaking in the context of sensework is to focus on seven key factors (Weick, 1995, pp. 17–63) that are proximal influences on the sense that is made. These seven factors include: Social context, Identity formulation, Retrospective interpretation, meaningful Cues, Ongoing events, Plausible stories and effortful Enactment (SIR COPE). A description that uses the language of SIR COPE allows the analyst to retain part of the psychological reality of safety-relevant work. For example, consider the employees at Bhopal who were working at 11:30 PM on a humid December night (2 December 1984) in a deteriorating Union Carbide chemical plant. As the runaway chemical reaction began to unfold, there was little communication among the six people on the crew (social context). There was also acceptance of low-status employment in a neglected plant (identity), unease that what had been occurring that evening was not right (retrospect), malfunctioning gauges (cues), rumbling sounds that grew louder and odours that became stronger (ongoing), explanations of the odours as insect spray (plausibility) and little immediate action other than a tea break to follow-up on the cues (enactment) (Weick, 2010, p. 545).

The interdependence of sensemaking and sensework is clearest to me in the leadership priorities of the wildland firefighter, Paul Gleason (1946–2003).

Reflecting on his experience in the Zig Zag hot shot crew, he told me, 'If I make a decision it is a possession; I take pride in it; I tend to defend it and not to listen to those who question it. If I make sense, then this is more dynamic, and I listen, and I can change it. A decision is something you polish. Sensemaking is a direction for the next period'. Gleason's ability to put decision making into perspective was again evident when he said (while reflecting on the Cerro Grande escaped fire that caused $1 billion in damage), 'When the wind blows, those lines from decisions to outcomes are meaningless'.

20.5 Conclusion

Abstractions, impermanence and sensemaking are about stabilising the flux of Chia's 'ceaseless change'. In this sense, stabilising in the service of safety is indistinguishable from stabilising through organisation. The fact that words are so prominent in this stabilising seems like a weak response to the challenge. But words – words that give meaning to other words – and redescriptions that remix words are much of what we have to work with.

Ironically, that is part of our strength. For example, redescription plays a crucial role in the balancing of impact and analysis. As Bergström's[6] shows, when the language of resilience engineering is introduced into air transport regulations, it now holds pilots responsible for both the behaviour of the system in which they are embedded, and for their beliefs in resilience theory. An accident within this context means that neither the system nor the theory seem to be significant contributors. Redescription focusses attention back on those contributors, and back toward analysis. This occurs because redescription favours descriptive curiosity and critical commentary, rather than a preoccupation with normative impact.

Each analysis in this book represents a new way to think about safety, but at the same time, each one is a mere fragment of safety science. Again, that in itself can be a strength depending on the uniqueness and overlap of the fragment. The context within which the claim holds true is Donald Campbell's (1969) 'fish-scale model of omniscience'. He describes the model this way: 'the slogan is collective comprehensiveness through overlapping patterns of unique narrowness. Each narrow specialty is in this analogy a fish-scale... (O)ur only hope of a comprehensive social science, or other multiscience, lies in a continuous texture of narrow specialties which overlap with other narrow specialties' (p. 328).

Redescriptions can create novel narrowness, especially when investigators live by this credo: 'Let my pattern of inevitably incomplete incompetence cover areas neglected by others. Make me a novel fish-scale' (p. 342).

Newer and older instances of novel narrowness improve our understanding of safety as long as we remain attuned to the overlap. The preceding authors are mindful both of that novelty and of that necessary overlap.

Notes

1. On Ignorance and Apocalypse: A Brief Introduction to 'Epistemic Accidents', John Downer.
2. Standardisation and Digitalisation: Changes in Work as Imagined and What This Means for Safety Science, Petter G. Almklov and Stian Antonsen.
3. Sensework, Torgeir Haavik.
4. Visualising Safety, Jean-Christophe Le Coze.
5. Safety and the Professions: Natural or Strange Bedfellows? Justin Waring and Simon Bishop.
6. The Discursive Effects of Safety Science. Johan Bergström.
7. Investigating Accidents. The Case for Disaster Case Studies in Safety Science, Jan Hayes.
8. Revisiting the Issue of Power in Safety Research, Stian Antonsen and Petter Almklov.
9. Drift and the Social Attenuation of Risk, Kenneth Pettersen Gould and Lisbet Fjæran.
10. Coping with Globalisation: Robust Regulation and Safety in High-Risk Industries, Ole Andreas Engen and Preben Hempel Lindøe.
11. The Interaction between Safety Culture and National Culture, Tom Reader.
12. Governance for Safety in Inter-Organisational Project Networks, Nadezhda Gotcheva, Kirsi Aaltonen and Jaakko Kujala.

References

Baron, R. M. and Misovich, S. J. (1999). On the relationship between social and cognitive modes of organisation. In S. Chaiken and Y. Trope (Eds.), *Dual-Process Theories in Social Psychology* (pp. 586–605). New York: Guilford.

Barton, M. A., Sutcliffe, K. M., Vogus, T. J. and DeWitt, T. (2015). Performing under uncertainty: Contextualised engagement in wildland firefighting. *Journal of Contingencies and Crisis Management*, Volume 23(2), 74–83.

Campbell, D. T. (1958). Systematic error on the part of human links in communication systems. *Information and Control*, Volume 1(4), 334–369.

Campbell, D. T. (1969). Ethnocentrism of disciplines and the fish-scale model of omniscience. In M. Sherif and C. W. Sherif (Eds.), *Interdisciplinary Relations in the Social Sciences* (pp. 328–348). Chicago, IL: Aldine.

Chia, R. (2003). Organisation theory as a postmodern science. In H. Tsoukas and C. Knudsen (Eds.), *The Oxford Handbook of Organisation Theory* (pp. 113–140). New York: Oxford.

Collins, H. (1985). *Changing Order.* London: Sage.

Dewey, J. (1922). *Human Nature and Conduct.* Mineola, NY: Dover.

Henriqson, É., Schuler, B., van Winsen, R. and Dekker, S. W. A. (2014). The constitution and effects of safety culture as an object in the discourse of accident prevention: A Foucauldian approach. *Safety Science,* Volume 70, 465–476.

Irwin, R. (1977). Notes toward a model. In *Exhibition catalog for the Robert Irwin exhibition, Whitney Museum of American Art,* April 16–May 29, 1977 (pp. 23–31). New York: Whitney Museum of American Art.

Maitlis, S. and Christianson, M. (2014). Sensemaking in organisations: Taking stock and moving forward. *The Academy of Management Annals,* Volume 8(1), 57–125.

Rochlin, G. I. (2011). How to hunt a very reliable organisation. *Journal of Contingencies and Crisis Management,* Volume 19(1), 14–20. (1996).

Rochlin, G. I., La Porte, T. R. and Roberts, K. H. (1987). The self-designing high-reliability organisation: Aircraft carrier flight operations at sea. *Naval War College Review,* Volume 40(4), 76–90.

Rorty, R. (1979). *Philosophy and the Mirror of Nature.* Princeton University Press, Princeton, NJ.

Snook, S. A. (2002). *Friendly Fire: The Accidental Shootdown of U.S. Black Hawks Over Northern Iraq.* Princeton University Press, Princeton, NJ.

Sutcliffe, K. M. (2011). High reliability organisations (HROs). *Best Practice and Research Clinical Anaesthesiology,* Volume 25, 133–144.

Taylor, J. R. and Van Every, E. J. (2000). *The Emergent Organisation: Communication as Its Site and Surface.* Mahwah, NJ: Erlbaum.

Thompson, E. (2007). *Mind in Life: Biology. Phenomenology and the Sciences.* Cambridge, MA: Belknap Press.

Weick, K. E. (1987). Organisational culture as a source of high reliability. *California Management Review,* Volume 29, 112–127.

Weick, K. E. (1995). *Sensemaking in Organisations.* Thousand Oaks, CA: Sage.

Weick, K. E. (2011). Organising for transient reliability: The production of dynamic non-events. *Journal of Contingencies and Crisis Management,* Volume 19(1), 21–27.

Weick, K. E., Sutcliffe, K. M. and Obstfeld, D. (1999). Organising for high reliability: Processes of collective mindfulness. In: B. Staw and R. Sutton (Eds.), *Research in Organisational Behavior,* (Volume 21, pp. 81–123). Greenwich, CT: JAI Press.

Weschler, L. (1982). *Seeing Is Forgetting the Name of the Thing One Sees: A Life of Contemporary Artist Robert Irwin.* Berkeley, CA: University of California.

21

Skin in the Game: When Safety becomes Personal

Ron Westrum

CONTENTS

21.1 National Culture Is Not Like Organisational Culture 311
    21.1.1 'They Won't Go When I Go' 315
    21.1.2 Is This My Responsibility? 317
    21.1.3 Engagement in Safety.. 319
Notes 319

21.1 National Culture Is Not Like Organisational Culture

In going through Tom Reader's discussion of national influences on safety culture (this volume, Chapter 2), some of the following points occurred to me; this is how the discussion in my head went: first, I asked myself, 'What is culture?' For instance, sociologist Emile Durkheim famously defined a 'social fact' as something external and constraining.[1] Well, that would certainly be true for organisational culture as well. It is external, and it constrains. However, it is not immovable, and that is how it differs from national culture, which – at least in the short term – is immovable. For many years, I thought, as others do, that organisational culture and national culture were both 'culture', and that it was simply a difference of scale. But I no longer think this is correct. Organisational culture is a living thing; part of an ongoing living system. It is not passive, it reacts to things that people in an organisation do. By contrast, national culture *is* passive, and might be seen as a kind of distant background. Instead, we might call it 'national tendency'. (There are also 'regional tendencies'.[2]) It does not react to what people do, except perhaps in the very long run. The *living* organisational culture is built by management, and it is with management that this brief discussion will deal. While I know little about how national cultures come about, I do have some strong views on how organisations are created and changed.

So, let us discuss management for a moment. Management – the higher officials in an organisation – sets the conditions for organisational culture by clarifying what it is they want to see happen, as well as defining which behaviours are to be encouraged or punished.[3] If management is indeed at the root of organisational culture, it may be the case that managements with different orientations will create systematically different cultures.[4] Yet one matter directly important to safety has been neglected. That is the *presence* of management. Some managers are simply more psychologically present than others and take a greater ownership of what happens. This presence is one of the factors shaping safety and success. When managers are present, they demand higher performance, and that, in turn, is what makes both system success and safety more assured.

For instance, at the low end of presence, we have 'Brownie' Brown, head of the Federal Emergency Management Agency during the Katrina Hurricane disaster in New Orleans (29 August 2005). Clearly, Brownie Brown was not very happy with his role as the maestro of disaster response, and on his first day managing the hurricane disaster recovery, he sent an email asking, 'Can I quit now?'[5] In many other ways he showed that he was not into, and not up to, the job. In part, this could be because he had scant training for it. His previous position had been as a Judges and Stewards Commissioner for the International Arabian Horse Association, and he had gotten the disaster job without any special training or aptitude.[6] In fact, he got the job through what amounted to having known (and been hired by) a pal. This same pal had previously held the job himself, but he had also seemed to have little training or aptitude for disaster management. Accordingly, during Katrina he mostly seemed to concentrate on his public image and personal needs. The unfortunate citizens of New Orleans and the Gulf Coast had to depend on the skills of Brown's subordinates, who, it seems, couldn't much depend on Brown himself. So, Brown was not very much 'into' his role. It might have seemed funny if so many people had not been depending on Brown being into his role. Instead, they died by the dozens as Brown 'went through the motions'. The final toll of deaths during Katrina for New Orleans and the Gulf Coast was about 1245.[7]

Shortly before Katrina hit the Gulf Coast in 2005, the space shuttle *Columbia* had exploded on 1 February 2003, in a mass of fragments that fell all over Texas and Louisiana. In part, this too represented the actions of a team that was less than entirely 'into' the critical job they had been chosen to fill. The shuttle's Mission Management team was not deeply into their role, and while the shuttle was in space and preparing to return to earth in January, they were engaging in 'business as usual'. The team's leader, Linda Ham, was overconfident, and did not experience the 'chronic unease' that some experts feel goes with high-reliability management (Fruhen et al., 2013).[8] During the mission, the team took two weekends off, including the three-day Martin Luther King, Jr. weekend; meanwhile, the astronauts aboard the shuttle could not know that a foam strike at launch on 16 January had badly compromised their ability to get down to safety, and that re-entry might kill them.

Others at NASA, however, were alarmed at the potential state of the shuttle – none more so than engineer Rodney Rocha, head of the shuttle's structural engineering team at Johnson Manned Spaceflight Center in Houston. Rocha had looked again and again at the films of the launch's foam strike, and considered additional evidence produced by computer models of the event. He still felt he did not have enough information to evaluate whether or not the strike had fatally damaged the shuttle. He thought the team needed some additional information that a close-up photograph would provide. He, along with some other members of the team, sent a request to the Air Force, asking for such a photograph. The Air Force then geared up to produce it. Rocha also tried to alert other senior managers at the Johnson Space Center of the need for additional facts, without success. But ultimately, Linda Ham disagreed. She wound up telling the Air Force that the photo was unnecessary (Cabbage and Harwood, 2004, p. 111).[9] The structural engineering team never got the photos that might have shown the danger. The compromised shuttle *Columbia* was destroyed upon re-entering the atmosphere. Its astronauts died.

The other end of the spectrum is well represented by the attitude and behaviour of Eugene Kranz, NASA Flight Director during the Apollo 13 flight (Kranz, 2000).[10] When an internal explosion on 13 April 1970 shook the moon mission vehicle in space, NASA knew it had a problem. The three astronauts had to be brought home safely in the damaged spacecraft. While Kranz was only one of the flight directors involved, he had taken up a major role in training the controllers who were involved with the rescue. Kranz had also been part of the team when the Apollo fire had killed three astronauts. Kranz now made clear that no effort would be spared to get the astronauts back to earth. The phrase 'failure is not an option' was not Kranz's words, but it well represented his attitude. In the event, the astronauts had to shift back and forth between the command capsule and the moon lander so that they could conserve what scarce oxygen and power they had. They made faultless decisions to align the craft properly and fire the engines in the right amount at the right time. They did it all perfectly, and the astronauts all came home. Kranz was not about to let those astronauts die.

The contrast between Kranz's attitude and that of Linda Ham could not be more extreme. Kranz felt he was in charge. It was his shuttle, his people and he was *personally* not going to let them down. To Linda Ham, this was only one flight among many; the exact details did not seem to matter so much. The astronauts don't seem to be present for her, and she assumed no identification, no ownership; in fact, she seemed to lack compassion. It appears impossible not to think that Kranz's will to protect his people had a powerful effect on the rescue effort itself. And, in contrast, the psychological distance of Linda Ham from a possible rescue was thus a real danger to safety. One wonders what Ham said to the astronauts' families when things fell apart and the entire crew perished.

Going back more than a century, we come across a situation with, again, some similarities to Apollo 13. In 1850, England was preparing to house the first world's fair in Hyde Park, London. The date of the exhibition was less than a year away, but the design for the building was still unsettled.[11] A gifted committee of engineers had meanwhile agreed to consider building designs from around the world for a contest that, in the end, had received 245 entries. However, that contest now failed to produce a workable design, so the committee set about designing the building itself. In spite of the brilliance of the committee, the resulting concept was not an acceptable solution. (One recalls the joke that, 'a camel is a horse designed by a committee'.) The location required a temporary housing, and the committee's design was both unpleasing and unlikely to be temporary. Fortunately for the event however, at the eleventh hour, a solution came from the gifted landscape architect Joseph Paxton, who submitted his bid to construct a totally different design, essentially based on his previous experience with iron and glass greenhouses. Pictures of the proposed building soon appeared in the *Illustrated London News*; were popular with the British public, and eventually with the committee itself. Paxton's design would be called the 'Crystal Palace', and would be constructed by the talented builder, James Fox, through his firm: Fox, Henderson. The building's plans were being drafted something like *seven months* before the actual opening, on 1 May 1851. The opening would involve the young Queen of England, Victoria, and basically the entire civilised world, whose representatives had been invited to attend and bring exhibits from their respective countries.

What is of interest to us here is James Fox's comments on the design process.

> The drawings occupied me about *18 hours each day for seven weeks*, and as they went from my hand Mr. Henderson immediately prepared the ironwork and other materials required in the construction of the building. As the drawings proceeded the calculations of strength were made, and as soon as a number of the important parts were prepared....we invited Mr. Cubitt to pay a visit to our works at Birmingham to witness a set of experiments in proof of the correctness of these calculations... (Tomlinson, xxiii, italics added)

> From this time I took the general management of the building under my charge, and spent all my days on the works – feeling that, unless the same person who had made the drawings was always present to assign to each part as it arrived upon the ground its proper position in the structure, it would be impossible to finish the building in time to assure its opening on the 1st of May [1851]. ... (Tomlinson, xxiv)

> ...I felt in some degree that the credit of England was in our hands; that we had it in our power to achieve that that would be eminently successful or failing to do so would become a source of the most serious disappointment and disgrace. ... (Tomlinson, xxiv)

So, James Fox, taking personal responsibility for the correctness of the design, subsequently saw to it personally that *every girder went into the right place.* These were key factors in the safety and durability of the Crystal Palace. This huge and novel structure stood 1850 feet long, 408 feet wide, and, in places, 64 feet high.[12] Moreover, in 1851, there were no internal combustion engines. Truly, the 'horsepower' that lifted the building's parts into place came from real horses. Contrary to both uneducated and highly trained nay-sayers, the building went up on time and functioned perfectly. Millions of people, British and otherwise, saw the exhibits. The reputation of England was not only saved, but embellished.

So some sort of personal engagement in the outcome is essential to getting good results. One wonders if this is why aviation safety advanced so much more quickly than other kinds of systems safety, because, as one writer put it, when things go wrong, 'the pilot always goes in first' (Jouanneaux, 1999).[13] Getting pilots involved in safety is not hard, because their lives are on the line if safety doesn't work out! But usually, to get the right safety decision, one has to involve those further up the line, so that they too have 'skin in the game', to use an American expression. James Reason has called the higher officials in the organisation 'the blunt end', as distinct from the 'sharp end' at the front.

A perfect example of this distinction was involved in an aircraft seat manufactured by the Hughes Aircraft company, which went into the American F-106 Delta Dart fighter plane. This ejection seat, it turned out, was a pilot killer, since when pilots had to eject from the F-106, the seat often failed to save them. In 1964, this problem ensnared Colonel Jack Broughton, who was commanding officer of the Minot Air Force Base in North Dakota. He was the person who had had to tell the spouses of fourteen dead flyers that, 'sorry, but your husband did not survive the crash' (Broughton, 2008).[14] Broughton thus wanted to ground the F-106, but the Commanding General of the Air Defence Command Herbert Thatcher (4 stars), did not want to ground the plane, which was important to continental defence. In 1964, the Delta Dart was the primary all-weather interceptor for the U.S. Air Force. Broughton then took the very risky step of telling his high-up boss that he himself would accept the plane continuing to fly if the general would accept personal responsibility for the next pilot killed. The general was furious, 'Damn you, Broughton!' but the plane was grounded until the seat got fixed. Broughton was very lucky not get demoted, but the general had understood his passion, and backed off.

By the same token, those whose lives are not on the line often do not care all that much for the safety of those less fortunate.

21.1.1 'They Won't Go When I Go'

This is the title of a song by the singer Stevie Wonder (from the album *'Fulfillingness' First Finale'*), and reflects the indifference or lack of care of those who make, but are unaffected by, dangerous decisions. Some years back, when I was in graduate school, I knew a classmate who had worked for the CIA in Vietnam. He once described the experience to me of sitting on a roof somewhere in Saigon, watching a multi-mini-gun-armed C-130 (often called 'Puff the Magic Dragon') pasting the ground below with its fearsome weapons – while he sipped a cocktail. He said it sounded like a buzzsaw. At the time, he was just enjoying his drink and feeling happy to not be on the receiving end of the aircraft's murderous fire. It was only later, after returning to the United States, that his conscience awakened; it devastated him to think about all of the events in that country with which his agency had been involved, resulting in death and destruction for so many of its citizens.

I remember reading an account by one of the American pilots who was shot down (cf. the late Senator John McCain) over North Vietnam, shortly before being captured. He said that it 'never occurred to him' until he was shot down, just what was happening on the ground as he dropped his bombs(!). And I also remember reading one of the Viet Cong (South Vietnamese rebels) describing the intense fear generated by the campaign called 'Rolling Thunder', which involved huge bombs being dropped from American B-52 bombers, trying to destroy the Viet Cong's infrastructure. He said that everyone in the tunnels below defecated when the bombs went off. It was very embarrassing, he said.

The importance of 'closing the loop' between causes and effects also brings to mind another example, from my own study of missile design at China Lake (Westrum, 1999). Frank Knemeyer, an Associate Technical Director at the facility, told me of the enormous sense of involvement that the weapon designers felt.

'If you wanted to make a lot of money at your job, you didn't work for the government out here. So, what you did get, you got the satisfaction you're talking about – that the thing worked the way it was supposed to. In Korea and Vietnam, we actually helped pilots accomplish their mission and *saved their lives*. Carrier skippers and carrier air group commanders would come here before they deployed. We would spend a day with them here, telling them all about our systems and everything, everything we knew about their idiosyncrasies, and in some cases, we helped train them here. We would also send people out to the carriers to help them whenever they ran into any problems, everywhere from the technical, even to helping them with their tactical problems. When those people came back from deployment, they made a point to stop here and tell us how they made out, how our systems worked, what were the bad points, and basically, how it saved their lives. That is not a monetary reward – that is one of the richest rewards you can get' (Westrum, 1999, p. 224).[15]

We can well imagine how serious the responsibility was on the shoulders of the weapons designers, because weapons that didn't work would inevitably cost pilots' lives. Many of the China Lake designers remembered that, in World War II, some of the weapons *didn't* work. This was true of the American aerial torpedoes at the battle of Midway, which did not go off after they hit the enemy vessels, and whose pilots were all shot down.

By the same token, others with strategic responsibilities in Vietnam during the 1960s fretted over the American losses in the air war. One of these was Frank Ault, who was the commanding officer on the aircraft carrier *Coral Sea*. Captain Ault obsessed over the number of American planes getting shot down over North Vietnam, so different from the 'exchange ratio' during Korea. He wrote letter after letter to the Navy, until one day in 1968, in Washington, he found himself in a conference room facing Vice Admiral Tom Connally, and Rear Admiral Bob Townsend. The two admirals now acknowledged the problems that Ault had complained about. And, they said, he – Frank Ault – would be given the resources and the responsibility to fix those problems (Westrum, p. 216).[16] He took on the job. Ault quickly rose to the task, putting together a committee of five experts, who, in turn, commissioned visits to carriers, factories, repair depots and design organisations. When (90 days later) the information was brought together, a report came forth that brought a raft of changes. These changes included the creation of the Naval Fighter Weapons School in Miramar, California, commonly known as 'Top Gun'. Ault's involvement led to a greatly superior U.S. fighter community (Wilcox, 1990).[17]

21.1.2 Is This My Responsibility?

Paul Ehrlich, a biochemist (1854–1914), worked on inventing a cure for syphilis until he found one. 'He was hard-working, kind, and a modest man with a sense of humor'. When in 1890 he developed a better cure for Syphilis, he was very proud of his accomplishment. But he did not become complacent with his invention or the profits that came from using it.[18] When the original formulation of Salvarsan (the trade name for his product) caused too many side effects, he developed a better formulation, Neosalvarsan, with fewer side effects. He was a tireless experimenter, and constantly checked and re-checked his experimental results.[19] One could wish all drug developers were this conscientious, but they aren't.[20] The opposite would seem to be true of Raymond Sackler and the Perdue family, developers and manufacturers of Oxycontin, an opioid painkiller. When early reports came to the company that the drug was being over-prescribed and used illegally on a large scale, it would seem that they ignored the misuse. The problem then grew more serious, as over-prescription continued. Even restricting the drug turned into a nightmare, as addicted users who were unable to afford street versions of the pill turned to cheaper heroin. Deaths in the United States then rose to the tens of thousands from heroin overdoses and heroin mixed with even deadlier drugs such as fentanyl. But still the Sackler and Perdue families took little responsibility for the devastation they had unleashed, because their products continued to make them billions of dollars.[21]

And finally, we have the Great Leap Forward, China's campaign to increase the productivity of its agricultural system, 1958–1962. The Great Leap Forward was one of the worst excesses of communism, because it resulted in 38 million deaths. It was supposed to work by forming communes from individual farms, but it was a huge failure. Predictions about greater productivity did not work out and eventually led to starvation. The problem was simply that there was not enough food to export, yet the Chinese government exported it anyway. The explanation for this failure was that party officials had miscalculated the harvest through bureaucratic exaggeration. When shortages appeared, leading to widespread starvation, local officials did not speak up, but continued to misrepresent the situation. And the reason for this was that the party cadres themselves had access to stores of food that the peasants did not. So, while the people they managed were starving and dying, the cadres lived on.

Not all the deaths were due to starvation. Many were due to coercion, punishment and torture. The party continued to encourage compliance with the plan. Officials who complained about or opposed the policies were brutally beaten, sanctioned, or killed. At the top of the bureaucratic pyramid, Mao Zedong learned about the shortages but did nothing. He later joked with others about the starvation and ensuing cannibalism. He was that kind of person. (He would have sex with the female members of his staff, even though he knew he had a venereal disease). He was capable of extreme brutality because he could get away with it. The great leader of Chinese communism viewed the deaths with indifference.

Such moral indifference is not a given quality of senior leaders. After World War II, when general George Marshall, chief of American military power, gave a speech at the Pentagon on 26 November 1945, he spoke with compassion for those faced with recovery from the 'storm of war'.

> Most of you know how different, how fortunate America is compared to the rest of the world... The world of suffering people looks to us for such leadership. Their thoughts, however, are not concentrated alone on the problem. They have the more immediate and terribly pressing concerns – where the mouthful of food will come from, where they will find shelter tonight and where they will find warmth from the cold of winter. Along with the great problems of maintaining the peace we must solve the problems of the pittance of food, of clothing and coal and homes. Neither of these problems can be solved alone. They are related, one to other. (Pogue, 1987, p. x)[22]

It is perhaps surprising that these words would have come from a military officer, but George C. Marshall was much more than simply a four-star general. Later (as American Secretary of State) he would aid in the creation of the European Recovery Plan, which is known to us today as 'The Marshall Plan'. This had much to do with the European recovery from World War II, addressing issues such as food, shelter and warmth. And it was later followed by the Berlin Airlift into that city, which again provided such things as food, clothing and coal. This too took place during Marshall's tenure as Secretary of State.

21.1.3 Engagement in Safety

The safety of others requires our engagement. Others have to be psychologically present to us to secure our attention, concern and expertise. When this concern is lacking, harm can come to those for whom we have responsibility. Without the personal acceptance of responsibility, secured by emotional or moral bonds, others can be left to suffer or perish. The example of Apollo 13 versus the *Columbia* shuttle is, to me, a powerful reminder that no matter how 'hard' the technology, human care and human 'presence' can make a difference. And obviously it was not *American* culture that changed between those two accidents, but the culture of NASA itself. Organisational culture is 'up close and personal' in a way that national culture is not. And it may even vary between different departments of the same organisation, with different leaders, and thus, different sub-cultures. We also see these variations when we look at effective performance in extreme environments generally, in war, in operation of complex technical systems and in medical practice. When we have 'skin in the game', we behave differently.

Notes

1. Durkheim, E. (1919). *The Rules of the Sociological Method*. Paris: Felix Alcan.
2. Putnam, R., with Leonardi, R. and Nanetti, R. (1993). *Making Democracy Work: Civic Traditions in Modern Italy*. New York: Princeton University Press.
3. Whyte, W. F. (1969). *Organizational Behavior: Theory and Applications*. Homewood, IL: Dorsey Press.
4. Westrum, R. (2004). "A Typology of Organizational Cultures". *BMJ Quality and Safety in Healthcare*, Volume 13, Supplement II, pp. ii22–ii27.
5. Anonymous (2005). "Can I quit now? FEMA chief wrote as Katrina raged." *CNN Special Report on Hurricane Katrina*, 4 November, posted 11:12 AM.
6. "Michael D. Brown," *Wikipedia* article accessed 20 February 2019.
7. "Michael D. Brown," *Wikipedia* article accessed 20 February 2019.
8. Fruhen, L.S., Flin, R. and McLeod, R. (2013). "Chronic Unease for Safety in Managers: A Conceptualization." *J. Risk Research*, Volume 17, 25 September, pp. 969–979.
9. Cabbage, M. and Harwood, W. (2004). *Comm Check: The Final Flight of Shuttle Columbia*. New York: Simon & Shuster.
10. Kranz, G. (2000). *Failure Is Not an Option: Mission Control from Mercury to Apollo 13 and Beyond*. New York: Simon & Shuster.
11. A Comprehensive discussion of the Crystal Palace and its history is given in Charles Tomlison's: *Encyclopedia of the Useful Arts and Manufactures*, London: George Virtue, ca. 1853, vol. 3, pp. i–lxxx.
12. Tomlinson, p. xxiii.
13. Jouanneaux, M. (1999). *Le Pilote est toujours Devant*. Paris: Editions Octarres.

14. Broughton, J. (2008). Rupert Red Two: *A Pilot's Life from Thunderbolts to Thunderchiefs.* St Paul: Zenith Press.
15. Westrum, R. (1999). *Sidewinder: Creative Missile Design at China Lake.* Annapolis, MD: Naval Institute Press.
16. Westrum, R. (1999). *Sidewinder: Creative Missle Design at China Lake.* Annapolis, MD: Naval Institute Press, p. 216.
17. Wilcox, R. K. (1990). *Scream of Eagles: The Creation of Top Gun and the U.S. Air Victory in Vietnam.* New York: John Wiley.
18. Bosch, P. and Rosich, L. (2008). "Paul Ehrlich's Contributions to Pharmacology," *Pharmacology (online journal),* p. 82.
19. Westrum, R. *Technologies and Society, The Shaping of People and Things.*
20. See my further remarks in Technologies and Society.
21. Meier, B. (2018). Pain Killer: *An Empire of Deceit and the Origin of America's Opioid Crisis.* New York: Random House.
22. Pogue, F. C. (1987). *George C. Marshall, Statesman 1945–1959.* New York: Viking, 1987.

Index

A

Accident, xv, xvii, xviii, xix, xxi, xxii,
        xxiii, xxv, xxvi, xxvii, xxviii,
        xxx, 22, 23, 25, 40, 41, 43, 45,
        48, 59, 65, 68, 71, 78–83, 87, 88,
        91–93, 96, 97, 99, 105, 108, 109,
        113, 119–124, 126–130, 152–154,
        156, 157, 160–162, 173, 178, 183,
        187–190, 192–194, 196–200, 204,
        210, 211, 213, 215, 226, 227, 235,
        250, 251, 254, 256–258, 264–266,
        269–271, 273, 275, 276, 281, 282,
        286, 290–293, 295, 300–302, 305,
        306, 308, 319
AcciMap, xxxii, 189, 204, 210–214, 219, 227
Action research, 206
Adaptation, xxxii, 8, 14–16, 42, 105,
        120, 166, 210, 211, 213, 219, 220,
        250, 259
Aircraft, xviii, 28, 41, 112, 153, 180, 188,
        257, 269, 270, 300, 315–317
Amalberti, René, xxiv, 162
Artefacts, xxii, 63, 107, 110, 151, 152, 157,
        159, 160, 164, 168, 193, 207, 224,
        304, 306
Audits, xxxii, 7, 8, 10, 16, 46, 141, 143, 274
Autonomy, 49, 119, 136–138, 255

B

Boundaries, xxv, 4, 21, 23, 46, 47, 49, 50,
        59, 93, 94, 123, 128, 134, 136, 137,
        146, 155, 156, 159, 160, 166, 167,
        178, 206, 256, 282
Bourrier, xxiv, 42, 92
Braithwaite, John, xix, xxi, xxvi, 64, 105,
        174, 175, 179, 227
Bridges, xxiv, 49, 160, 169, 205, 224

C

Causes, 62, 79, 91, 126, 143, 188, 189, 198,
        200, 213, 257, 265, 266, 270, 281,
        305, 316

Cognition, xxiii, xxviii, xxxi, 107, 110,
        152, 154, 156, 160, 161, 263
Cognitive engineering, xix, xxvii, xxviii,
        152, 175, 182
Collective mindfulness, xxi
Communication, xxx, 23–26, 30, 31, 40,
        44, 93, 97, 176, 206, 237, 239, 264,
        266, 281, 284, 294, 296, 305, 307
Complexity, xvi, xvii, xxii, xxv, 39, 40, 49,
        56, 59, 63, 66, 70, 90, 99, 100, 104,
        140, 152, 153, 162, 164, 168, 169,
        175, 176, 181, 183, 197, 268, 269,
        271, 284, 289, 292, 293, 304
Compliance, 14, 16, 23, 46, 56, 57, 61, 65,
        69, 89, 97, 139, 141, 146, 192, 194,
        233, 234, 256, 293, 301, 318
Consultants, 174, 205, 208, 227, 228, 230,
        231, 291, 292
Controversies, xxii, 75, 76, 106, 108, 109,
        113, 273
Cooperation, xxx, xxxii, 71, 72, 94, 121,
        130, 251
Coordination, xxx, 5, 9, 11, 13, 41, 42, 47,
        66, 125, 284, 293
Crew resource management, xxiii, 176,
        178, 255, 291
Critical, xix, xxi, xxviii, xxix, xxxi, xxxii,
        4, 9, 11, 22, 24, 33, 41, 42, 44, 46,
        71, 80, 83, 92, 96, 99, 100, 119,
        130, 136, 137, 173, 181, 182, 188,
        190, 191, 195, 211, 238, 259, 307,
        308, 312

D

Dams, 153
Decision-making, xxii, xxiii, 14, 42, 45,
        47–49, 81, 89, 95, 100, 110, 112,
        127, 178, 190, 196, 284
Deep Water Horizon, xxvii, 276
Deregulation, 9, 274
Digitalisation, xxvii, xxviii, xxix, xxxiii,
        4, 8, 13, 14, 16, 56, 98–100, 258,
        259, 295, 300

Disasters, xviii, xx, xxi, xxii, xxiv, 22,
    40, 45, 59, 76, 80, 87–89, 93, 95,
    98, 119–121, 123, 126, 128, 133,
    134, 152, 156, 157, 159, 162, 166,
    168, 176, 178, 182, 188–191, 193,
    194, 198–200, 215, 269, 271–273,
    276, 312
Discourse, 3–5, 13, 16, 68, 92, 100, 110, 138,
    158, 173, 174, 182, 271, 295, 305
Drift, xxx, xxxiii, 7, 119–130, 175, 181,
    219, 254, 256, 265, 269, 275, 294,
    302, 303

E

Edmondson, Amy, xix, xxiii, xxvii
Emergency, xxiii, 94, 129, 135, 141, 147,
    276, 296, 305
Engineering, xvii, xviii, xix, xxiii, xxv,
    xxvi, xxvii, xxviii, 68, 76, 78, 80,
    105, 133, 152, 154, 157, 174–177,
    182, 183, 189, 193, 194, 196, 231,
    250, 255, 271, 272, 277, 282, 291,
    294, 296, 297, 308, 313
Engineers, xvii, xviii, xxii, 41, 80, 83,
    110, 134, 136, 154, 175, 180, 183,
    193, 196, 231, 271, 272, 274, 300,
    304, 314
Epistemic accidents, xv, xxiii, xxviii,
    xxx, 79, 80–83, 258, 265, 300
Ethnography, 189, 190, 193
Experts, 77, 79, 83, 120, 122, 134, 136, 147,
    191, 215, 301, 304, 312, 317

F

Failure, 40, 67, 71, 79–83, 105, 107–109,
    113, 121, 125, 126, 134, 139,
    156, 173, 175, 178, 182, 187, 188,
    191–194, 196, 197, 250, 253, 257,
    259, 270, 271, 273, 276, 290, 294,
    300, 301, 304, 313, 317, 318
Fieldwork, xxi, 190, 274
Financialisation, xxvii, xxxiii, 275
Flin, Rhona, xvi, xix, xxiii, xxvi, xxvii,
    24, 28, 252, 255, 256
Fukushima, xxvii, 59, 79, 95, 98, 178,
    215, 276

G

Global, xxxiii, 7, 39, 50, 56, 60, 63, 66, 67,
    72, 113, 129, 138, 275, 276, 281,
    284, 286, 293, 294
Globalisation, xxvii, xxviii, xxx, 5, 6, 15,
    55, 56, 62, 66, 68, 71, 72, 98, 152,
    162, 164, 168, 169, 263, 274–277,
    283, 286, 289, 293
Governance, 8, 10, 13, 39–44, 46–50,
    56–58, 66, 69, 128, 162, 265, 276,
    284, 294, 305
Graphical, 152, 157, 159, 295, 296

H

Hale, Andrew, xix, xxiii, xxvi, 15, 66, 68,
    70, 109, 176, 194, 218, 264
High-reliability organisations (HROs),
    xix, xx, xxvi, xxvii, xxviii, xxv,
    41, 42, 144, 174, 250, 256, 271,
    272, 291
Hollnagel, Erik, xvi, xix, xxiii, xxvi, xxxiii,
    24, 104, 105, 107, 109, 110, 162, 164,
    174, 175, 178, 197, 250, 259
Hopkins, Andrew, xix, xxi, xxiv, xxvi,
    44, 45, 59, 61, 154, 166, 176,
    188–191, 198, 272, 275, 282
Hospitals, xxviii, 7, 8, 134, 141, 142
Human error, xix, xxiii, xxvi, xxvii, xxxi,
    82, 96, 97, 100, 129, 154, 156, 157,
    159–161, 189, 250, 271, 272

I

Identity, 23, 93, 136, 140, 144, 272, 275, 307
Incentive, 45, 46, 69, 126, 144, 189, 230,
    234–236, 238, 239, 241, 252, 255,
    284
Incidents, 10, 23, 24, 29, 30, 32, 41, 47, 48,
    65, 76, 120, 122, 128, 133, 134,
    142, 143, 146, 193, 197, 228, 234,
    250, 264, 270, 273, 274, 290, 292,
    302, 305
Incubation, xix, xx, xxiv, 182, 188, 271
Interaction, xvii, xix, xxiii, xxviii, xxxi,
    13, 15, 21, 22, 28, 29, 32, 33, 40,
    46, 49, 50, 62, 63, 76, 88, 104, 121,
    136, 137, 151, 152, 154, 156, 163,

166, 168, 181, 183, 190, 193, 199, 205, 208, 210, 211, 213, 224, 257, 294, 296, 304, 306

Interface design, xix, xxiii, xxvii, 169

Investigation, xviii, 22, 30, 33, 40, 76, 77, 79, 91, 92, 95, 143, 144, 159, 160, 166, 188, 190, 192, 197–200, 205, 211, 227, 233, 234, 251, 254, 265, 282, 286, 295, 305

Investigators, 82, 122, 160, 226, 227, 273, 274, 304, 308

J

Just culture, xxxi, 174, 178–180, 182, 273

K

Klein, Gary, xix, xxiii, 112, 191

L

Lapses, 133

Learning, xix, xx, xxiii, xxiv, xxxiii, 14, 24, 27–30, 32, 41, 46, 48, 64, 65, 68, 71, 82, 105, 108, 126, 140, 142–145, 182, 187–189, 191, 198–200, 250, 269, 273, 291, 302

Leveson, Nancy, xix, xxiii, xxvi, 63, 110, 175

M

Macro, xv, xxv, 59, 60, 98, 154, 167, 174, 283, 295, 305

Major events, xxxi, xxxiii

Management, xvii, xviii, xix, xxii, xxiii, xxiv, xxvi–xxviii, xxxiii, 3–10, 16, 21–26, 28–32, 39–41, 47, 48, 50, 51, 61, 64, 65, 71, 88, 93, 95, 98–100, 120, 121–123, 125–130, 135, 138, 139–142, 145–147, 157, 163, 166, 173–183, 188, 189, 195–197, 199, 204, 207–211, 215, 218, 227, 228, 231–234, 238, 250, 255, 259, 267, 269, 271, 272, 289–293, 296, 305, 311, 312

Managers, xxii, xxx, 4, 8, 10, 32, 41, 45, 46, 93, 97, 100, 123, 124, 128, 129, 137, 138, 143–146, 154, 193, 196,

199, 206, 208, 228, 232, 251–253, 258, 273, 283, 293, 312, 313

Meso, xxv, 93, 167

Micro, xv, xxv, 59, 60, 75, 93, 167, 174, 181, 283, 295, 299

Mines, 163

Mistakes, xx, 120, 133, 139, 140, 143, 147, 189, 266, 290

Models, xx, xxiii, xxiv, xxvi, xxxiii, 23, 25, 26, 28, 30, 33, 40, 41, 47, 77, 78, 80, 83, 87, 88, 92, 96, 100, 110, 113, 141, 152, 155, 157, 159–161, 164, 166, 189, 193, 196, 203, 204, 207, 208, 210, 211, 218, 227, 250, 254, 255, 257, 273, 283, 284, 291–296, 307, 308, 313

N

Narratives, xv, xxxii, xxxiii, 22, 69, 91, 94, 154, 188, 198, 253

Near misses, xxxi, 48, 121, 124

Networks, xv, xvi, xix, xxv, xxvii, xxviii, xxx, xxxiii, 9, 10, 12, 39–43, 45, 48–50, 65, 89, 106, 112, 121, 130, 159, 162, 164, 168, 195, 263, 264, 284, 294, 295, 305, 306

Normal Accident, xviii, xix, xxi, xxvi, xxviii, xxx, 81–83, 99, 153, 154, 250, 271, 300

Norman, Donald, xix, xxiii, 235, 241

Nuclear power plants, xvii, 43, 160, 213, 215, 258

O

Offshore, 15, 40, 45, 59, 66, 68, 71, 110, 112, 215, 258

Organisation, xxi, xxv, xxix, xxx, 8, 9, 15, 22, 24, 33, 40, 43, 46, 89–91, 97, 110, 122, 125, 129, 130, 134–136, 138, 141, 142, 147, 158, 163, 166, 181, 183, 190, 193, 196, 198, 228, 229, 238, 250, 257, 258, 272–274, 276, 283, 285, 286, 292, 301–305, 311, 315, 319

Outsourcing, xxviii, 5, 9, 56, 190, 274, 277, 284

P

Perception, 23, 24, 32, 33, 67, 95, 100, 119,
    121, 123–125, 127, 129, 275, 282,
    299, 301–304, 307
Performance, xviii, xxiii, 10, 23–28, 40,
    41, 44–50, 61, 71, 77, 80, 83, 119,
    120, 134, 138, 141, 146, 155, 156,
    166, 173, 178, 188, 197, 198, 251,
    253, 257, 284, 300–303, 312, 319
Perrow, Charles, xviii, xix, xxi, xxvi, 81,
    82, 88, 90, 98–100, 103, 104, 119,
    133, 134, 152–154, 157, 159, 160,
    162, 183, 189, 209, 250, 266, 271
Persuasion, xxii, 105, 180
Pidgeon, Nik, xvi, xix, xx, xxxiii, 22, 88,
    119–121, 182, 191, 194, 271–275
Pinch, Trevor, xix, xxii, 63, 77, 106
Power, xv, xvii, xxx, xxxi, xxxiii, 3, 5,
    10–12, 14–16, 25, 26, 31–33,
    41, 43, 48, 56, 57, 59, 62, 69,
    88–100, 111, 120, 129, 130,
    134–136, 138, 141, 147, 154,
    160, 168, 173, 180–183, 213,
    215, 218, 255, 257–259, 265,
    272, 295, 305, 313, 318
Power, Michael, 8, 10, 57, 141
Practical, xxi, xxiii, xxvi, xxviii, xxix,
    xxx, xxxi, xxxii, 12, 14, 28, 33,
    40, 48, 61, 68, 119, 124, 125, 129,
    142, 160, 190, 198, 203–206, 211,
    238, 240, 242, 256, 283, 289, 292,
    297, 302, 307
Practices, xv, xvii, xxi, xxii, xxiii, xxviii,
    xxix, xxx, xxxi, xxxii, xxxiii, 3,
    8, 10, 12, 13, 16, 23–26, 28, 29,
    32, 33, 42, 44, 47, 56, 58–66, 76,
    83, 94, 96, 107, 109, 110, 120,
    122, 123, 126, 136, 138, 140–148,
    158, 166, 169, 173, 174, 179, 181,
    183, 187, 190, 194, 199, 200,
    203–211, 219, 220, 224–229, 231,
    235, 237–242, 251–256, 259, 271,
    274–276, 283, 289, 292, 294, 295,
    301, 303, 306, 307, 319
Practitioners, xviii, xxiii, xxvii, xxviii,
    xxxii, xxxiii, 88, 92, 160, 169,
    179, 197, 203, 213, 215, 218–220,
    224–242, 250–252, 256, 264, 265,
    270, 272, 281, 282

Pressure, 22, 28, 65, 67, 77, 122, 125, 126,
    144, 192, 193, 234–236, 238, 301
Privatisation, 190, 274, 277
Procedures, 5, 6, 9, 10, 12–15, 25, 27, 46,
    61–65, 69, 92, 95, 110, 112, 122,
    125, 134, 139, 143, 178–180, 190,
    193, 238, 258, 272, 273, 276, 284,
    300, 301, 306, 307
Professionalisation, 135–137, 145, 146
Professionals, xxvi, xxxi, xxxii, 6, 58–62,
    65, 66, 70, 103, 109–110, 122,
    134–148, 191, 193, 195–196,
    198–200, 207, 226, 231, 239–240,
    252, 257–258, 264, 270, 272, 285,
    305
Professions, xxviii, xxxi, xxxiii, 8,
    134–137, 139–141, 143, 145–147,
    178, 195, 196, 224, 231, 257, 272,
    283, 289, 292, 305
Projects, xv, xvi, xvii, xxx, xxxiii, 11, 25,
    28–30, 39–51, 133, 137, 145, 228,
    232, 238, 240, 251, 257, 269, 270,
    274, 284, 289, 294, 305
Psychology, xiv, xx, xxiii, 127, 174, 189,
    225, 226, 231, 238, 250–254, 270,
    272, 273, 292

R

Rasmussen, Jens, xviii, xxiii, xxvi, 63, 93,
    96, 119, 146, 154–157, 159, 160,
    163–167, 175, 181, 189, 211, 250,
    254–256
Reason, xxiii, xxvi, 22, 27, 47, 97, 119, 133,
    142, 146, 157, 159, 160, 164, 166,
    174, 178, 179, 182, 188, 189, 194,
    250, 256, 265, 273, 315
Redundancy, xx, 166
Refineries, vii
Regulation, vii, xix, xxi–xxii, xxiv, xxvii,
    xxx, xxxi, xxxiii, 50, 55–72, 91,
    94–95, 99, 122, 123, 127, 129, 140,
    141, 146, 147, 174, 178, 180, 183,
    194, 218, 224, 232, 233, 238, 255,
    257, 265, 272, 293, 294, 308
Regulation regimes, xxii, 56, 61, 70, 71, 257
Reporting, 4, 9, 23, 24, 27, 29, 30, 46, 48,
    88, 105, 122, 142–143, 145, 146,
    174, 179, 180, 232, 234, 252,
    273, 305

Reports, xx, 6, 10, 30, 32, 46, 59, 143, 145,
180–182, 188, 190–194, 215, 227,
228, 233, 235, 236, 239, 250, 253,
273, 274, 317
Research traditions, vii, xv, xvi,
xviii, xix–xxv, xxvi, xxvii,
xxviii–xxxii, 106, 166
Resilience, xxiv, xxv, xxvi, xxxi, 3–5, 11,
15, 41, 68, 156, 174–176, 178–183,
255, 259, 264, 271, 272, 284, 291,
292, 301, 304, 308
Resilience engineering, xxv, xxvi, 41,
97, 105, 174–178, 180, 182, 250,
271–273, 294, 308
Responsibilities, 40, 41, 47, 48, 61, 62, 64,
72, 90, 93, 95, 123, 129, 143, 192,
195, 206, 254, 255, 257, 274, 275,
296, 314–319
Risk, xvii–xviii, xxii, xxxi, xxxiii, 8, 10,
22, 26, 39, 41, 45, 46, 55–69, 71,
81, 83, 91–94, 98–100, 104, 108,
109, 112, 119–130, 133–135,
139–148, 154, 162, 163, 166, 168,
173, 175, 176, 180, 183, 189, 193,
196, 210, 211, 232, 233, 250–252,
254–259, 265, 269–277, 290, 293,
294, 296, 297, 302, 303, 305, 306
Roberts, Karlene H., xvi, xx, xxii, xxvi,
xxvii, xxxiii, 42, 133, 144, 174,
271, 300, 306
Rochlin, Gene, xx, 174, 250, 300–302

S

Safety climate, xix, xxii–xxiii, xxiv, xxvi,
xxvii, xxviii, 24, 215, 218, 253
Safety-critical systems, vii, xix, xxxi, 62,
109, 159, 166
Safety culture, vii, xix–xx, xxiv, xxvi,
xxix, xxx, xxxii, xxxiii, 10, 21–
33, 41, 45, 69, 88–92, 97, 100, 142,
144–145, 156, 174, 199, 204, 210,
213–219, 228, 233, 256–258, 265,
271, 284, 289, 291, 292, 294, 311
Safety leadership, xxii, 199
Safety management systems (SMS),
xviii, xxii, xxiii, 10, 28, 30,
50, 142, 228, 232, 290
Safety regulations, vii, xix, xxi, xxiv,
xxvii, xxxiii, 72, 94–95, 232, 272

Safety research, xv, xvi, xvii–xviii,
xxv–xxvii, xxviii, xxix, xxx,
xxxi–xxxiii, 16, 41, 48, 87–100,
103–109, 112, 113, 121, 129, 151,
152, 154, 157, 159–161, 168, 180,
187, 198, 219, 223–242, 249–259,
263, 264, 266, 267, 270, 271, 274,
276, 283, 289–297
Safety science, vii, xv, xix, xxv, 3–16, 39–43,
49, 88, 92, 96, 97–98, 100, 119,
120, 135, 138, 139, 142, 145–147,
173–183, 187–200, 203–220,
249–255, 259, 263, 274, 275, 277,
281–286, 290–297, 308
Sagan, Carl, xxi, xxvi, 81
Sanctions, xxii, 66, 67, 69, 71, 89, 90, 273
Schulman, Paul, xvi, xx, xxxiii, 14, 111,
119, 123–127, 129, 130, 162, 176,
282, 291, 294, 295
Science and technology, xix, xxii, xxviii,
78, 103, 159, 263, 282
Self-regulation, xxvii, 43, 46, 61, 62, 71,
72, 137, 139, 141, 142, 147, 257
Sensemaking, xxx, xxxiii, 105, 111,
191, 250, 274, 300, 301, 304,
306–308
Sensework, xv, 103–113, 265, 295, 300,
306–308
Sharp-end, xxxiii, 14, 180, 181, 183, 299
Snook, Scott, xix, xxi, 119, 124, 125, 182,
188, 198, 219, 302
Society, xvii, xxvi, 8, 10, 26–28, 39, 58, 59,
61, 65, 95, 99, 126, 129, 133, 136,
138, 141, 275, 276, 285
Socio-legal, xix, xxi, xxiv, xxvii, 134
Sociology, xxiv, xxviii, 40, 107, 108, 134,
135, 147, 166, 189, 263, 272, 275
Sociotechnical systems, xvii, 103–105,
107–110, 112, 113, 121, 152,
154–155, 157, 159–161, 163, 164,
168, 187–189, 211, 213, 271, 290
Speak up, xxiii, 306, 318
Standardisation, xxvii, xxix, xxxiii, 4–9,
11, 13–16, 56, 72, 100, 112, 207,
215, 258, 265, 291, 295
Standards, xvii, 5–9, 13, 15, 42, 58–66,
69, 71, 72, 126, 136, 137, 139–141,
143, 144–147, 200, 215, 227,
231–233, 251, 253, 274, 291
Subcontracting, 190

Systemic, xxiii, 67, 91, 100, 105, 112, 113, 129, 142, 144, 153, 162, 163, 166, 168, 181, 227, 275, 300, 304
System safety, xix, xxiii, xxviii, 126, 134, 218, 229
System view, xxiii, xxvii, 154, 163

T

Teamwork, xxii, 176
Theories, xxviii, 4, 9, 31, 63, 64, 97, 103, 105, 106, 110, 113, 119, 120, 134, 135, 158, 173, 174, 178, 181–183, 188, 189, 204, 215, 218, 240, 241, 251, 252, 264, 271, 284, 295, 308
Tools, xvii, xviii, xxxiii, 47, 98, 104, 108, 110, 113, 138, 204, 207, 209, 210, 218–220, 229, 230, 233, 234, 242, 252, 253, 264, 307
Turner, Barry, xix, xx, xxvi, 119, 121, 157, 182, 188, 189, 215, 269–272

U

Unanticipated, 82, 126, 141, 275, 300
Uncertainties, xxii, 25–27, 31–33, 56, 59, 63, 65, 78, 80, 83, 104, 112, 113, 128, 134, 142–144, 147, 148, 195, 196, 257, 271, 284, 291–293, 296, 304
Unintended effects, xxxi, xxxiii

V

Values, xvi, xxi, xxii, xxxi, xxxii, 22, 24–26, 30, 33, 42, 44, 47, 58, 61, 63, 65, 69, 70, 77, 80, 89, 121, 123, 125–127, 129, 140, 145, 157, 195, 196, 198, 205, 224, 230, 233–235, 240, 250, 254, 275, 282, 294
Vaughan, Diane, xix, xxii, xxvi, 40, 119, 121, 125, 126, 133, 134, 182, 188–191, 198, 219, 256, 300
Visualising, 152, 168, 283, 295

W

Weick, Karl, xvi, xx, xxiv, xxvi, xxxiii, 46, 49, 64, 103–105, 109, 111, 122, 174, 187, 191, 199, 219, 250, 256, 265, 267, 271, 274, 301, 303, 306, 307
Westrum, Ron, xvi, xxii, xxxiii, 316, 317
Woods, David, xxiii, xxvi, 48, 110, 175, 176, 178, 181, 182, 256, 291
Work as done, 105, 110
Work as imagined, 3–16, 110, 301
Wynne, Brian, xxii

Z

Zohar, Dov, xxiii, 24, 215

Printed in the United States
by Baker & Taylor Publisher Services